58 Springer Series in Chemical Physics
Edited by V.I. Goldanskii

Springer Series in Chemical Physics

Editors: Vitalii I. Goldanskii Fritz P. Schäfer J. Peter Toennies

Managing Editor: H. K. V. Lotsch

40 **High-Resolution Spectroscopy of Transient Molecules**
By E. Hirota

41 **High Resolution Spectral Atlas of Nitrogen Dioxide 559–597 nm**
By K. Uehara and H. Sasada

42 **Antennas and Reaction Centers of Photosynthetic Bacteria**
Structure, Interactions, and Dynamics
Editor: M. E. Michel-Beyerle

43 **The Atom-Atom Potential Method**
Applications to Organic Molecular Solids
By A. J. Pertsin and A. I. Kitaigorodsky

44 **Secondary Ion Mass Spectrometry SIMS V**
Editors: A. Benninghoven, R. J. Colton, D. S. Simons, and H. W. Werner

45 **Thermotropic Liquid Crystals, Fundamentals**
By G. Vertogen and W. H. de Jeu

46 **Ultrafast Phenomena V**
Editors: G. R. Fleming and A. E. Siegman

47 **Complex Chemical Reaction Systems**
Mathematical Modelling and Simulation
Editors: J. Warnatz and W. Jäger

48 **Ultrafast Phenomena VI**
Editors: T. Yajima, K. Yoshihara, C. B. Harris, and S. Shionoya

49 **Vibronic Interactions in Molecules and Crystals**
By I. B. Bersuker and V. Z. Polinger

50 **Molecular and Laser Spectroscopy**
By Zu-Geng Wang and Hui-Rong Xia

51 **Space-Time Organization in Macromolecular Fluids**
Editors: F. Tanaka, M. Doi, and T. Ohta

52 **Multiple-Photon Laser Chemistry**
By. R. V. Ambartzumian, C. D. Cantrell, and A. Puretzky

53 **Ultrafast Phenomena VII**
Editors: C. B. Harris, E. P. Ippen, G. A. Mourou, and A. H. Zewail

54 **Physics of Ion Impact Phenomena**
Editor: D. Mathur

55 **Ultrafast Phenomena VIII**
Editors: J.-L. Martin, A. Migus, G. A. Mourou, and A. H. Zewail

56 **Clusters of Atoms and Molecules**
Salvation and Chemistry of Free Clusters, and Embedded, Supported and Compressed Clusters
Editor: H. Haberland

57 **Radiationless Transitions in Polyatomic Molecules**
By E. S. Medvedev and V. I. Osherov

58 **Positron Annihilation in Chemistry**
By O. E. Mogensen

59 **Soot Formation in Combustion**
Mechanisms and Models
Editor: H. Bockhorn

60 **Ultrafast Phenomena IX**
Editors: P. F. Barbara, W. H. Knox, G. A. Mourou, and A. H. Zewail

Volumes 1–39 are listed at the end of the book

Ole Erik Mogensen

Positron Annihilation in Chemistry

With 44 Figures and 13 Tables

Springer-Verlag
Berlin Heidelberg New York London Paris
Tokyo Hong Kong Barcelona Budapest

Dr. O.E. MOGENSEN (†)

ISBN-13:978-3-642-85125-4 e-ISBN-13:978-3-642-85123-0
DOI: 10.1007/978-3-642-85123-0

CIP data applied for

Softcover reprint of the hardcover 1st edition 1995

Typesetting: Macmillan India Ltd, Bangalore-25

SPIN: 10069941 54/3140/SPS - 5 4 3 2 1 0 - Printed on acid-free paper

Preface

The electron, positron, and positronium (Ps, the electron-positron bound state) are the only light particles which participate in normal low-energy processes of physics and chemistry. Because of their small masses quantum mechanical phenomena play a great role in their behavior. Good accounts of the excess electron behavior are available. Similarly, several good books treat the behavior of the positron in solid state and gas physics. However, the research on the positron and Ps in chemistry-oriented fields of science is not well treated in the literature. In liquids Ps is normally in the "bubble" state, and hence, its chemical reactions are very exceptional, indeed. The states and reactions of the positron in insulating liquids can profitably be compared to those of the only other charged light particle, the excess electron. Furthermore, Ps is formed by a reaction between the positron and excess electron in the positron spur, and Ps formation studies give, therefore, information on excess electrons, too. Consequently, to understand "light particle chemistry" it is necessary to have a good understanding of the chemistry of the positron and Ps.

This book is a critical review of the research on the behavior of the positron and Ps performed within chemistry-oriented positron annihilation. In particular, I have emphasized the interpretation of selected, typical, experimental results in the field. The book does not give an exhaustive discussion of all the published results. This is impossible because of lack of space anyway. Discussions of experimental results and theoretical explanations believed to be outdated, erratic, or misleading are mainly omitted, with the exceptions of those which have been used extensively in the literature. The omission of certain works might just indicate that the accepted ideas and concepts in the field have changed very much recently.

Partly because of lack of space positron annihilation in several fields on the borderline of chemistry, physics, and/or materials science is not discussed. For example, research on metals, semiconductors, ionic crystals, ceramic materials, high T_c superconductors, gases, and liquid noble gases is not discussed, except if the results are of general interest.

The writing of this book would have been impossible without many years of very good collaboration with M. Eldrup and N.J. Pedersen at Risoe. Furthermore, I spent seven months doing fruitful work at IMFUFA, Roskilde University

Center, and I have been finishing the manuscript during my stay at the Chemistry Department, JAERI, Japan. I am very thankful to both these institutions.

Tokai-mura, Japan, December 1991 O.E. Mogensen

Sadly, the author of the present book, Ole Erik Mogensen, suddenly passed away in April 1993. He therefore never saw this book in which he had invested so much effort.

The editorial procedure was in progress at the time of his demise, and the reviewer's comments on the manuscript were not in time for him to consider them. The book, as it now appears, is thus the original version from Ole Mogensen's hand apart from the corrections of some misprints and an appended subject index.

Risø, August 1994 M. Eldrup

Contents

1 Introduction

The chemical physics of the positron and positronium (Ps) is concerned with the chemical aspects of the interaction of low-energy positrons with matter. Positrons injected into matter thermalize rapidly. In many insulating compounds some of the positrons form Ps. The positron and Ps might react with solutes, etc. before they ultimately annihilate into gamma photons. The properties of these photons, measured by gamma spectroscopy techniques, give information about the positron and Ps reactions before the annihilation.

This chapter starts with a short general discussion of the basic principles of the chemistry-oriented positron annihilation research. The next section gives a brief description of the history of this subdivision of low-energy positron annihilation. A short discussion of the physics-oriented disciplines of the field is included in Sect. 1.3. In the last section a discussion of the content of this book is given.

1.1 Positron Annihilation in Chemistry

The positron is the antiparticle of the electron. Hence, it has the same mass as the electron, and its charge is numerically identical to, but of opposite sign of that of the electron. Energetic positrons injected into condensed matter are slowed down to thermal energy in a very short time of about one ps. They penetrate 10–1000 μm into condensed matter. Thermalization occurs by creation of excess-electron positive-ion pairs along the track. The last part of this radiation damage along the track will take place at a short distance from the thermalized positron. It is called the positron spur.

In many insulating solids and liquids some of the injected positrons form positronium (Ps) with an electron of the sample. Positronium is a bound state of an electron and a positron. It is the lightest known atom, which can formally be obtained from a hydrogen atom by replacing the proton by the positron. The thermalized positron cannot form Ps with one of the electrons bound in the molecules, since the binding energy of an electron in Ps is only 6.8 eV compared to that in ordinary molecules of 9–12 eV. However, Ps can be formed by a reaction between the thermalized positron and one of the excess electrons in the positron spur. This occurs in competition with other spur processes, such as excess-electron positive-ion recombination, excess-electron and positron diffusion out of the spur, and excess-electron and positron reactions with the molecules, ions, etc. of the sample, free radicals, excited states, and other spur species.

Figure 1.1 shows a flow chart for some of the most important positron processes which occur in the time interval between the emission of the positron from a β^+ nucleus, e.g., ^{22}Na, and the detection of the annihilation photons emitted when the positron ultimately annihilates with an electron. Normally the lifetime of the positron and the angular correlation of the annihilation photons are measured. Typical results are shown schematically at the bottom of the flow chart.

In a lifetime (LT) experiment the time interval between the detection of a 1.3 MeV photon, emitted from ^{22}Na almost simultaneously with the positron, and an annihilation photon of 0.511 MeV is measured by use of gamma spectroscopy techniques. The main parameters of a lifetime spectrum are the lifetimes (actually the average lifetimes) and the relative intensities of its exponential terms. The lifetime is determined by the density of electrons, having opposite spin direction of that of the positron, at the positron.

In the usually found two photon annihilation, the two photons, of about 0.511 MeV each, are emitted almost back to back. The angular deviation from π, θ, is measured by use of a linear slit angular correlation (AC) apparatus, which measures the number of annihilations, where a component of the total momentum is given by $k_z = mc\theta$, as a function of θ. m is the electron mass, and c the velocity of light. Except for weak influence of third bodies, the angular correlation spectrum is given by the distribution of the total momentum of the electron-positron pairs participating in the annihilations. The positive ions repel the positron, and hence, only the valence and the outer core electrons can annihilate with the positron.

After thermalization the positrons not forming Ps diffuse in the sample for 100–500 ps before they annihilate with an electron of the molecules into normally two γ-rays, each having an energy of about 0.511 MeV. The AC spectrum is broad because the annihilating electron is bound in the molecules.

Positronium is formed in two states: 25% as para Ps (p Ps) in which the spins of the two particles have opposite directions, and 75% as ortho Ps (o Ps) in which the spins are parallel. The lifetime of p Ps is about 0.125 ps. It annihilates mainly intrinsically into two γ-rays, of 0.511 MeV each, i.e., the positron in p Ps annihilates with the electron in p Ps. The AC spectrum, determined by the zero-point motion of Ps, is narrow. In vacuum o Ps annihilates into three γ-rays with a mean lifetime of 142 ns. However, in condensed matter the positron in o Ps annihilates mainly with an electron of the molecules having opposite spin direction to that of the positron into two γ-rays of about 0.511 Mev each. Hence, the AC spectrum is broad. The mean o Ps lifetime is about 1 to 5 ns. This is called pick off annihilation.

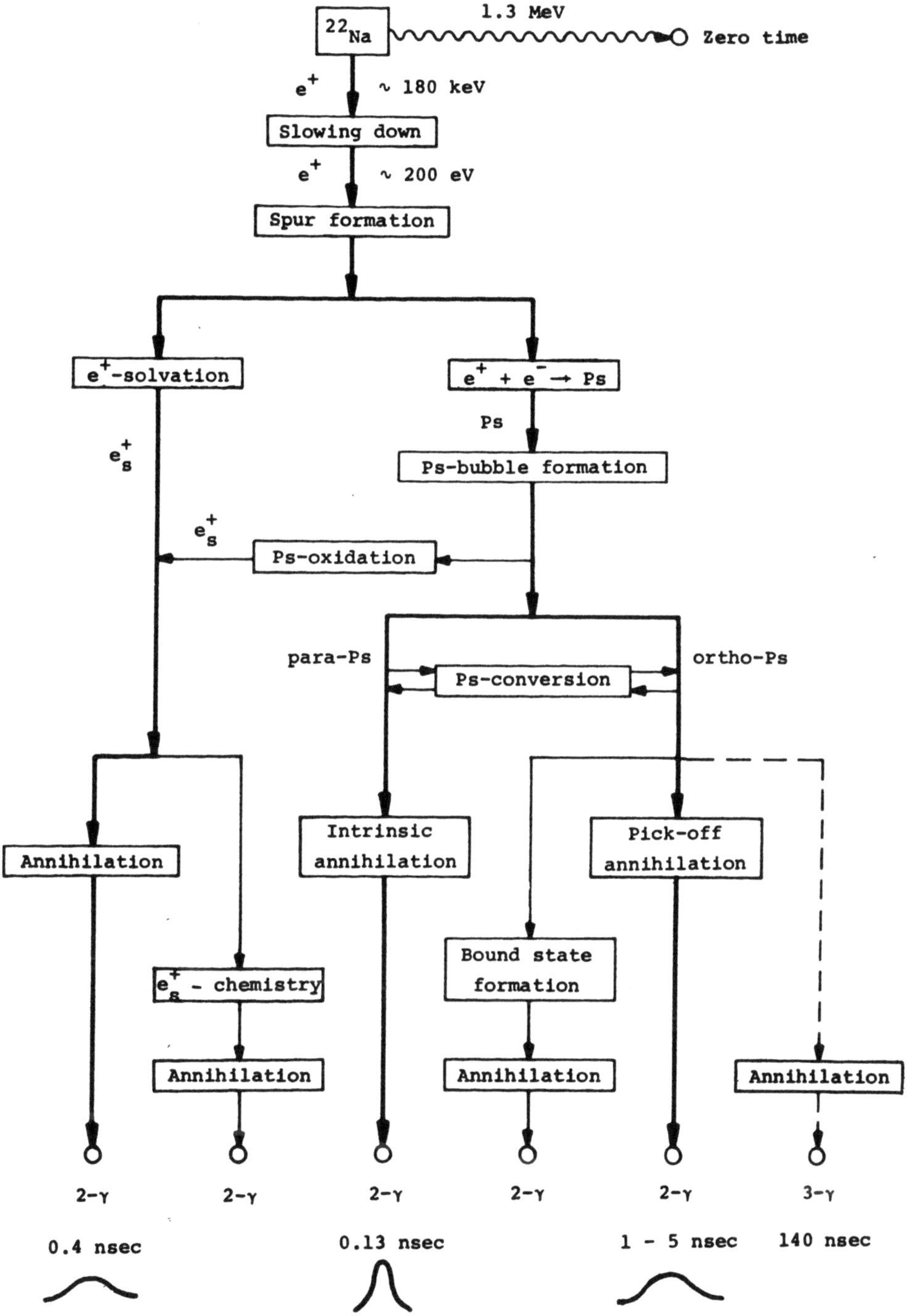

Fig. 1.1. Flow chart for some of the positron and Ps processes in liquids. Only the most important processes are shown

During their lifetimes the free positron, p Ps, and o Ps can react with atoms, ions, and molecules of the sample and/or become trapped in defects in solids. This causes normally changes in the properties of the emitted annihilation photons. By measuring these changes one can obtain information on the processes which have influenced the states of the particles before the annihilation. Because the particle processes depend in a detailed way on the structure of the atoms, ions, molecules, defects, etc. one can thereby obtain information on the properties of these reactants. Furthermore, the properties of the annihilation photons depend on the states of the particles in the bulk of the sample if the concentration of the reactants is so low that the particles annihilate before the reaction. Hence, the states of the positron and Ps in the solid or liquid can be investigated, too. This is the basic idea of usual low-energy positron annihilation studies of matter.

In the latest three decades the excess electron, i.e., the electron in the "conduction band", in liquids has attracted great attention, in particular in radiation chemistry. As it is the mostly investigated light particle, a comparison of its behaviour with that of the positron, the only other charged light particle, is very useful. Another reason for the inclusion of a discussion of the excess electron in this book is that Ps is formed by a reaction between the thermalized positron and one of the thermalized excess electrons created in the positron spur by the slowing down processes of the positron. The Ps yield can be strongly influenced by the addition of compounds which react with the excess electron and/or positron before Ps is formed. Positronium formation measurements can, therefore, give information on the chemical reactions of the excess electron.

The electron, positron, and Ps are the only light particles participating in low-energy physics and chemistry. Because of their low masses quantum mechanical phenomena, such as tunnelling, delocalization, and zero-point energy, play a great role in their behaviour. Only for these three particles quantum mechanics effects are not just perturbations of the behaviour, as, e.g., for the muon, Mu, proton, and H, but decisive for their reactions, diffusion, mobility, lifetime, etc. The reactions of the light particles can usefully be denoted "Light Particle Chemistry".

In liquids the presence of a light particle influences the position and orientation of the nearby molecules. If the molecules are strongly attracted by the light particle, as is the case for the electron and positron in polar liquids, the particle is strongly solvated, and hence, slowly moving. For example, the excess electron in thermal equilibrium in water, denoted e^-_{aq}, is hydrated by an energy of about 2 eV. If a strong repulsion exists between the light particle and the molecules the particle is in a bubble state, and also in that case it moves slowly. The molecules around the particles are pushed away, and a cavity, the bubble, is formed, in which the zero-point energy of the particle is balanced by the sum of the surface and pressure-times-volume energies of the bubble. Positronium is in a bubble state in most liquids, while the bubble state for the excess electron is found only in liquid He, Ne, and H_2. If the light particle has roughly the same energy in the liquid as in vacuum it is partly delocalized in the liquid, and hence, it moves much faster than the heavy molecules and ions. In particular, the excess electron has a very high mobility and diffusion constant in liquids where it has a slightly lower energy than in vacuum (e.g., −0.55 eV in tetramethylsilane and −0.35 eV in neopentane). The rates of reactions of the light particles depend, of course, very strongly on the structure of the particle states.

Let us briefly discuss some examples of light particle reactions, which have been studied by positron annihilation. The free positrons, i.e., the positrons not bound in Ps but normally solvated, react with Cl^-, Br^-, and I^- in water and alcohols, and with CCl_4, CBr_4, and some crown ethers in nonpolar liquids. Positronium in the bubble state is oxidized by $MnO_4{}^-$ and $IO_4{}^-$, and spin converted by Mn^{++}, in water, while it reacts with nitrobenzene and other molecules of high electron affinity in most liquids. With few exceptions Ps formation is inhibited by all compounds reacting with Ps by oxidation and/or bound state formation, because these compounds react with the excess electrons in the positron spur. However, some weakly electron-accepting molecules and ions, not reacting with Ps, inhibit also Ps formation by reaction with the positron-spur excess electrons. Examples are $NO_3{}^-$ in polar and CCl_4 in nonpolar liquids. Cl^-, Br^-, and I^- partly inhibit the Ps yield in water because they react with the positron. Positronium reacts with hole-type defects, such as vacancies and vacancy clusters, in molecular crystals, which results in longer o Ps lifetimes than in the bulk. Similarly, the positron is trapped in hole-type defects in many metals and in F-centers in ionic crystals.

Monographs on positron and Ps chemistry are [1.1–4]. The reviews in the collection of articles [1.1] have been written with fairly different aims than this book. The reviews in [1.5] give good accounts of the physics-oriented positron solid-state research. The bibliography [1.6] seems to be very complete. However, its index is poor, and hence, the search for literature on particular topics is very difficult in [1.6]. The proceedings of some of the International Conferences on Positron Annihilation (ICPA) are [1.7–12]. About 15–30% of the papers in these proceedings belong to the topics discussed in this book. [1.13–18] are proceedings of recent workshops etc. on various special fields of positron annihilation. The 9ICPA took place in Hungary in 1991, and the proceedings will be published. A fourth positron and Ps chemistry symposium will be arranged in 1993 by the Strasbourg group. Workshops on positron annihilation in gases and on slow positron technique will probably continue to be arranged in the future.

1.2 History of Chemistry-Oriented Positron Annihilation Studies

The positron was introduced into physics by *Dirac* in 1930 [1.19]. He proposed in the theory of holes that the negative – energy states of the Dirac equation (see, e.g., [1.20]) are normally completely filled by electrons. Hence, a transition to the negative-energy states of the positive-energy electrons is prevented by the

Pauli principle. The infinite number of negative-energy electrons were assumed not to cause electric field and gravitational effects. A negative-energy electron can be excited to positive-energy states resulting in the creation of a hole in the "sea" of negative-energy electrons. The presence of the hole means that the "sea" behaves like a positively charged particle, which Dirac assigned to the proton. It was subsequently identified as a new particle, the positron. A discussion of the history is given by *Dirac* [1.21]. Quantum electrodynamics (QED), i.e., the theory of interacting electrons, positrons, and the quantized field of radiation, was developed (except for higher order corrections) at about the same time (1930). It predicts that an electron-positron pair can annihilate into photons and can be created by photon absorption, if certain conditions are prevailing [1.19,20], as discussed in Chap. 2.

On the 2nd of August 1932 *Anderson* [1.22] observed the positron for the first time in cosmic-ray tracks studied by use of a Wilson cloud chamber. It was soon confirmed that the positron behaves as predicted by QED. The positron was the first discovered antiparticle in physics.

The first written discussion of positronium (Ps) is that of *Mohorovicic* [1.23] in 1934. However, Ps seems to have been discussed before 1934 in several institutes. The basic features of the physics of Ps were calculated before Ps was experimentally discovered by *Deutsch* [1.24], who produced Ps in gases in 1951. *Rich* [1.25] discusses the history of Ps physics in details.

The experimental techniques of positron annihilation are based upon the techniques of nuclear spectroscopy, in particular gamma spectroscopy, which was strongly developed in the two decades after 1945. Especially, the invention of the scintillation detector in 1944–1953 was of great importance for positron annihilation studies. Somewhat later the introduction of high-quality NIM electronic modules, in particular the time-to-amplitude converter (TAC), strongly improved the efficiency of data collection.

The first reaction of Ps, namely spin conversion by NO in gases, was also reported by *Deutsh* [1.24] in 1951. *Pond* [1.26] observed the effect of addition of DPPH to benzene in 1954. This is probably the first report on a Ps reaction in a liquid. The now standard LT method of Ps chemistry seems first to have been employed by *Berko* and *Zuchelli* [1.27], who studied DPPH in benzene in 1956. Apparently, *Stewart* [1.28] observed first the narrow p Ps component (in teflon in 1955), while *De Zafra* [1.29] used first the AC method in Ps reaction studies in 1959. It is remarkable that the first (of the now standard) methods of Ps chemistry were invented by physicists and published in Physical Review.

The first general theory of Ps formation was published in 1949 by *Ore* [1.30], who assumed that Ps is formed during the slowing down of the positron at energies, E_+, in the "Ore gap", $V - 6.8$ eV $< E_+ < V$, where V is the ionization energy of the molecules and 6.8 eV the Ps binding energy. The Ore model of Ps formation is discussed in detail in Sect. 7.7. However, the measured Ps yields for liquids could not be well explained in terms of the Ore model. This prompted *Jackson* and *McGervey* [1.31] to propose the "Ore + hot-Ps-reaction" (OhPr) model, in which it is assumed that Ps is formed by the Ore model process, but that Ps can undergo "hot Ps" reactions before it thermalizes. In the general case, Ps is, of course, formed with an excess kinetic energy in its center-of-mass motion in the Ore model. Also the OhPr model is discussed in Sect. 7.7. The OhPr model became the mostly used model to explain the Ps formation in condensed matter at around 1968, although other models were used, too (see Sect. 7.7).

Ferrel [1.32] introduced the bubble model of the Ps state in liquids in 1958. Actually, it was used in Ps physics before it was realized that the excess electrons are also in bubble states in liquid He, Ne, and H_2. In 1963 *Lee* and *Celitans* [1.33] discovered that the oxygen content of most liquids exposed to air at atmospheric pressure reduces the o Ps lifetime by up to 40% because oxygen reacts with Ps. It was realized that only the degassed liquid gave an o Ps lifetime, which could be explained in terms of pure liquid properties. *Gray* et al. [1.34] measured the o Ps lifetime and yield for 193 degassed, pure liquids in 1968. This comprehensive work was of great importance for the later works in Ps chemistry.

The spur model of Ps formation was proposed by *Mogensen* [1.35] in 1973. The proposal was triggered by the strong developments in radiation chemistry, in particular concerning the excess electrons and the structure of charged particle tracks (spurs) in liquids, taking place after 1960. At first the spur model was met with fairly strong opposition. However, around 1982–1985 it became the accepted model amongst the majority of scientists studying the chemistry aspects of positron annihilation, while several physicists interested in slow-positron-beam investigations still have used the Ore model without "hot-Ps" reactions recently (see Sects. 7.7 and 9.5). Many Ps yield effects, predicted and/or explained by use of the spur model, were measured from 1974 to 1985, when the Ps formation was the mostly studied topic of Ps chemistry.

The spur model was later used to explain the muonium (Mu) formation by *Percival* et al. [1.36] in 1978. Also in that case the spur model was met with strong opposition, but it seems now to be the accepted model of most (but certainly not all) scientists interested in Mu formation.

The first rate constant of a solvated positron reaction, namely that of the hydrated positron, e^+_{aq}, reaction with Cl^-, was measured *Mogensen* and *Shantarovich* [1.37] in 1974. It was discovered that only the AC (but not the LT) spectra changed significantly on the positron bound state formation with Cl^-. The solvated positron reactions are discussed in Chap. 6. As the positron is short lived in liquids ($\sim$ 400 ps), fairly high concentrations of the reactant are necessary, and hence, time-dependent rate constants and initial close encounters must be taken into account in the interpretation of the reactions.

The molecular solids, i.e. polymers, molecular crystals, etc., were studied by positron annihilation methods already in 1950–1960. It was soon realized that Ps is formed in many molecular solids. In 1960 *Brandt* et al. [1.38] proposed the "free volume model", which has been used very much to explain the measured Ps lifetimes. It was assumed that Ps moves in the free volume, defined as the volume not occupied by the van der Waals spheres around each atom in the molecules in molecular solids and liquids. However, Ps is a "seeker" and a "digger". It "seeks", and gets trapped, in the hole-type defects and open structures in the liquids and solids, and after the trapping the repulsion between most molecules and Ps causes an enlargement of the trap in the softer solids and in all liquids. In nearly all liquids a bubble state in equilibrium is formed before Ps decays. Hence, the "free volume model", as it is used in many papers, is incorrect. This is further discussed in Sect. 10.2 on polymers.

A new development occurred in 1969 when *Brandt* et al. [1.39] discovered delocalized Ps in a Bloch function state in single crystals of quartz. The AC spectra determine the numerical value of the Fourier-transform of the center-of-mass wave function at the reciprocal-lattice points (Ps is thermalized), which cannot be measured directly for the two other light particles. Ps is the only atom shown to be in a Bloch function state.

Delocalized Ps was also found in single crystals of ice [1.40]. Subsequently, it was shown (see e.g. *Eldrup* [1.41]) that Ps is trapped in vacancies, i.e., missing water molecules, in ice above −100°C, and in irradiated ice and HF doped fast-cooled ice below −180°C. The results were used to determine directly the vacancy formation and migration energies, and to get information on small vacancy clusters, for the first time for any molecular crystal (ice). Furthe more, the vacancy concentration in ice just below the melting point was found to be about 5 ppm, i.e., 50.000 times higher than assumed previously.

The Ps states and possible trapping in vacancies in ten molecular single crystals were studied in a continuation of the ice work, see Sect. 9.2. An important result of these studies was a curve of the o Ps lifetime versus hole size for solids and liquids, Fig. 4.1. It has been used much to interpret the o Ps lifetimes in many compounds recently. Simultaneously, *Goworek* [1.42] studied Ps formation and lifetime in many aromatic molecular crystals.

In the latest five years the studies of Ps in liquids have mainly been concerned with Ps states and reactions. An anomalous magnetic quenching of Ps in nitrobenzene/hexane, but not in nitrobenzene/benzene and nitrobenzene/cyclohexane, solutions [1.43] has attracted much attention. It seems to be well explained in terms of a general model of the temperature and pressure dependence of the rate constants of Ps reactions with electron-acceptor molecules in liquids [1.44]. Apparently, the anomalous magnetic quenching is found when Ps is just stable energetically, as is also the case in naphthalene at room temperature. Several cases of a continuous change from a Ps state to a non-Ps state at decreasing temperature, e.g., for liquid CS_2 and some aromatic hydrocarbon solids, have been reported. The same general model explains well the strong dependence of the Ps rate constant on external pressure for certain reactant/solvent combinations, e.g., for nitrobenzene in hexane but not in benzene, reported by *Kobayashi* [1.45,46]. Roughly speaking, the energy of the bubble state of Ps can be increased either by increasing the surface tension, and thereby the surface energy, by lowering the temperature, or by imposing an external pressure, thereby increasing the pv-energy of the bubble. For certain solute/solvent combinations this energy increase results in a change from an endothermic to an exothermic reaction, i.e., the Ps reaction changes from being much slower than diffusion-controlled to a mainly diffusion-controlled reaction. Hence, "squeezed" Ps has attracted attention recently.

Concerning the techniques of positron annihilation, in particular the LT apparatus has been improved within the last two decades. The use of the so-called constant fraction discriminator has permitted a much higher counting rate at a reasonable time resolution. Furthermore, the full-width-at-half-maximum (fwhm) of the time resolution used in practice has decreased from about 400 ps to roughly 200 ps. However, the slow-positron beams have not been extensively utilized to solve chemistry problems, as yet. The normally used computer programs for the analyses of LT and AC spectra (PATFIT) were developed at Risoe around 1975 [1.47]. They have also been of great importance for the efficiency of information collection.

1.3 Review of Physics-Oriented Positron Research

This book discusses positron annihilation in chemical physics. Furthermore, some topics in the borderland between physics and chemistry discussed in details in other publications are excluded, or are only treated very briefly. In this section a brief review of the physics-oriented disciplines of positron annihilation, not discussed elsewhere in the book, will be given.

Positron annihilation in pure metals has now been investigated in details for four decades [1.5–12]. In metals Ps is not formed, mainly because the positron is bound to many electrons simultaneously. If, by chance, a kind of pseudo-Ps is formed it will immediately form Ps^- with, and/or get spin converted by, one of the valence electrons of the metal, and hence, it will be extremely short-lived. Only the valence electrons and some of the outer core electrons participate in the annihilation as the positron is repelled by the positive core of the atoms. The positron polarizes the freely moving electrons in metals resulting in a much higher density of electrons at the positron than the average density of valence electrons. Consequently, one short positron lifetime (τ, between 100 ps for transition metals and 400 ps for Rb and Cs) is measured. Theoretically, τ is normally calculated by use of many-body theories for a "Fermi sea" of free electrons adapted to the given metal. Electron Bloch functions are normally not used. On the other hand, the AC curve is generally calculated by use of Bloch functions for the electrons and the positron obtained in the independent-particle approximation. Surprisingly, these theoretical procedures give reasonably good results for both the LT and AC cases.

For non-defected metals AC measurements are mainly used to study the Fermi surfaces in single crystals. They can be used in metals and alloys in which the mean free path of electrons is short, and hence, where most of the other methods of investigating Fermi surfaces cannot be used. Apparently, this is one of the best methods of getting information on Fermi surfaces in alloys. Before about 1978 only the linear-slit one-dimensional AC set-ups (Chap. 3) were used. Recently, however, the expensive, two-dimensional AC apparatus has been build in several institutes. They are particularly useful for Fermi surface studies. However, with few exceptions they have not been applied on chemistry-oriented problems. The "Fermi surface" of high T_c materials have been much studied lately. However, the results were not promising [1.7].

At about 1964–1967 it was realized that positrons are trapped in hole-type defects in solids, and defect studies became rapidly one of the largest topics of positron annihilation after 1970. In particular, vacancies and small polyvacancies in metals were extensively investigated. Vacancy formation energies for metals of high melting temperatures could be determined, apparently for the first time. Furthermore, information on the small, submicroscopic polyvacancies was first obtained by use of positron trapping measurements. Recently, gas bubbles in metals has been much studied. Voids (polyvacancies) and gas bubbles constitute very important materials science problems in the fast-breeder fission, and in the first wall of fusion, reactors. Today, the positron trapping method is recognized as being an important technique in metal defect research [1.5–11].

Positrons are also trapped in some color centers in ionic crystals and in hole-type defects in semiconductors. However, positron trapping seems not, as yet, to be a very important method of studying defects in the bulk of these materials [1.5–12]. The reason is partly that other techniques (EPR, infrared and optical absorption, etc) of studying defects can profitably be applied in these solids. One difficulty is that a "reference" ionic crystal with such a low concentration of color centers that the positron is not trapped, seems not to have been studied by use of positrons, yet. Furthermore, the radiation damage due to positron slowing down influences the resulting AC and LT spectra in many cases.

At around 1970 the possibility of making a slow positron beam of a useful intensity was demonstrated. Up to 1970 the high-energy positrons of average energy of about 200 keV emitted from nuclear sources were mainly used. Recently, the use of slow positrons to investigate surface phenomena of solids has been developed into being the most important topic in positron annihilation [1.5–11,17,48]. Positrons having energies of around 1–30000 eV are injected into the solid, and, for example, the yield of Ps escaping through the surface and/or the Doppler broadened annihilation energy of the positrons annihilating in the solid, are measured as a function of energy of the injected positron. In other studies the energy of Ps, and/or the yield and energy of positrons, escaping from the solids are determined. In particular, thin layers deposited on semiconductors, and the near-surface distribution of defects in materials, have been investigated very much.

The theory of the behaviour of slow positrons in solids is fairly similar to that discussed in Sect. 6.2. The positron is assumed to diffuse randomly, and to drift in the electric fields, in the solid (see (6.7)), and the decay (trapping) of the positron is described approximately by an annihilation (trapping) probability per second. Hence, a term similar to the left-hand side of (6.3) is substituted into the diffusion/drift equation (6.7), for example. This is, of course, only an approximation as (6.3) is usually obtained by solving (6.7). Normally, only one spatial dimension is assumed to be important (see e.g., p. 737 in [1.48]).

Generally speaking, the slow positron work seems not to have caused a strong impact on the applied science of the surfaces of materials, although some spectacular results have been obtained. However, very important new information on the basic science of the behaviour of slow positrons in solids has resulted. Very few results of slow positron studies of the chemistry-oriented problems are available. Certainly, the potential of the slow positron methods in chemistry is not fully investigated at present. Some slow positron results are discussed in Sect. 9.5.

Positron annihilation in gases has been well reviewed in several publications [1.49–52] and in the proceedings of international workshops [1.15,16]. The sixth workshop was held in Sydney in 1991. These workshops have been made partly as a reaction to the fact that the bigger ICPAs have become dominated by the larger fields, e.g., positron annihilation in solid state physics and slow-positron surface studies. A similar situation is found in the field of chemistry-oriented positron annihilation [1.13,14], for which the ICPAs have also become less profitable. Except for a few remarks, positron investigations in gases (and in liquid inactive gases) are not discussed outside the brief review in this section.

Before about 1970 slow positrons were unavailable, and hence, fast positrons from mainly nuclear sources had to be used to study the positron and Ps interaction with gases. In particular, positron annihilation under the variation of the density, temperature, and structure of the gas molecules was studied. In general, thermalized positrons tend to become trapped in clusters of gas molecules at higher densities and lower temperatures, in particular in polar gases. This decreases the positron lifetime. For example, the lifetime of positrons in NH_3 gas is roughly equal to that in liquid NH_3 at gas densities above 0.4% of the liquid NH_3 density at 22°C due to NH_3 cluster formation around the positron [1.53]. On the other hand, Ps tends to form bubble states in gases at higher densities [1.13,49,50], whereby the o Ps pick off lifetime is increased above that obtained by extrapolating the low density results to higher densities. In general, the self-trappings of the positron in clusters, and of Ps in bubbles, are very interesting physical processes both in theory and experiments. Of course, a comparison with the behaviour of the third light particle, the excess electron, in gases might be very fruitful. Opposite to the excess electron case, the mobility of the positron has been studied very little in gases (and in condensed matter) [1.54,51]. Furthermore, only few and scattered results on Ps chemical reactions, and the inhibition of Ps formation, on the addition of compounds to gases are available. For example, no data on Ps reactions and inhibition at temperatures above the critical point is available. Also the excess electron reactions in high pressure gases seem to have been very little studied.

The Ps formation in gases at various pressures is discussed briefly in Sect. 7.7. Three recently studied effects attract attention [1.7,8,13,49–52]. First, the rather few reported results indicate that the transition between the Ore model and the spur model processes seems to take place at densities about 5–10 amagat in polyatomic gases. Second, the positron cannot thermalize completely in xenon, which influences strongly the measured Ps yields. Third, the cross sections for Ps formation in very low density gases have significant values even at positron energies of about 100 eV.

The introduction of slow positron beams after 1970 had a very strong impact on positron studies of gases. The cross sections for scattering of slow positrons could now be measured and compared with those of the other, charged, light particle, the electron. At present, the total cross sections for positron scattering on most monoatomic and several of the simple polyatomic molecules have been measured [1.49–51]. Furthermore, the determination of partial cross sections for various inelastic processes, such as excitation, ionization, and Ps formation, has attracted attention recently [1.52]. In theory the experimental cross sections for positron scattering have been applied for the testing of the approximations used previously in the calculations of electron cross sections. Slow positrons have also been used to study positron attachment to various organic molecules, and the positive ions occurring as a result of positron annihilation on the molecules [1.7,15].

A small, but important, field of positron annihilation is the test of QED by measurements of the physics of Ps [1.25,15,7]. In particular, the annihilation rates and the low energy atomic physics of Ps have been studied [1.25]. The conclusion seems to be that all the results obtained until now are in full agreement with the predictions of QED. Annihilation rates and binding energies of Ps negative ions have been calculated and experimentally studied, too [1.15,25].

Positron annihilation plays also a small role in astrophysics. A 511 keV annihilation line source in the galactic center region has attracted attention recently [1.15,55].

Positron Emission Tomography (PET) appears to be the best high resolution technique for the determination of the distribution of radioactive tracer compounds in nuclear medicine. Apparently, PET has, by far, the lowest detection limit of any of the imaging techniques used in medicine [1.56]. Especially, the distribution of compounds, marked by use of short-lived positron emitters (e.g., ^{11}C, ^{13}N, ^{18}F) in the human brain has been studied. Maybe PET will play an important role in the measurements of the distribution of compounds in other samples, e.g., lubricants in machines and reactants in chemical reactors, in the future. At present, PET is a fairly expensive imaging technique, though.

1.4 The Content and Intention of this Book

This book is intended to be a critical review of chemistry oriented positron annihilation research. Several topics in the borderland between chemistry and neighbour disciplines, which are well discussed in other publications, have been excluded from the discussions, with the exception of certain remarks treating phenomena of fundamental importance.

Some of the fundamental theoretical problems are briefly discussed in Chap. 2, and the experimental techniques are described in Chap. 3. Both topics are well treated in the available literature. The purpose of the present discussion is, therefore, only to give a guide to the fields for newcomers.

A very important part of the book concerns the reactions of light particles with solutes in liquids and solids, and with defects in solids. In principle, the following topics can be discussed for each reaction:

A) The states of the light particles before the reactions.
B) The formation of these states.
C) The properties of the particle states, solutes, and defects making a reaction possible.
D) The rate of the reactions.
E) The products of the reactions.
F) The possibility of further reactions of the products.

In the following brief review of the contents of the Chaps. 4–10 the points A to F are indicated in parentheses.

In Chap. 4 the states of light particles, the excess electron, positron, and Ps, in insulating matter (A) are described in some details. A discussion of the basic processes of Ps formation in pure liquids is given in Chap. 5. Hence, the formation of the positron and excess electron states (B) in the positron spur is described, and the spur reaction of the positron and an excess electron (C,D,E,F) into Ps and the roughly simultaneous formation of the Ps bubble state (B) are discussed. Furthermore, the measurements of the Ps yields and typical yields for various types of liquids (E) are described. Chapter 6 treats the reactions of the positron in liquids (C,D,E,F). A brief review of selected topics of reaction kinetics (D) is included.

A "classical" field of positron annihilation in chemistry is the investigation of the influence of added solutes on the formation of Ps in liquids. This is discussed in terms of the spur model in one of the larger chapters of the book, Chap. 7. Hence, the changes in the measured Ps yields caused by the spur reactions of the solutes with the excess electrons and/or the positron (C,D,E,F) are treated. The older models of Ps formation are mainly treated in Sect. 7.7. Another great, "classical" topic is the study of Ps reactions with solutes in liquids (C,D,E). This is discussed in Chap. 8, which is also a large chapter. In both chapters the fundamental problems are treated in the first sections, while a fairly detailed survey of the measured results for various solute/solvent combinations is given in the following sections.

Chapter 9 is a discussion of positron annihilation in selected molecular solids (C,D,E). The criteria for the selection are mainly that the basic principles of positron annihilation in insulating solids are well illustrated by the described results. Important topics are the selftrapping of Ps (B), and the reaction of Ps with defects (C,D,E), in solids. The chapter closes with a description of the results of certain slow positron beam measurements of positron annihilation in the surface regions of solids.

Chapter 10 deals with positron annihilation in polymers and miscellaneous molecular compounds of complex structures. Annihilation in polymers and liquid crystals is treated in Sect. 10.2 and Sect. 10.4, respectively, while zeolites, small grain oxide powders, and samples cut out of living systems are discussed in Sect. 10.5. Furthermore, several special effects, such as transitions from glassy to normal liquid state (Sect. 10.1) and the dependence of the Ps yields on electric fields (Sect. 10.3), are discussed.

Conclusions and future developments are discussed in Chap. 11.

The book is intended to be a comprehensive treatment of the fundamental principles of positron annihilation in chemistry. In particular, the reasons why the measured results are obtained will be discussed in details. Hence, the experimental results used in the discussions have been strongly selected in order to illustrate the fundamental questions: What are the properties of the light particles in the samples, which particle reactions take place, and how can the measured results be explained in terms of these properties and reactions?

A complete review of all the published results is not given in the book. Certainly, the amount of data and the available space make this task impossible. Furthermore, a discussion of the interpretations, etc. in published papers, which are incorrect to the best of my knowledge, is normally not included, either. Such discussion would necessarily occupy much of the available space and make the book very difficult to read and understand. Only those incorrect interpretations which seem to have been used extensively in the literature have been critically commented.

The 9ICPA (August, 1991) took place when the book was almost finished. The data presented at the 9ICPA are not discussed in the book (see the proceedings). The main reason is that no new results which would imply that the main interpretations given in the book had to be changed, were advanced. A complete inclusion of the 9ICPA data in the discussions would have delayed the book significantly.

2 Theory

To provide the basis for the detailed discussion of the experimental results in the following chapters we shall in this chapter discuss briefly the theories of positron annihilation; Ps atomic physics; and the positron and Ps states in condensed matter. Of course, detailed calculations of the measurable quantities are normally impossible or very difficult indeed to perform, except in the simplest cases. As is the case in most of condensed matter science the use of judicious approximations is essential! However, an understanding of the fundamental theoretical principles, and the fairly qualitative results which can usually be derived, must be achieved before the positron annihilation results can be interpreted, in general.

2.1 The Annihilation Probability

The probability of positron-electron annihilation is calculated in quantum electrodynamics (QED), as briefly discussed in Chap. 1. In QED the dynamical equation can be written

$$i\hbar(\partial\psi/\partial t) = (H_{\mathrm{par}} + H_{\mathrm{ph}} + H_{\mathrm{int}})\psi, \tag{2.1}$$

similar to the Schroedinger equation of nonrelativistic quantum mechanics. H_{par} is the Dirac-Hamiltonian of the particle system consisting generally of many electrons and positrons. Normally, H_{par} is formulated in the occupation number representation (second quantization, field quantization). H_{ph} is the Hamiltonian of the photon system obtained by field quantization (second quantization, occupation number representation) of the classical Maxwell equations for a system without charged particles. H_{int} is the Hamiltonian of the interaction between the particles and photons. Equation (2.1) is mainly solved by use of time-dependent perturbation theory as H_{int} can be considered to be a "small" operator. The resulting probabilities are given as power series in the fine structure constant, $\alpha = e^2/4\pi\varepsilon_0\hbar c = 1/137$, where e, $\hbar$, and c are the electron charge, Planck's constant divided by 2π, and velocity of light, respectively. Normally, the first nonvanishing term in the series is finite, while the higher orders in α give infinite values, from which small corrections of the first term can be derived by use of the so called renormalization procedure. For example, the annihilation probability into a certain number of photons, described by a final photon wave function, of a system of electrons and positrons, described by an initial particle wave function, can be calculated. The Dirac equation and QED are briefly discussed in most textbooks on intermediate quantum mechanics and discussed in details in monographs on QED (e.g. [1.20]), to which we refer.

At first, let us discuss 2γ-annihilation, which is, by far, the most important annihilation process in condensed matter, to illustrate important aspects of the derivation of annihilation probabilities in QED. This will be followed by a general discussion of positron-electron annihilation.

By use of the nonrelativistic approximation in QED the following probability per second for the annihilation into two photons, of which one is emitted into the space angle $d\Omega$, and for which the sum of the wave vectors $\mathbf{k} = \mathbf{k}_1 + \mathbf{k}_2$ is within $d\mathbf{k}$, has been derived [2.1]

$$dw = (r_0^2/(2\pi)^3)d\mathbf{k}d\Omega \sum_{j=1}^{n} \sum_{\sigma_1} \int d\mathbf{r}_1 \ldots$$
$$\times \sum_{\sigma_{j-1}} \int d\mathbf{r}_{j-1} \sum_{\sigma_{j+1}} \int d\mathbf{r}_{j+1} \ldots \sum_{\sigma_n} \int d\mathbf{r}_n$$
$$\times \left| \sum_{\sigma_j} \sum_{\sigma_p} \int d\mathbf{r}_j \varepsilon_{\sigma_p \sigma_j} e^{-i\mathbf{k}\cdot\mathbf{r}_j} \psi(\mathbf{r}_1\sigma_1, \ldots, \mathbf{r}_j\sigma_j, \ldots, \mathbf{r}_n\sigma_n, \mathbf{r}_j\sigma_p) \right|^2, \quad (2.2)$$

where $\psi(\mathbf{r}_1\sigma_1, \ldots, \mathbf{r}_n\sigma_n, \mathbf{r}_p\sigma_p)$ is the initial nonrelativistic Schroedinger wave function of a system of n electrons (variables $\mathbf{x}_j = \mathbf{r}_j\sigma_j$) and one positron ($\mathbf{x}_p = \mathbf{r}_p\sigma_p$). We use units where $\hbar = c = 1$. $r_0 = \alpha/m$, where m is the electron mass. $\varepsilon_{11} = \varepsilon_{22} = 0$ and $\varepsilon_{12} = -\varepsilon_{21} = 1$. This is the first nonvanishing term in the perturbation series. By use of the antisymmetry in the electron coordinates (2.2) can be reduced somewhat. However, in its present symmetric form it can also be used if the antisymmetry is disregarded. Clearly, dw depends only on the initial wave function, and it is independent of the space angle Ω. Furthermore, dw is determined by a) the density of electrons at the positron (as $\mathbf{r}_p = \mathbf{r}_j$ in (2.2)), b) the Fourier transform with respect to the coordinates $\mathbf{r}_j$ and $\mathbf{r}_p$, and c) the fact that the electron spin direction must be opposite to that of the positron (assured by $\varepsilon_{\sigma_p\sigma_j}$).

Apparently, there is no complete and thorough derivation of the 2γ-annihilation probability for a reasonably general many-electron positron system in the literature. My derivation in [2.1] is incomplete partly because some of the approximations used in the calculations were uncontrolled [2.2]. Part of the wording in [2.1] is unsatisfactory, too. However, [2.1] seems to give the most general derivation available, as discussed in [2.2]. Expression (2.2) seems to be equivalent to the expression used in many-body calculations of the positron lifetime in metals (e.g., [2.3]), except for the additional use of the adiabatic hypothesis on the Coulomb potential between the particles in the metals.

To a fairly good approximation we may assume that the state of a system consisting of Ps and $n-1$ electrons is described by the wave function

$$\psi = c_1 A\{\psi_p(\mathbf{x}_1, \mathbf{x}_p)\psi_1(\mathbf{x}_2, \ldots, \mathbf{x}_n)\}, \tag{2.3}$$

where c_1 is a normalization constant ($c_1 = n^{1/2}$, if the Ps-electron does not overlap the other electrons), and A is the antisymmetry operator. $\psi_p(\mathbf{x}_1, \mathbf{x}_p)$ is the Ps wave function and $\psi_1(\mathbf{x}_2, \ldots, \mathbf{x}_n)$ is the antisymmetric wave function for the $n-1$ electrons. For this wave function (2.2) gives the following 2γ-annihilation probability, defined as for (2.2)

$$dw = (r_0^2/(2\pi)^3)d\mathbf{k}d\Omega(|c_1|^2/n)\left[\left|\sum_{\sigma_1}\int d\mathbf{r}_1 \sum_{\sigma_p}\varepsilon_{\sigma_p\sigma_1}\right.\right.$$
$$\left.\times\, \mathrm{e}^{-\mathrm{i}\mathbf{k}\cdot\mathbf{r}_1}\psi_p(\mathbf{r}_1\sigma_1, \mathbf{r}_1\sigma_p)\right|^2 + (n-1)\sum_{\sigma_1}\int d\mathbf{r}_1 \ldots \sum_{\sigma_{n-1}}\int d\mathbf{r}_{n-1}$$
$$\left.\times\left|\sum_{\sigma_n}\int d\mathbf{r}_n \sum_{\sigma_p}\varepsilon_{\sigma_p\sigma_n}\mathrm{e}^{-\mathrm{i}\mathbf{k}\cdot\mathbf{r}_n}\psi_p(\mathbf{r}_1\sigma_1, \mathbf{r}_n\sigma_p)\psi_1(\mathbf{r}_2\sigma_2, \ldots, \mathbf{r}_n\sigma_n)\right|^2\right] \tag{2.4}$$

The next approximation is to assume that

$$\psi_p(\mathbf{x}_1, \mathbf{x}_p) = \psi_2(\mathbf{r}_1, \mathbf{r}_p)\psi_s(\sigma_1, \sigma_p), \tag{2.5}$$

where $\psi_s(\sigma_1, \sigma_p)$ is either the p Ps or one of the three o Ps spin wave functions (Sect. 2.2). Similarly, we assume that $\psi_1(\mathbf{x}_2, \ldots, \mathbf{x}_n) = \psi_3(\mathbf{r}_2, \ldots, \mathbf{r}_n)\psi_4(\sigma_2, \ldots, \sigma_n)$, and that the state of the $n-1$ electron spins could be described by a uniform ensemble (i.e., the equal probability and random phase assumption). This gives the "spin-averaged" 2γ-annihilation probability defined as for (2.2)

$$dw = (r_0^2/(2\pi)^3)d\mathbf{k}d\Omega(|c_1|^2/n)\left[\left\{2\left|\int d\mathbf{r}_1\mathrm{e}^{-\mathrm{i}\mathbf{k}\cdot\mathbf{r}_1}\psi_2(\mathbf{r}_1, \mathbf{r}_1)\right|^2\right\}_{\text{para}}\right.$$
$$+ ((n-1)/2)\int d\mathbf{r}_1 \ldots \int d\mathbf{r}_{n-1}\left|\int d\mathbf{r}_n\mathrm{e}^{-\mathrm{i}\mathbf{k}\cdot\mathbf{r}_n}\right.$$
$$\left.\left.\times\, \psi_2(\mathbf{r}_1, \mathbf{r}_n)\psi_3(\mathbf{r}_2, \ldots, \mathbf{r}_n)\right|^2\right] \tag{2.6}$$

where the subscript para indicates that the term only differs from zero if $\psi_s(\sigma_1, \sigma_p)$ is equal to the p Ps spin function.

The total annihilation rate w, (i.e., the reciprocal lifetime) is obtained by integration over $\mathbf{k}$ and Ω. Hence,

$$w = 2\pi r_0^2(|c_1|^2/n)\left[\left\{2\int d\mathbf{r}_1|\psi_2(\mathbf{r}_1, \mathbf{r}_1)|^2\right\}_{\text{para}}\right.$$
$$\left.+((n-1)/2)\int d\mathbf{r}_1 \ldots \int d\mathbf{r}_n|\psi_2(\mathbf{r}_1, \mathbf{r}_n)|^2|\psi_3(\mathbf{r}_2, \ldots, \mathbf{r}_n)|^2\right] \tag{2.7}$$

The two terms in (2.4,6,7) are the intrinsic and pick off terms, respectively. Apart from these terms two other terms, the intrinsic-pick-off-exchange and pick-off-exchange terms, are obtained [2.4,2]. Use of the wave function (2.3) without the antisymmetry operator would only give the first two terms. Correspondingly, the two exchange terms are zero if we use the approximation that the Ps-electron does not overlap the other electrons. The exchange terms seem only to have been published in [2.2,4] and, apparently, they have not been used and their values have not been estimated. Consequently, they are not given here. As the normally found repulsion between Ps and most molecules is mainly caused by the repulsion between the Ps electron and the electrons of the molecules, we expect that the Ps-electron overlap with the other electrons is smaller than that of the positron. Hence, we expect that the exchange terms are much smaller than the pick off term, and that they are more important in cases where Ps is strongly "squeezed" (as in ionic crystals and ice).

Of course, (2.3) is only an approximation of the wave function in condensed matter, and the more the two particles in Ps overlap the other electrons, the worse is the approximation. On the other hand, (2.3) seems to be the best of possible approximations in which Ps can be associated with a wave function of its own. In a higher approximation the total wave function would not be separable as given in (2.3).

The "spin averaged" probabilities (2.6,7) are calculated under the assumption that the spins of the Ps electron and positron are not correlated with the spins of the other electrons. This is not necessarily a good approximation. In particular, the density at the positron of electrons with the spin parallel to the Ps-electron spin is expected to be lower than that of electrons with the opposite spin direction because of the Pauli-principle. This will reduce the pick off 2γ-annihilation rate of p Ps compared to that of o Ps, an effect which seems not to have been estimated.

Furthermore, the spins of the positrons in β^+ decay are oriented for most part parallel to the positron momentum vector. The degree of spin polarization is given by v/c the ratio of the positron velocity to the velocity of light. This will influence the annihilation probability if a direction in space is characterized in the experiment, e.g., if the electron spins are polarized, too (ferromagnetic materials) or in the presence of a magnetic field. This effect, discussed briefly in Sect. 2.2, is, of course, disregarded if spin-averaged probabilities are used. The conditions in most experiments discussed in this book make the effects of spin polarization of the positron very small or insignificant.

If we assume that $\psi_p(\mathbf{x}_1, \mathbf{x}_p) = \psi_1(\mathbf{x}_1)\psi_p(\mathbf{x}_p)$ in (2.4) and average over the spin directions we get

$$dw = (r_0^2/2(2\pi)^3)d\mathbf{k}d\Omega \left| \int d\mathbf{r}\phi_-(\mathbf{r})\phi_+(\mathbf{r}) \exp(-i\mathbf{k}\cdot\mathbf{r}) \right|^2 \tag{2.8}$$

This is the expression used normally to calculated the angular correlation curve for solids in the independent particle approximations.

Let us assume that the Ps wave function in (2.5) is

$$\psi_2(\mathbf{r}_1, \mathbf{r}_p) = \psi_c(\mathbf{r}) f(\mathbf{R}) \tag{2.9}$$

i.e., a product of a function $\psi_c(\mathbf{r})$ of the center-of-mass coordinate $\mathbf{r} = (\mathbf{r}_1 + \mathbf{r}_p)/2$ and a function $f(\mathbf{R})$ of the relative coordinate $\mathbf{R} = \mathbf{r}_1 - \mathbf{r}_p$ (e.g., an s-state wave function). Hence, the intrinsic p Ps term in (2.6) gives

$$dw \propto \left| \int d\mathbf{r} \psi_c(\mathbf{r}) \exp(-i\mathbf{k} \cdot \mathbf{r}) \right|^2 d\mathbf{k} d\Omega \tag{2.10}$$

This expression is used in Sect. 4.4 on the Ps states in liquids and in Sect. 9.1 on Ps in Bloch function states in solids.

On assuming that we have p Ps in vacuum we get from (2.7) the first approximation of the p Ps lifetime $\tau = 1/w = 125$ ps. Furthermore, a system consisting of one free positron in a sea of free electrons of average density n gives an annihilation lifetime of $\tau = 1/\pi c n r_0^2$ by use of (2.8). Experiments on metals give a much shorter lifetime than that given by the average density of valence electrons, because the valence electron density at the positron is much higher than the average valence electron density.

The above given discussion illustrates how QED calculations of annihilation probabilities are performed, in general. At first one calculates the probability of transition from a certain initial state to a certain final state of the particles and photons. As initial state we selected a wave function for n electrons, one positron, and no photons, using our knowledge about the actual physical systems we wished to study. The final state was selected to give a probability of one of the processes detected by our apparatus. We chose a final state of $n - 1$ electrons and two photons with given momenta $\mathbf{k}_1$, $\mathbf{k}_2$ and polarization vectors $\mathbf{e}_1$, $\mathbf{e}_2$. We then summed all the probabilities for transitions to any final state of the $n - 1$ electrons. Furthermore, we summed the probabilities for the possible photon polarizations, which cannot normally be detected by our apparatus. At last we transformed the obtained probability for the detection of the variables $\mathbf{k}_1$ and $\mathbf{k}_2$ to that for the detection of $\mathbf{k}(\mathbf{k} = \mathbf{k}_1 + \mathbf{k}_2)$ and $\mathbf{k}_1$ and used the deltafunction, which ensures energy conservation, to integrate over $|\mathbf{k}_1|$ using polar coordinates for $\mathbf{k}_1$ (i.e. $d\mathbf{k}_1 = |\mathbf{k}_1|^2 d\Omega d|\mathbf{k}_1|$). This explains the particular probability given by (2.2).

Expressions (2.3–10) illustrate how (2.2) can be used to calculate 2γ-annihilation probabilities for specific initial wave functions, how to use statistical assumptions (spin averaging), and how to sum all the transition probabilities to get the total decay rate (2.7). The calculations of other annihilation processes are fairly similar to that of 2γ-annihilation.

After thus having reviewed the most important 2γ-annihilation probabilities in condensed matter in some detail, we have to limit the treatment of the other annihilation processes to a brief discussion of the main results.

The 2γ-annihilation cross sections, i.e., the annihilation probabilities per second divided by the flux of the incoming particle, for a relativistic, free electron and a relativistic, free positron in the center-of-mass and the electron-at-rest coordinate systems are given in [1.20], Chap. 33. Cross sections for the emission of polarized photons and for incoming polarized particles are discussed, too. Similarly, 3γ-annihilation and 1γ-annihilation in the presence of a third body are also discussed in [1.20], Chap. 33. Other special annihilation cross sections (probabilities) can be found in the monographs and journals.

From the law of conservation of charge parity in QED (e.g., [1.20], Sect. 23.3) we know that an arbitrary neutral system, if it has a definite charge parity, can decay either only into an even or only into an odd number of photons, as a system consisting of N photons has the charge parity $(-1)^N$. Positronium in vacuum is neutral, and in the absence of a magnetic field the eigenstates of the energy can be chosen to have a relative angular momentum of the electron and positron, l, and the total spin, $s(s = 0, 1)$. The charge parity is then definite, namely $(-1)^{\ell+s}$. Hence, for example, the ^1S $(l = 0, s = 0)$ state can only decay into an even $(N = 2, 4, \ldots)$, while the $^3S(l = 0, s = 1)$ state can only decay into an odd $(N = 3, 5, \ldots)$, number of photons. These states are two out of the four unexcited Ps states found experimentally. Why this is the case will be discussed in Sect. 2.2.

A high-energy positron is normally thermalized before it annihilates. Theoretically, this can, in principle, be studied by a comparison of the annihilation probability with that of slowing down of the positron at all the energies above thermal. At energies above about 100 eV it is fairly easy to show by use of the Bethe formula [2.5] that annihilation during slowing down is insignificant. However, the slowing down processes of the positron (and electron) in condensed matter below about 100 eV are not well understood, in particular below 10 eV. On the other hand, experiments show that the positron is thermalized before significant annihilation occurs in condensed matter. In some monoatomic gases annihilation during slowing at very low energies ($\lesssim 1$ eV) is important. Furthermore, annihilation of nonthermalized p Ps ($\simeq 0.5$ eV) occurs in a xenon gas in silicagel (Sect. 10.5).

Consequently, it is only necessary to use the 2γ-annihilation probabilities derived by use of the nonrelativistic approximation in the interpretation of experimental results in this book, as in almost all of low-energy positron annihilation research. The 1γ-annihilation rate in the presence of a third body is also insignificant mainly because of the fast slowing down of the positron. Hence, it will not be discussed further.

The first non-vanishing term in the perturbation series of the 3γ-annihilation probability is of third order in α. Hence, it is roughly two orders of magnitudes smaller than that of 2γ-annihilation. Clearly, only the nonrelativistic limit is important for 3γ-annihilation, too. The terms of order n in α for $n > 3$, i.e., the so called radiation corrections [1.20,25], are insignificant for all annihilation processes except in the interpretation of very special experiments, for example, in certain tests of QED [1.25].

The total 3γ-annihilation probability in the nonrelativistic approximation averaged over particle spins and summed over photon polarization is [2.6,1.3,20]:

$$w_{3\gamma} = (4/9\pi)(\pi^2 - 9)\alpha w_{2\gamma} = (1/371.3)w_{2\gamma}, \tag{2.11}$$

where $w_{2\gamma}$ is the similar 2γ-annihilation probability. Both are calculated by use of the first non-vanishing term in α, i.e., without radiation corrections [1.20,25]. This result has been confirmed by experiments. The probability of the emission of a γ-photon of energy E under the same conditions as for (2.11), is essentially linear in E [2.6,1.3,20]. The particle spin and photon polarization effects in the 3γ-annihilation probability are discussed below and in [2.7,8].

Further theoretical discussion of positron annihilation can be found in the literature, see, e.g., [1.20,25,2.6–8] and refs. therein. For example, it is, in principle, possible to stimulate the emission of a certain annihilation photon by use of a high-density of that photon [1.20]. However, 2γ-annihilation photons have a very high energy for which efficient γ-mirrors cannot be made, and hence, a high photon density cannot be obtained, at present. Furthermore, in 3γ-annihilation the probability of photon emission at low energies, where lasers can be used to make the high density of photons, is very low indeed. Stimulated annihilation may be of importance in astronomy [2.9] (e.g., for black holes and neutron stars) and in the future.

2.2 Atomic Physics of Positronium (Ps)

The annihilation of a positron-electron pair in a Ps state depends on the properties of the state. Hence, a brief discussion of the Ps atomic physics is appropriate here. Only the Ps states of importance for the understanding of annihilation in condensed matter and high density gases will be treated.

In the simplest nonrelativistic approximation the Schroedinger equation for Ps is that of the hydrogen atom in which the reduced mass M_{H} is replaced by $M_{\mathrm{P}} = m/2$ (see, e.g., [1.20,3]). Consequently, following (2.5) and (2.9) the eigenfunctions can be determined from

$$-(\hbar^2/4m)\nabla^2\psi_{\mathrm{c}}(\mathbf{r}) = E_r\psi_{\mathrm{c}}(\mathbf{r}), \tag{2.12}$$

$$-(\hbar^2/2M_{\mathrm{P}})\nabla^2 f(\mathbf{R}) - (e^2/4\pi\varepsilon_0|\mathbf{R}|)f(\mathbf{R}) = E_R f(\mathbf{R}), \tag{2.13}$$

for

$$\psi_{\mathrm{p}}(\mathbf{x}_1, \mathbf{x}_{\mathrm{p}}) = \psi_{\mathrm{c}}(\mathbf{r})f(\mathbf{R})\psi_s(\sigma_1, \sigma_{\mathrm{p}}). \tag{2.14}$$

(2.12) is the eigenvalue equation for the center-of-mass of Ps moving as a free particle. (2.13) is the hydrogen eigenvalue problem for $M_{\mathrm{P}} = m/2$. Hence, the eigenvalues E_R are very close to half those of hydrogen, e.g., the ionization energy of Ps is 6.80 eV. $\psi_s(\sigma_1, \sigma_{\mathrm{p}})$ can be chosen as eigenvectors of S^2 and S_z, where $S = \sigma_1 + \sigma_{\mathrm{p}}$, the total spin. The four eigenvectors are

$$\chi_{11} = \alpha(\sigma_1)\alpha(\sigma_{\mathrm{p}}) \tag{2.15}$$

$$\chi_{1-1} = \beta(\sigma_1)\beta(\sigma_{\mathrm{p}}) \tag{2.16}$$

$$\chi_{10} = (2)^{-1/2}\{\alpha(\sigma_1)\beta(\sigma_{\mathrm{p}}) + \beta(\sigma_1)\alpha(\sigma_{\mathrm{p}})\} \tag{2.17}$$

$$\chi_{00} = (2)^{-1/2}\{\alpha(\sigma_1)\beta(\sigma_p) - \beta(\sigma_1)\alpha(\sigma_p)\} \tag{2.18}$$

in the usual notation. (2.15–17) are the triplet (o Ps) states and (2.18) is the singlet (p Ps) state.

In this nonrelativistic approximation we get from (2.4,7) the 2γ-annihilation probability of p Ps

$$w_{2\gamma} = 4\pi r_0^2 c|f(0)|^2 = 4\pi r_0^2 c/\pi a^3 = 0.803 \times 10^{10}\ \mathrm{s}^{-1}. \tag{2.19}$$

Here $a = 2a_0$, where a_0 is the Bohr radius. Furthermore, the 3γ-annihilation probability of o Ps is

$$w_{3\gamma} = (4/9\pi)(\pi^2 - 9)\alpha w_{2\gamma} = w_{2\gamma}/1113.9 = 7.211 \times 10^6\ \mathrm{s}^{-1}. \tag{2.20}$$

by use of (2.11), taking into account that (2.11) gives the spin-averaged annihilation rate. Hence, the lifetimes are

$$\tau_S = 124.49\ \text{ps and } \tau_T = 138.67\ \text{ns}. \tag{2.21}$$

The fine structure of the Ps energy levels can be calculated (see, e.g., [1.20]) by use of the eigenvalue equation

$$H\psi_p(\mathbf{x}_1, \mathbf{x}_p) = (H_0 + V_1 + V_2 + V_3 + V_4 + V_5)\psi_p(\mathbf{x}_1, \mathbf{x}_p). \tag{2.22}$$

Here H_0 is the sum of the Hamiltonians of (2.12) and (2.13), while V_i are the five terms describing the relativistic corrections of H to the order of c^{-2}. They are similar to the relativistic corrections to the hydrogen atom in the same order of c. However, in the Ps case two extra terms appear, namely the spin-spin term, V_4, which is very small in hydrogen, and an exchange interaction term, V_5, unique for Ps. The operators V_i contain only operators acting on the spin variables in the form of the total spin S. Hence, the Ps levels can still be divided into triplet and singlet states. Similarly, the presence of a constant electric field does not mix the singlet and triplet states in the given approximation.

The operators V_i are small compared to H_0. The first-order relativistic corrections of the Ps energy levels can, therefore, be calculated by use of perturbation theory; the V_i being the perturbations. The shifts in the energy levels are then expressed in terms of the matrix elements of V_i calculated by use of the eigenfunctions (2.14) [1.20].

The higher order corrections of the energy levels and annihilation rates of Ps are calculated by use of the theory of radiation corrections in QED [1.20,25].

The radiation corrections give small changes in the lifetimes and energies. Apparently, the best theoretical value of τ_S is $\tau_S = 123.91$ ps [1.25]. The small radiation correction of τ_S seems not to have been verified yet. The best theoretical value of τ_T is $\tau_T = 142.08$ ns, while the latest experimental value (1990) is $\tau_T = 141.88$ ns [2.10]. There seems to be good reasons to believe that the small difference between theory and experiment is mainly caused by yet uncalculated higher order terms in the perturbation series. Hence, as usual QED explains the measured results.

The energy difference between the ground states of o Ps and p Ps calculated from (2.22) by use of the first term in the perturbation series is [1.20]

$$\Delta = (7/12)\alpha^4 mc^2 = 8.453 \times 10^{-4}\ \text{eV}. \tag{2.23}$$

It can be shown that $\Delta \propto |f(0)|^2$ in that approximation. *Ritter* et al. [2.11] measured (1984) $\Delta = 8.41150 \times 10^{-4}$ eV, while the theoretical value (1984) is $\Delta = 8.41195 \times 10^{-4}$ eV. This difference between experiment and theory can be explained as caused by higher order radiation corrections not yet calculated. Clearly, the radiation corrections of the first order Δ-value (2.23) are small. The energies of the excited Ps states, not needed in this book, can be found in the literature [1.25].

The next step is to discuss the energies of the ground states of Ps in a magnetic field **B**. In the first order in **B** the magnetic part of the Hamiltonian is [1.20,25]

$$H_M = (g'/2)\mu_B|\mathbf{B}|\{\sigma_{1,z} - \sigma_{p,z}\}, \tag{2.24}$$

where g' is the g-factor ($g' \simeq 2$), and μ_B is the Bohr magneton. The z-axis is along **B**. The magnetic field mixes the χ_{10} and the χ_{00} states but does not influence the χ_{11} and χ_{1-1} states. Diagonalization of the Hamiltonian including (2.24) gives the following eigenstates, energies, and annihilation rates [1.20,25]

$$\psi'_T = (1+y^2)^{-1/2}\{\chi_{10} + y\chi_{00}\}, \tag{2.25}$$

$$\psi'_S = (1+y^2)^{-1/2}(\chi_{00} - y\chi_{10}), \tag{2.26}$$

$$E'_T = \Delta\{1 + (1+x^2)^{1/2}\}/2, \tag{2.27}$$

$$E'_S = \Delta\{1 - (1+x^2)^{1/2}\}/2, \tag{2.28}$$

$$\lambda'_T = (1+y^2)^{-1}\{w_{3j} + y^2 w_{2j}\}, \tag{2.29}$$

$$\lambda'_S = (1+y^2)^{-1}\{w_{2j} + y^2 w_{3j}), \tag{2.30}$$

for $x = 2g'\mu_B|\mathbf{B}|/\Delta \simeq |\mathbf{B}|/36.5$ ($|\mathbf{B}|$ in kG), and $y = x/\{1+(1+x^2)^{1/2}\}$. The subscripts T and S indicate the quantities which reduce to those associated with χ_{10} and χ_{00}, respectively, in the limit of **B** equal to zero. The zero of energy is taken to be E_S, the energy of the singlet state at $\mathbf{B} = \mathbf{0}$, i.e., $E_T = \Delta$.

Expressions (2.29,30) follow from the fact that [1.20]

$$\lambda = |c_1|^2 w_{2j} + |c_2|^2 w_{3j} \tag{2.31}$$

for a spin wave function

$$\psi = c_1\chi_{00} + c_2\chi_{10} \tag{2.32}$$

($|c_1|^2 + |c_2|^2 = 1$). For example, $(\lambda'_T)^{-1} = 142$, 117, 32.5, 6.7, and 2.0 ns at $|\mathbf{B}| = 0$, 1, 4, 10, 20 kG, respectively. Hence, λ'_T is markedly reduced at normal magnetic field strengths.

Δ can be measured by applying a static magnetic field in the region where Ps is formed, and then applying a rf magnetic field $B_x \cos(2\pi f t)$ perpendicular to **B**. The 2γ-annihilation is increased at the resonance, where hf is equal to the energy difference between the $\psi'_T(m = 0)$ and $\psi_T(m = \pm 1)$ states. This determines x from (2.27) and hence Δ.

The state of two electrons and one positron, i.e., Ps^-, has been studied, too. *Ho* [2.12] has calculated the annihilation rate of Ps^- as 2.08613 ns^{-1}. The AC spectrum of Ps^- is narrow, as is that of p Ps. *Mills* [2.13] has measured the Ps^- decay rate $\lambda = 2.09(9)$ ns^{-1}.

It is important to note that the until now given expressions for the annihilation probabilities can be applied for the explanations of only the simplest, but also the normally encountered, experiments. Several effects which have been disregarded must be taken into account if more complex experimental conditions prevail.

First, in 3γ-annihilation of Ps in a magnetic field, **B**, the probability of the detection of the three photons depends on the angle, θ, with the direction of quantization, i.e., the direction of **B**, apart from the dependence on the energy, E, of the photons and on which state of Ps is annihilating [2.7]. This is also the case even if only one photon is detected (e.g., in LT experiments).

Second, the positrons emitted in β^+-decay (e.g., from ^{22}Na) are polarized along the direction of their momentum. The degree of polarization is given by v/c, the ratio of their velocity to that of light. The positrons are only partly depolarized during the slowing down in normal matter. Apparently (see below), the polarization influences the annihilation probabilities only if the annihilating positrons have a momentum distribution which is anisotropic with respect to a specially emphasized direction in the sample, e.g., if a magnetic field is present, or if the spins of the annihilating electrons are polarized, too (ferromagnetic materials).

If an external positron source is used (e.g., in most AC [2.14,15] and in special LT [1.25,2.16,17] studies) with the sample in a magnetic field polarization effects must be taken into account. These effects cause the magnetic field effects on the narrow components of the AC spectra to change in magnitude when the field is changed from **B** to $-$**B**. Furthermore, the relative intensities of the lifetime components (2.29,30) in the LT spectra change on reversal of the direction of **B** at intermediate field strengths. These effects have mainly been used to measure the degree of polarization of positrons from β^+-decay [2.14,16,17]. In [2.15] the polarization effect in AC spectra for single crystals of quartz in a magnetic field was used to assign unambiguously the narrow AC peaks to p Ps decay.

Third, in 3γ-annihilation with the sample in a magnetic field and if an external positron source is used, counting rate oscillations at a frequency of the triplet Ps level splitting can be observed [2.8,18]. The amplitude of this beat effect depends on the direction to the photon detectors.

In addition, I do not understand the way the annihilation probabilities of Ps are calculated. In the absence of a magnetic field the probabilities of decay for each of the wave functions (2.15–18) are calculated separately. For example, it is assumed that either o Ps or p Ps are formed, and hence, no linear combinations are considered. According to the interpretation rule of quantum mechanics this is a correct procedure if the states of Ps were measured just before the annihilation. However, this is never done. Consequently, the nature of the Ps states is always unknown, in principle, and hence, statistical quantum mechanics should be used for the correct choice of the initial Ps states by use of the knowledge about the way Ps is formed and its interaction with surrounding matter if present. In the presence of a magnetic field the energy eigen functions (2.15,16,25,26) are used similarly in simple calculations [1.20,25]. If the positron polarization is taken into account in the calculations the following procedure is used [1.25,2.16,17]. The wave functions for the assumed initial Ps states (e.g., for given positron and electron spin directions) are expressed in terms of (2.15,16,25,26), which are chosen to decay exponentially by the rates w_{3j}, w_{3j}, λ'_{T}, and λ'_{S}, respectively. The sum of the squared wave functions is then differentiated with respect to time to give the theoretical LT-spectrum. The AC-spectrum is obtained in a similar way [2.14].

Clearly, the theoretical results agree with experiments within the experimental uncertainties, which are very different from one experiment to another, though. In particular, the very good agreement between the theoretical and experimental w_{3j}-values for o Ps in vacuum [2.10] implies that the o Ps wave functions contain a very small, if any, p Ps component, i.e., the constants c_1 in (2.32) are very small indeed. Hence, my problem of understanding is not whether the theoretical calculations are in agreement with nature; they are! I do not understand where in quantum mechanics, etc. it is stated that the used calculation procedures are the only correct ones. I have discussed this problem with several wellknown scientists, but have not, as yet, obtained satisfactory (in my opinion) solutions.

2.3 Application of the Theory

We have until now discussed the theory of positron annihilation and positron atomic physics for vacuum. If the annihilation takes place in liquids, solids, and high-density gases the main problem in the use of the theory discussed in Sects. 2.1,2, is the calculation of the positron many-electron wave function. As mentioned above, it is normally impossible or very difficult to perform detailed calculations even for fairly simple systems. However, the theory can be used in semi-quantitative calculations in which rather crude approximations are used. Chaps. 4–10 give many examples of such applications. We shall here only discuss selected, general examples.

The positron-electron annihilation probability for a typical condensed matter system can be calculated in two steps. First, the positron many-electron wave function(s) are derived by use of normal nonrelativistic quantum mechanics. Second, the derived wave function(s) are inserted into one of the annihilation expressions given above. As usual in QED, the particle-photon interaction influences the particle wave function very slightly, and it can be disregarded in the first step.

Let us first discuss shortly the annihilation of "free" positrons, i.e., positrons not forming Ps but normally solvated in liquids and/or partly bound to the molecules, ions, etc. of the system. The positron is repelled by the positive core of the atoms, and hence, only the valence and some of the outer core electrons annihilate with the positrons. This will be one of the results of the calculation. We discussed positron annihilation in metals, and positron trapping in defects in metals, semiconductors, and ionic crystals, in Sect. 1.3. In general, the understanding of the positron many-electron states in metals and semiconductors is much better than in typical chemical systems, such as liquids, solutions, and insulating molecular solids. Actually, the states of the positron, both the states formed directly at thermalization and those formed as a result of Ps chemical reactions, have been studied very little both theoretically and experimentally. Exceptions are positron bound states at some halides in water, discussed in Chap. 6. A restricted Hartree-Fock calculation of the AC spectrum for Cl^-, Br^-, and I^- agreed well with experiments. However, the derived lifetimes were much too long, because the enhancement of the electron density at the positron could not be well described without "correlation" included in the calculation. The agreements between the theoretical AC spectra for bound states of the positron at polyatomic, negative ions (e.g., pseudohalides) and experiment were much poorer. In general, realistic calculations of AC and/or LT results for positrons in liquids and/or bound to molecules seem not to have been published. However, theoreticians seem to have treated fairly complex positron-electron systems recently, and hence, the simpler, theoretical problems of positrons in insulating, polyatomic matter might be solved in the near future. For calculations of properties of "free" positrons in (mainly monoatomic) high-density gases see, e.g., [1.13–15].

In summary, the best way to get an overview of the present and future theoretical possibilities of explaining "free" positron annihilation in condensed matter is to consult the papers on calculations done on (intrinsic or defective) metals and semiconductors, high density gases, and positron-ion bound systems.

The calculation of Ps annihilation in condensed matter is even more difficult than that of the "free" positron annihilation. In principle, one way of calculating the wave function would be to calculate ψ_1 in (2.3) taking into account the presence of Ps (including its influence on the atomic positions, e.g., bubble formation in liquids). ψ_p in (2.3) might then be calculated under the assumption that the rest of the system constitutes an external potential (including approximate electron exchange terms, etc.) for the particles in Ps. This external potential must be derived from the atomic positions and ψ_1. The calculation may, of course be done in a self-consistent way. However, such reasonably advanced calculation of the Ps state has not been done at present. It seems to be a very difficult task. On the other hand, many simple approximation schemes are possible.

The most important difference between Ps in condensed matter and in vacuum is the pick off annihilation. Hence, p Ps decays by the two processes given in (2.4,6,7), while o Ps decay is described by the sum of the pick off term in (2.4,6,7) and the 3γ-annihilation terms (2.11,20). In the presence of a magnetic field the pick off terms calculated by use of ψ'_{T} (2.25) and ψ'_{S} (2.26) are added to λ'_{T} (2.29) and λ'_{S} (2.30) respectively. A typical Ps pick off lifetime of o Ps is 2–10 ns, while the "free" positron lifetime is 0.3–0.45 ns.

As mentioned in Sect. 2.1 the pick off 2γ-annihilation rate of p Ps is expected to be lower than that of o Ps. The o Ps ($m = 0$) state becomes increasingly p Ps like as the magnetic field is increased. Disregarding a possible decrease in the pick off rate of o Ps ($m = 0$) for increasing magnetic field causes that the magnetic field effect is estimated to be too large. The difference in pick off rates has not been taken into account in interpretations, as yet. Of course, pick off 3γ-annihilation of Ps occurs, too, but with a much lower probability (see (2.11)). The Fermi (Pauli) hole around the Ps electron will cause this rate to be lower for o Ps than for p Ps, an effect which has not been measured yet.

As mentioned in Chap. 1, Ps is always in a bubble state in a liquid, if it is formed. This is discussed in details in Sect. 4.4. The Ps bubble state is a result of the repulsion between Ps and the molecules caused mainly by the electron in Ps being repelled by electrons in the molecules. The positron is only repelled by the positive core of the atoms, not by the outer valence electrons. The positron tends, therefore, to penetrate more into the molecules than the electrons, except maybe if the molecules are electron acceptors. Consequently, Ps is normally "swollen" in condensed matter, i.e., the density of its electron at the positron is lower than that of Ps in vacuum.

In a first approximation this can be described by

$$|f(0)|^2 = \alpha'(|f(0)|^2)_{\mathrm{v}}, \tag{2.33}$$

where α' is a constant (e.g., $\alpha' \simeq 0.8$ for linear hydrocarbons) As shown by (2.19), (2.20), and (2.23) w_{2j}, w_{3j}, and Δ are proportional to $|f(0)|^2$ in the simplest approximation. Hence, in condensed matter $w_{2j} = \alpha'(w_{2j})_{\mathrm{v}}$, $w_{3j} = \alpha'(w_{3j})_{\mathrm{v}}$, and $\Delta = \alpha'(\Delta)_{\mathrm{v}}$, where the subscript v refers to the vacuum values (see, e.g., [2.19]). In particular, Ps annihilation in a magnetic field is normally interpreted by use of α'. Hence, α' in w_{2j}, w_{3j}, and x in (2.29) and (2.30) is used as a fitting parameter to get agreement with experiments [2.19]. For very "squeezed" Ps low values of α' have been obtained (e.g., $\alpha' \simeq 0.13 - 0.25$ for naphthalene, see Fig. 9.7). α' describes an approximate correction to the wave function $f(\mathbf{R})$ for the relative coordinate $\mathbf{R}$.

In condensed matter the center-of-mass wave function $\psi_{\mathrm{c}}(\mathbf{r})$ may be approximated by the addition of a potential term on the left hand side of (2.12). Hence, it is assumed that Ps center-of-mass moves like a single particle in a potential caused by the surrounding molecules. The intrinsic p Ps annihilation is determined by the Fourier transform of $\psi_{\mathrm{c}}(\mathbf{r})$ (2.10). For example, delocalized Ps in single crystals is well described by (2.12) with an added potential having lattice translational symmetry. $\psi_{\mathrm{c}}(\mathbf{r})$ is then a Bloch function and hence, the resulting AC spectrum has narrow peaks at the reciprocal lattice points (Sect. 9.1, Figs. 3.4 and 9.1). In Sect. 4.4 localized Ps is discussed in terms of square-well and harmonic potentials for the center-of-mass wave function $\psi_c(\mathbf{r})$. Such models may be used to get reasonable approximations for the zero-point energy, and the AC spectrum due to p Ps intrinsic annihilation, of localized Ps. The pick off lifetime of localized o Ps is more difficult to approximate. Several ways of doing this are discussed in Sect. 4.4 (e.g., Fig. 4.1).

Linear combinations of localized and delocalized Ps have been treated in [2.2]. Effects of lattice vibrations on delocalized Ps are discussed in [2.2], too. Several other methods of calculating Ps properties can be found in [1.5] (solids) and [1.13–15] (dense gases). The effects of chemical reactions on the measured Ps properties are discussed in details in Chap. 8. Transitions from Ps states to non-Ps states are described in Sects. 5.5 (CS_2), 8.3, and 9.3 (aromatic crystals). In Sect. 5.5 the "fourth positron state" is discussed.

3 Experimental Techniques

The basic experimental technique, used in the studies of positron annihilation in chemical physics, is the detection of γ-radiation. The main exceptions are the detections of positrons and Ps in the slow positron beam investigations. Gamma-spectroscopy is well treated in several books, e.g., in the monograph [2.5] and the textbook by *Knoll* [3.1]. We shall refer to these books, in particular [3.1], in the discussions of this chapter. Only the techniques of general importance for the chemistry-oriented part of positron annihilation will be treated in some details. The less used techniques will only be briefly mentioned.

The three most important techniques in positron and Ps chemistry are the lifetime (LT), the angular correlation (AC), and the Doppler broadening (DB) techniques. The DB spectrum is essentially identical to the AC spectrum, except it is obtained with a poorer resolution. It is very important to realize that both methods, LT and AC (DB), must be used in the investigation of most chemical problems. For example, Ps inhibition cannot be separated from Ps quenching by use of only AC (DB) measurements. If Ps quenching occurs the spin conversion reaction cannot be well separated from the other Ps reactions by use of only the LT technique. Furthermore, only the AC (DB) method has given information on the solvated positron reactions until now, and AC is the only technique to use in practice if delocalized Ps in single crystals is studied.

Several reviews of the three techniques are available. Recently, a fairly elementary discussion was given in [3.2], while [3.3,4] describe particular experimental problems in detail. An "open discussion" took place at the 8ICPA [3.5]. Original papers are mainly published in Nucl. Instr. Meth. and in the proceedings of the positron annihilation meetings [1.7–18]. Furthermore, valuable information can be found in the catalogues and instrument manuals published by the manufacturers of devices (e.g., [3.6]).

3.1 Positron Lifetime (LT) Technique

The LT measurements are probably the most important ones used in the chemistry-oriented field of positron annihilation. The time resolution of a LT system limits the information obtainable in many cases. Hence, the LT technique is fairly difficult, and consequently, it is not an easy task to discuss the LT systems here. At first we shall give a fairly elementary description of a typical LT apparatus. Some special problems will then be discussed, and the section is finished by a treatise of the computer analyses of LT spectra.

3.1.1 Elementary Description of a LT Spectrometer

The principles of a modern LT spectrometer are shown in Fig. 3.1, and Fig. 3.2 shows several measured LT spectra for CCl_4/benzene solutions [3.7]. They show that CCl_4 inhibits the o Ps yield in benzene.

Normally, the β^+-emitter ^{22}Na is used as positron source. ^{22}Na emits a 1.28 MeV photon almost simultaneously with emission of the positron. One of the detectors is adjusted to detect this photon. The other detector detects one of the annihilation photons (0.511 MeV). The time interval between the detections of the 1.28 MeV and the 0.511 MeV photons is the positron lifetime. Both detectors are scintillation gamma-ray detectors [3.1], each consisting of a scintillation crystal (SC), a photomultiplier (PM), and a PM base unit. A high-voltage supply powers the PM tubes. The SC is a fast plastic (or BaF_2) crystal emitting light when a gamma ray is stopped or Compton scattered in it. This light is detected by a photo-cathode of the PM, and the PM amplifies the electrical signal from the cathode.

Both detectors deliver a fast voltage pulse (0.3–1 V) from the PM, signalling the detection of the gamma ray, to the input of a NIM module [3.1] (e.g., Ortec 583 [3.6]), consisting of a delay (D), a constant-fraction discriminator (CF), a single-channel analyzer (SCA), and a coincidence circuit (COIN). The CF determines the time of arrival of the voltage pulse from the PM. If

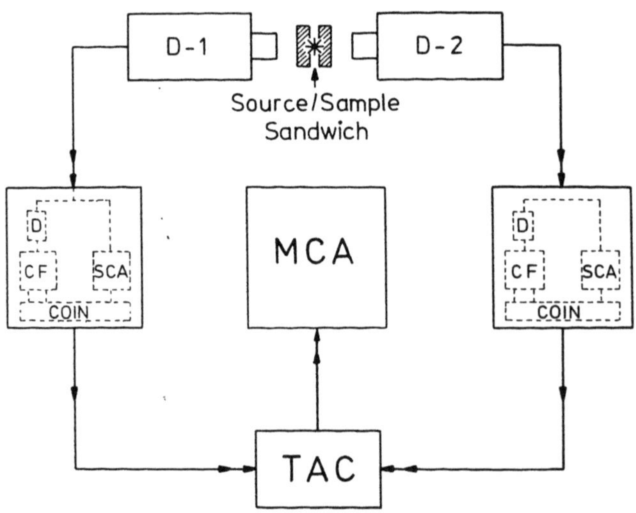

Fig. 3.1. A typical positron lifetime (LT) spectrometer using the fast-fast coincidence method. The position of the source/sample is discussed in the text

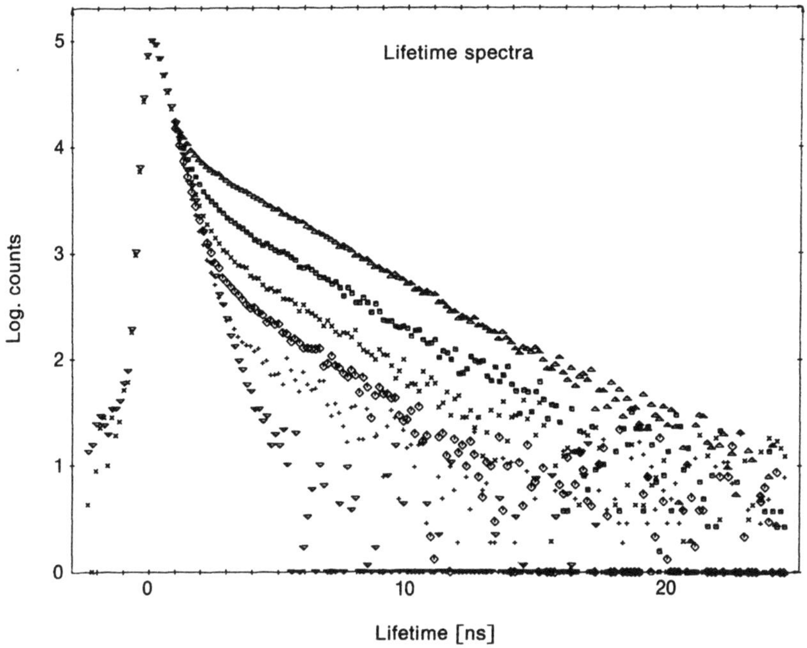

Fig. 3.2. Lifetime (LT) spectrum for CCl_4/benzene solutions of concentrations: △ pure benzene, □ 0.06 M CCl_4, × 0.2 M CCl_4, ◇ 0.4 M CCl_4, + 1.0 M CCl_4, and ▽ pure CCl_4 [3.7]. The constant background was subtracted, and each mark shows the average of two measured points. Counts at or below one are shown at the x-axis. CCl_4 inhibits the Ps yield in benzene

the height of the pulse is within preselected limits, tested by the SCA, a standard pulse is send to the time-to-amplitude-converter (TAC). Hence the time between the standard pulses from the two D-CF-SCA-COIN modules is the positron lifetime.

The TAC delivers to the multi channel analyzer (MCA) a voltage pulse of amplitude proportional to the time between the two standard pulses from the 1.28 MeV and the 0.511 MeV detector branches. The MCA contains at least 512 channels. Each channel corresponds to a preselected small interval of voltage amplitudes. The MCA measures the amplitude of the pulse from the TAC, and adds one to the channel which corresponds to the amplitude of that pulse. This is a registration of one positron lifetime. Hence, the contents of the MCA channels accumulated over a preselected time (hours) constitute a histogram of the positron lifetime spectrum. In that way about a million lifetimes can be measured and stored within some hours. The full width at half maximum (fwhm) of the time resolution spectrum is normally 200–400 ps. If the spectrum contains only one exponentially decaying component of 100–200 ps lifetime the lifetime can be determined with an uncertainty of about 3 ps.

The LT spectra of Fig. 3.2 are such histograms measured with a resolution of fwhm $\sim$ 400 ps. For example, the slope of the longest lifetime component of benzene in Fig. 3.2 gives the average o Ps lifetime of 3.15 ns. The relative area of that lifetime component is the relative yield of o Ps (43%) in benzene. The histogram of positron lifetimes is normally fitted by a sum of exponentially decaying components, given by the average lifetimes, τ_i, and intensities, I_i, and a constant background (see below).

This is the simple description of the LT measurements. All parts of the apparatus are described elementary in, e.g., [3.1], to which non-specialists are referred.

3.1.2 Special Problems of LT Measurements

We shall now discuss certain selected problems of the LT technique. The methods described have been used in practice with fair success. However, it must be realized that they are not necessarily the best methods. Almost every LT specialist has his own ideas of how to get the best performance for a LT system in the details. Furthermore, new components and methods will appear in the future. Hence, read this section with a pinch of salt.

The ^{22}Na LT positron source is often made by drying a droplet of ^{22}NaCl aqueous solution on a thin foil ($\sim$ 1 mg/cm^2). The ^{22}NaCl spot is then covered by another similar foil. If Kapton foils are used they can be glued together by epoxy glue. The source is placed in the center of the ampoule containing a solution, or sandwiched between two pieces of a solid sample. The sample on both sides of the source should be thick enough to ensure that all positrons are stopped in it. The glued Kapton foil source can be used in most solutions, but it is advisable to keep it in the solution only for the time of the actual measurements.

Kapton seems to be the only polymer with no Ps yield and hence no longlived component ($\tau \sim 382$ ps). Other polymer (glued or welded) and metal (soldered or just folded) foils can be used as a source cover [3.4]. The amount of annihilation in the source, i.e., the intensity of the source correction, can be determined by use of the source in a LT measurement on a suitable sample with a known LT spectrum. Pure CCl_4 and saturated $K_2Cr_2O_7$ or $KMnO_4$ aqueous solutions can be used in liquid studies, and a metal or benzophenone if solid insulators are studied. All these samples have no longlived components. The shape of the source correction spectrum is obtained by a LT measurement of the foil material. Because of backscattering of the positrons in the sample the source correction depends somewhat on the sample, in particular for high-Z materials.

The LT source strength, Q, is normally $Q = 5 - 50$ μCi (1 Ci $= 3.7 \times 10^{10}$ Bq). As the random background (LT spectrum) is proportional to Q^2 (Q) the background-spectrum ratio is proportional to Q. Hence, the best source strength is a compromise. High-quality determinations of longlived components necessitate a low source strength. A very long, low-intensity lifetime ($\tau \gtrsim$ 20 ns) can easily be missed if the background is too large!

Scintillation detectors are discussed in great details in, e.g., [3.1], Chaps. 8–10. The fast plastic scintillators (SC) used in LT systems (e.g., Pilot U or NE 111) contain too light atoms and are too small to give photopeaks. The energy spectrum contains essentially only a Compton continuum. The energy window accepted by the SCA in the start (stop) branches is normally about the upper 30% of the 1.28 MeV (0.511 MeV) Compton continuum. Apparently, the fastest collection of the light created in the SC is obtained for conically shaped crystals. Crystals of 3.8 cm height, 3.8 cm (2.5 cm) diameter at the PM (in the top) seem to be a reasonable compromise between high counting efficiency and good time resolution. The background-spectrum ratio is roughly independent of the efficiencies of the two detector branches. The higher density and Z-value of the BaF_2 SC result in a reasonably large photopeak, and consequently, narrow energy windows at the photopeaks are practically useful. A crunched Al foil around the SC increases the amount of light hitting the PM.

The Philips XP2020 (XP2020Q) PM has been used for fast plastic (BaF_2) SC. BaF_2 emits uv light, and hence, quartz windows are necessary. XP2020 and XP2020Q have 12 dynodes. The best results are obtained if the cathode is grounded. The highest permitted voltage difference between the cathode and the first dynode(s) should be used. Normally, the signal from the 9'th dynode is used as the timing signal fed into the discriminator. The timing performance of the PM is strongly dependent on the presence of a magnetic field. The earth magnetic field of about 0.5 G is strong enough to reduce the time resolution appreciably. Magnetic shielding can be used, but has sometimes given bad results, too. By moving the detectors better resolution can usually be obtained. Having the magnetic field (earth, building, etc.) along the axes of the PM seems to be best. In [3.8] the BaF_2/Hamamatsu R2083Q SC/PM combination gives very good time resolution if the anode signal is used. This example illustrates that other SC/PM systems give good timing results, too.

The adjustment of the D-CF-SCA-COIN module (e.g., Ortec 583) is described in the instrument manual and catalogue [3.6]. *Knoll* [3.1] discusses pulse timing systems in his Sect. 17.7. In the CF timing systems an output signal is produced when the leading edge of the pulse from the PM reaches a constant fraction of the peak pulse amplitude [3.1]. This method eliminates partly the uncertainty of the arrival time of the pulses caused by differences in the pulse amplitude. In practice, CF timing allows that a much wider range of the pulse amplitudes can be used for the timing than in other timing methods [3.1] (e.g., leading edge timing). A range consisting of the highest 30% of the pulse amplitudes in the Compton distribution (selected by the SCA) is a reasonable compromise in both channels in a CF LT-system, if plastic SC are used (see below).

At Risoe the performance of the Ortec 583 module ($\sim$ 1985 version) was improved significantly by use of three changes in the electronics. They cannot by discussed here. Nonspecialists in electronics cannot perform such changes except if detailed diagrams are available. Newer version of the 583 module might also respond differently to the changes.

Apparently, it is best to use the pulses from the PM directly, i.e., without preamplification, as input for the CF module. The anode signal is a negative fast pulse suitable as input for CF timing. The dynode pulse must be inverted. This can be done by conducting it through about 60 cm of a 50 Ω small-diameter, coaxial cable. At the dynode end of the cable the signal goes to the center wire, while the shield is grounded. At the CF end the signal is taken from the shield and the center wire is grounded. The inverted, bipolar dynode pulse can also be input directly for CF timing, i.e., the small positive leading pulse can be neglected. It can also be input for zero-crossover timing [3.1] in the module as it is shaped correctly. Apart from these comments the tuning of the module follows from the manual.

The time-to-amplitude converter (TAC, e.g., Ortec 566 or 567) is normally used as received. The connection to the other modules and the adjustment follow from the manuals. The same can be said about the multi channel analyzer (MCA). A combination of certain NIM modules and a PC can replace the MCA.

The geometry of the source-sample-detector system can be the cause of important distortions of the LT spectrum [3.3]. For example, the collinear geometry, where the detectors are separated by the source-sample, the detectors and source-sample being on the same line, should not be used. The reason is that, in an otherwise normal case of a detected 1.28 Mev start-photon and 0.511 MeV stop-photon, the other 0.511 MeV photon may be detected by the 1.28 MeV detector. The 1.28 MeV and 0.511 Mev signals from the start PM tube will then be summed because both signals are 5–10 ns wide, i.e., they are long compared with most positron lifetimes. Such events will distort the time resolution. One way of avoiding this situation is to place the detectors facing one another and move the source-sample so far away from the detector symmetry axes that the second 0.511 MeV photon cannot be detected by the 1.28 MeV detector if the first 0.511 photon has been detected by the 0.511 MeV detector.

Backscattering of the photons might cause problems, too. Photons backscattered from material surrounding the source-sample-detector system, such as cryostats, shielding, etc., might cause the detected photons to take a longer path. Hence, the time resolution might be broadened. The signal from a backscattered photons might also be added to a "good"signal, as discussed above. The photons might be scattered by the SC themselves, too. For example, if a Compton scattering of the 1.28 MeV photon occurs in the start SC and the scattered photon is detected by the stop SC, a contribution to the LT spectrum of zero lifetime will be recorded. A reasonably high energy window in the stop channel SCA will make this effect smaller. In particular, if a BaF_2 SC is used in the 0.511 MeV detector this "prompt peak" effect tends to be important because of the high detection efficiency of the BaF_2 SC.

A source of 10 μCi ^{22}Na gives about 40 counts per second for energy windows selecting the upper 30% of the Compton distribution in both channels. A resolution fwhm of 200–250 ps can be obtained.

The LT system described above is the most elementary one. More complicated systems are used. For instance, a spectrum stabilizer [3.1] can be used to stabilize the position of the time zero of the LT spectrum. A recording of the "feedback" output of the stabilizer is a very useful test of the working conditions of the main part of the system. To permit a higher counting rate a "pile-up" rejecting module can be implemented. Furthermore, a stabilization of the temperature of the room in which the LT system is placed seems to be necessary for high quality measurements.

3.1.3 General Remarks on the Data Analysis

The measured LT, AC, and DB spectra are normally analyzed by use of least-squares-fitting programs. It is assumed that the spectra are composed of several components, and the purpose of the analysis is to extract the parameters characterizing the components. We shall here discuss general principles which are common for such fitting analyses, performed by use of, e.g., the programs: POSITRONFIT, RESOLUTION, PAACFIT, and PARAFIT [1.47] ([3.9], latest version, also on PC).

In a typical analysis of a LT, AC, or DB spectrum we do not have an "ideal" model. We have ideal model if we know before the analysis the number of and type of terms (exponentials, gaussians, etc.) in the measured spectra and that the uncertainties of the spectra are in good agreement with the general assumptions used in the programs. Hence, we normally have to select a hypothetical model by use of our knowledge about the positron processes, and test this model by use of the programs. Four types of results of such test are usually found: 1) The "variance of the fit" is large, and the standard deviations of the fitting parameters are small. Normally few iteration steps are necessary. 2) The variance of the fit is close to one (or acceptable), and the standard deviations of the fitting parameters are small (or acceptable). A "normal" number of iteration steps are necessary. 3) The variance of fit is close to one (or acceptable), and the standard deviations of the fitting parameters are very large. Many iteration steps are necessary. 4) The fitting does not converge. In many fitting cases, but not always, case 1 is associated with too few fitting parameters, while in case 4 too many fitting parameters are used. Sometimes cases 2 and/or 3 are absent if the number of components is increased starting from one component only. In general, the models are so versatile and contain so many fitting parameters that a good fit can be obtained in most cases provided suitable components are selected.

Two facts should be kept in mind when the programs are used: 1) A fitting analysis is a mathematical exercise and the resulting parameters are at first mathematical quantities. They get a physical meaning only if they can be associated unambiguously with the physical processes of the positron. 2) An analysis can answer the question: Is a particular model in agreement with the measured spectrum? A good fit is a necessary but not a sufficient condition for the validity of a particular model; other models might give good fits, too. In general, extreme care should be exercised in the interpretation of the parameters in terms of physical quantities. An intelligent use of constraints in the analysis might solve the interpretation problems partly in many cases.

3.1.4 Data Analysis of LT Spectra

We shall here discuss briefly the normally used analysis of the LT spectra. The problems of LT spectra analysis are discussed in, e.g., [3.3,5].

In particular, the programs: POSITRONFIT and RESOLUTION [1.47,3.9] have been used in the analysis of the LT spectra, in general. In these programs the model function consists of a sum of decaying exponentials convoluted with the time resolution function of the LT spectrometer, plus a constant background.

In POSITRONFIT the resolution function given by a sum of gaussians is assumed to be known. Each gaussian is characterized by its position, fwhm, and relative intensity. To perform the necessary source correction the time-zero, t_0, of the spectrum must be known. Hence, the spectrum is fitted by a sum of exponentials folded with the resolution function plus background in a first iteration cycle to determine t_0. The source contribution given by a sum of exponentials folded with the resolution function is then subtracted. At last the corrected LT spectrum is fitted by a sum of exponentials folded with the resolution function plus the background in the second iteration cycle. In these fittings the number of gaussians is fixed. In general, the fitting parameters are the relative intensities and lifetimes of the exponentials, the background, and the time-zero. However, several constraints can be imposed on the analysis. For example, some or all the fitting parameters may be fixed to chosen values. Furthermore, one or more linear combinations of the intensities may be put equal to zero. These constraints may be used to fix ratios of intensities. Finally, the total area of the spectrum may be fixed, and the background may be fixed to the mean value between two channels. It is also possible to change the number of lifetime components after the source correction has been applied.

In the RESOLUTION program the mathematical model is the same as in POSITRONFIT. The purpose of RESOLUTION is to extract the shape of the time resolution function. Hence, the widths and displacements of the gaussians are included as fitting parameters. However, in order not to have too many fitting parameters the intensities of the gaussians are fixed parameters. The resolution function is normally determined by fitting simple LT spectra, e.g., spectra containing only one major lifetime component. RESOLUTION does not include as many features as POSITRONFIT, e.g., there are no source correction, no constraints on the time-zero and on the area, and the background is always fixed to a mean value between certain channel limits.

Both programs run fast on mainframe computers. For the new PC versions [3.9] a memory of minimum 640 kbyte is needed. A math coprocessor (8087, 80287, or 80387) is necessary to get reasonable run times. Without the coprocessor computing times increase by about 10 times, e.g., 20–40 minutes for fairly normal fits on an 8086 or 80286 PC.

The experience with the programs has been discussed in many papers. Brief reviews are given in [1.47,3.9]. POSITRONFIT in various versions has been used since 1972 by many groups. RESOLUTION is more than 10 years old, and it is also used routinely in the field. It is impossible to give a reasonable discussion of the use of the programs here. We have to refer to the general discussions in the papers (see also Sect. 3.1.3). Most of the LT spectra discussed in this book have been analyzed by use of POSITRONFIT, and the discussions illustrate several of the methods applied in typical analyses. A RESOLUTION analysis of LT spectra for samples giving essentially only one lifetime, such as benzophenone (311 ps) or well annealed copper (112 ps), normally gives the correct shape of the resolution function with the correct lifetime. The lifetime can also be fixed in the analysis. It is important to realize that the resolution shape determined directly by use of the two photons emitted by ^{60}Co differs somewhat from the correct shape. Furthermore, the shape depends somewhat on the geometry of the source-sample system. High quality LT measurements require that the resolution is carefully determined at regular intervals in a series of LT measurements.

3.2 Angular Correlation (AC) Measurements

3.2.1 Experimental

The measurements of the angular correlation (AC) of the annihilation radiation (also called ACAR) are very important in the studies of many chemical problems. In fact, several fields of positron annihilation in chemistry can only be studied in any detail by use AC measurements (see, e.g., Chap. 6). However, AC studies have been applied by very few groups. Because of its poor resolution DB- cannot replace AC-measurements.

The principles of a modern AC spectrometer are shown in Fig. 3.3, while Fig. 3.4 shows the AC spectra for water and single crystals of ice. The narrow peaks in the ice spectra caused by p Ps annihilation from Bloch function states are discussed in Sect. 9.1.

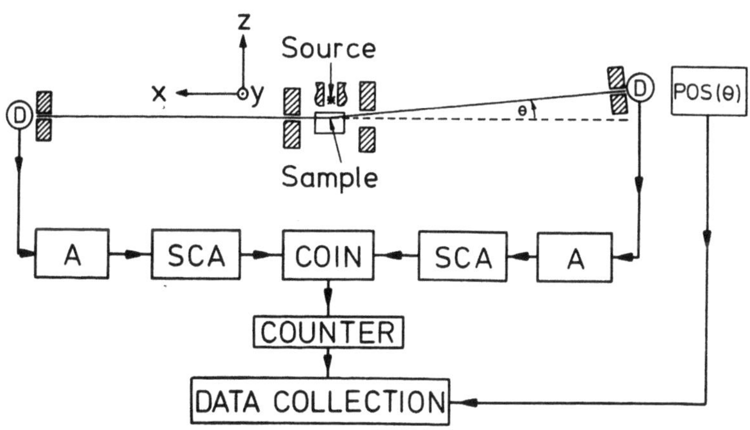

Fig. 3.3. A typical 2γ angular correlation (AC) apparatus

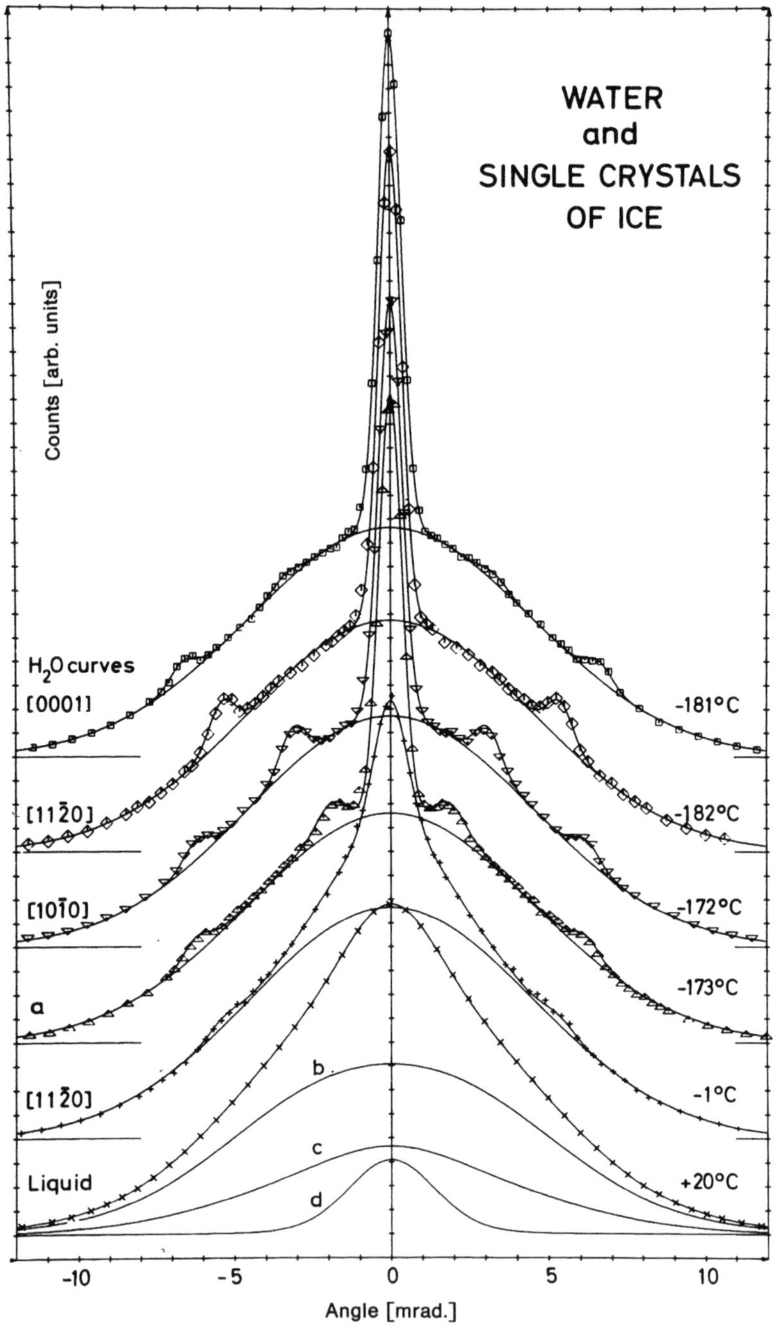
WATER
and
SINGLE CRYSTALS
OF ICE
Counts [arb. units]
H2O curves
[0001]
[11$\bar{2}$0]
[10$\bar{1}$0]
a
[11$\bar{2}$0]
Liquid
b
c
d
-181°C
-182°C
-172°C
-173°C
-1°C
+20°C
-10
-5
0
5
10
Angle [mrad.]

In 2γ-annihilation the angle between the two photons is π except for a small deviation, θ ($\theta \lesssim 15$ mrad). The linear-slit AC apparatus measures in a good approximation the number of annihilations, where a component of the total momentum is given by, $k_z = mc\,\theta$, as a function of θ. m is the electron mass, and c the velocity of light. Let us first describe the typical apparatus shown in Fig. 3.3, and then discuss the normally used approximations, the angular resolution, and the use of the theoretical expressions of Chap. 2.

The positron source (e.g., 30–50 mCi ^{22}Na) is placed roughly 1 cm above the sample, and some of the emitted positrons hit the sample. The range of ^{22}Na positrons in water and samples of similar density is about 0.2 mm. The positrons annihilate with the valence and outer core electrons of the sample molecules into normally two 0.511 MeV photons. Both detectors (D) (Fig. 3.3) give an electrical pulse if they are hit by one of these photons. The coincidence module COIN gives only an output pulse to the counter if both detectors have detected a 0.511 MeV photon within its coincidence resolution time (e.g., 50 ns). This time is so short that two 0.511 MeV photons have a reasonable chance of being counted only if they originate from the same annihilation event. The two identical detectors (e.g., 38 cm long, 5 cm diameter NaI(Tl) SC each optically coupled to two PM tubes) are placed at a distance, L, (e.g., 2 m) from the sample, behind lead collimators which define a long slit of narrow width (e.g., 1 mm). Hence, only those annihilation photons which hit the space angles defined by the slit length, l, and slit width, b, are detected for both detectors. The collimators at the sample ensure that only the sample surface can be seen through the slits. Furthermore, one of the detectors can by moved whereby the angle, θ, is varied. Consequently, the number of annihilation events resulting in simultaneously detected photons in the two slits is measured as a function of the angle θ.

The SCA modules may be adjusted to accept pulses corresponding to energy windows around the photopeak. Sometimes larger energy windows which include the upper part of the Compton distribution are used. The increase in the windows results in more counts, but also in a greater difference in the pulse shapes. This enforces the use of a larger coincidence time, as the pulses from the PM are not shaped as standard square pulses. Hence, a larger random coincidence background results. The detectors and the electronics of the AC set-up are of standard type which are described in manuals and in [3.1].

The main basic problems in AC measurements are related to the geometry of the set-up and sample. We want the measured counting rate versus angle, θ, to be a very good approximation of the distribution of annihilations versus the z-component, k_z, of the total photon momentum, $\mathbf{k} = \mathbf{k}_1 + \mathbf{k}_2$ (see Sect. 2.1).

Fig. 3.4. The measured points and the fitting curves of the angular correlation (AC) distributions in water and single crystals of ice [2.2]. The measured spectra are normalized to equal area. The statistical uncertainties are smaller than or equal to the size of the points. Curves b, c, and d are the "free" positron, o Ps, and p Ps components of the water spectrum. The z-direction of spectrum a is in the plane of the a- and c-axes and tilted 20.6° from the c-axis. A small constant background was subtracted

This is ensured if $b \ll r \ll l \ll L$, where r is the maximum breadth of the distribution of second photons at the slit for a fixed direction of the first photon at the other slit. For $L = 2$ m and for a maximum of θ equal to 15 mrad $r = 3$ cm. Hence, the typical values of $(b, r, l, L) =$ (0.1 cm, 3.0 cm, 38 cm, 200 cm), mentioned above, fulfill the required inequalities. Of course, the most correct way of calculating the measured counts versus θ is to start from the theoretical annihilation probability for $\mathbf{k}_1$ to be in $d\mathbf{k}_1$ and $\mathbf{k}_2$ to be in $d\mathbf{k}_2$ (see Sect. 2.1) and sum up all photons hitting both slits, taking into account the detection efficiency of the detectors. Furthermore, all the annihilations in the samples must be added, taking into account the distribution of positron flux, and the scattering probability of the photons, in the sample. Actually, such detailed calculations have not been published. Normally, a rough semiquantitative estimation of the possible errors of the AC measurements will show that the errors are very small.

For example, an estimation of the errors caused by the use of finite slit lengths showed that they were much smaller than expected at first sight, partly because the sample length in the direction of the y-axis (Fig. 3.3) is usually not smaller than r [3.10].

Several effects contribute to the angular resolution. First, the contribution of the finite slit widths is a triangular resolution curve of fwhm $= b/L$. Second, there are two contributions due to the finite distribution of annihilations in the z-direction (Fig. 3.3) in the sample: A) The sample has a depth in the z-direction, as seen through the slits and collimators, mainly for the following reason. Some of the annihilation photons are scattered before escaping from the sample. To reduce the dependence of this effect on the angle the flat sample surface is normally tilted slightly around the y-axis, to ensure that the photon to the movable detector is passing through the side of the sample for all angles, while the other photon is passing through the upper flat surface. The tilt angle, v_t, must be larger than the maximum angle studied. This causes that the flat sample surface has a depth of $d = b_s v_t$ in the z direction, where b_s is the sample breadth in the x direction. The corresponding resolution function is a rectangular curve of width $2d/L$, if the sample surface is rectangular of breadth b_s. A circular surface of diameter d_c, and hence $d = d_c v_t$, gives a curve of fwhm $< 2d/L$ and shaped more like a gaussian. B) The stopping profile of the positrons is roughly an exponential function, $n(z = 0)\exp(z/\lambda)$ for $z < 0$, in the sample ($\lambda \simeq 0.2$ mm for water), giving a resolution contribution equal to $(L/2\lambda)\exp[(L/2\lambda)\theta]$ for $\theta < 0$. The total resolution function is obtained by folding the three contributions [3.11].

For example, for almost all the AC spectra for liquids measured at Risoe the sample surface hit by positrons was circular and the folding was assumed to result in roughly a gaussian of $\text{fwhm}_t = [(b/L)^2 + (2d_c/L)^2]^{1/2} \simeq 1.15$ mrad, where the depth of the sample, d_c, was estimated. The effect of the positron stopping profile was approximately included in d_c. The fwhm of the narrow peaks in Fig. 3.4 could not be fully explained by a folding of the total resolution curve ($b/L = 0.48$ mrad) with the annihilation curve for p Ps, assuming thermalized Ps and a vacuum effective mass of Ps. The asymmetry in the central narrow peak of the ice AC spectra caused by the positron stopping profile effect, was only significant in high-quality spectra.

The two resolution contributions caused by the finite depth of the sample can be reduced by increasing the detector-sample distance, L ($L > 10$ m have been used). However, a larger L implies that longer slits must be used to prevent a too low counting rate and a too strong increase in the finite slit length error.

Linear-slit AC set-ups with horizontal and vertical slits have been applied. However, the horizontal slits are much better if liquids are studied. The normal, horizontal surface of the liquid can then be irradiated by the positrons. The movable detector slit is placed so far below the plane of the liquid surface that the detected photons are always emitted through the side of the liquid container. For any sample, air at atmospheric pressure above the sample stops so many positrons that a significant air contribution to the AC spectra is measured. Vacuum around the sample solves that problem. The stopping power of helium is so low that helium around the sample can be used, too. Helium is, in particular, a good solution if the evaporation from a liquid or solid sample is too large in vacuum. At Risoe, charging of the sample surface due to positron annihilation and secondary electron emission in AC measurements has only been measured for one solid, namely solid neopentane. Radiation damage due to positron stopping in the sample might result in fairly strong effects on the AC spectrum within few hours for some solids, e.g., certain ionic crystals and polymers.

3.2.2 Data Analysis of AC Spectra

The data analyses of the AC spectra have, in particular, been performed by use of the programs PAACFIT and PARAFIT [1.47,3.9] (joined into ACARFIT in [3.9]). PAACFIT fits the AC spectra by a sum of central gaussians and sidepeak pairs of gaussians. The curves in Fig. 3.4 are such fitting curves. In PARAFIT the fitting curve is a sum of gaussians and curves made by folding a free-electron parabola with a gaussian. In principle, the analyses of AC spectra are fairly similar to those of LT spectra. However, the resolution curve is not corrected for in the AC analyses, because the effect of the finite resolution is normally much smaller in the AC case. The source correction is not needed either. Furthermore, the number of points in an AC spectrum is lower than in a LT spectrum. Hence, a fitting of the AC spectra is about ten times faster than that of the LT spectra.

In a typical AC spectrum analyzed by PAACFIT the number of points was 50–70 between ± 15 mrad, and the statistical uncertainty in the top of the spectrum was about 0.4%. In a three gaussian fit of such spectrum the "variance-of-the-fit" was between 1 and 4, typically around 1.7. The reason for such large values is that the fitting model is not ideal because the spectra cannot always be fitted by gaussians alone, and because the spectrum contains small experimental errors. The "variance-of-the-fit" for fittings of LT spectra are much closer to 1, partly because the average statistical uncertainty of the points is much larger than the errors due to an imperfect model and the experimental errors in the LT case.

In particular, in the analysis of the AC spectra the constraints have been used to fit a spectrum in terms of other spectra the fitting parameters of which have been obtained in other analyses. For example, the AC spectrum for the hydrated positron, e^+_{aq}, annihilation could be well fitted by the three gaussians given in Table 5.3, 1. In a fit of another spectrum by a sum of three or more gaussians, three of these gaussians can be constrained to give a curve of a shape identical to that of the e^+_{aq} curve if their fwhm's are constrained to the same values as those of the gaussians of the e^+_{aq} curve, and two linear-combination-of-intensities constraints are used to fix the ratios of their intensities equal to those of the e^+_{aq} curve (see Chap. 6). As the program can handle both positive and negative intensities, spectra of many different shapes can be analyzed.

PARAFIT has been made to fit metal AC spectra containing a free-electron parabola, and hence, it has not been used much in analyses of spectra of chemical interest.

3.3 Doppler Broadening (DB) Measurements

The information on the positron annihilation processes obtained by a measurement of the Doppler broadening (DB) of the 0.511 MeV annihilation photons by use of a high-resolution semiconductor detector is essentially identical to that of an AC measurement. However, the DB resolution is much poorer than that of an AC setup, and a source contribution is included in the measured DB spectrum. On the other hand, in a DB measurement the sample geometry and the detector placement are less constrained, and the source strength is much lower, than in an AC measurement. The counting rates are roughly identical in the two techniques if the slit widths of the AC set-up are increased to a value giving the same poor resolution as in the DB case.

On neglecting higher order terms in v/c we can easily show that the energy of the annihilation photon is

$$E = mc^2 - E_b/2 + (ck_z)/2, \tag{3.1}$$

where m is the electron mass, c the velocity of light, E_b the binding energy of the electron-positron pair, and k_z the component of total momentum of the two photons parallel to the sample-detector direction. In AC measurements $k_z = mc\theta$ with the z-axis shown in Fig. 3.3. If the two directions are identical we have $\theta = 2(E - mc^2 + E_b/2)/mc^2$ ($2/mc^2 = 3.91390$ mrad/keV).

Nowadays a high-quality intrinsic-germanium detector with an energy resolution for the 0.511 MeV photon of 1.1–1.6 keV is normally used in DB measurements. The efficiency is roughly 10% of that of a 7.5 cm diameter and 7.5 cm high NaI(Tl) SC/PM detector. The electronics of a DB apparatus can be obtained from several manufactures. The tuning of the electronics and the detector to obtain the best possible resolution is described in details in the manuals and, e.g., in [3.1]. The output signal from the detector is fed into a preamplifier placed at the detector, from where it goes to a high quality spectroscopy amplifier. To get a reasonable energy dispersion (50–100 eV per channel) in the final DB spectrum a constant voltage must be subtracted from the pulses in a biased amplifier or, if possible, by digital off-setting in the multi channel analyzer (MCA). The pulses are then fed to the MCA, which is tuned to give the final DB spectrum as a histogram of the pulse heights in a narrow energy interval around the annihilation photopeak.

The positron sources used in DB studies are normally identical to the LT positron sources. However, other positron emitting isotopes are frequently used [3.4,5], such as ^{68}Ge which has a larger positron endpoint energy and less background under the 0.511 MeV photopeak than ^{22}Na and ^{58}Co. The source/sample can be placed close to the detector or at larger distances (e.g., 1 m). The energy resolution of the detector depends on this distance, however. Furthermore, it depends on the presence of other structures, such as collimators, cryostat, and/or oven, surrounding the source/sample and detector, and on the direction of incidence of the photons. Hence, it is important to keep the conditions of the experiments fixed if the results are to be compared in a series of measurements.

It is useful to tune the DB system carefully to get the best possible energy resolution. For example. the resolution fwhm of a detector at Risoe was 1.33 keV, while an annealed aluminum sample gave fwhm = 2.68 keV, after the first, fairly brief parameter adjustment. However, a resolution fwhm = 1.17 keV was obtained as a result of a following comprehensive adjustment. The manuals and [3.1] give instructions on the tuning of the electronics.

The DB system described above can be extended by a pile-up rejecter [3.1], which rejects those events where two pulses are piled up with one another in the electronics. Pile up occurs, in particular, at high counting rates, and it will distort the energy resolution. Another important extension of the DB system is a spectrum stabilizer [3.1], which may be used to stabilize the position of the 0.511 MeV photopeak, but it can also be used for stabilization of a photopeak of a gamma ray from another isotope, e.g., ^{85}Sr, of energy close to 0.511 MeV, placed at the detector. In addition, it is important to stabilize the temperature of the system.

In the analyses of the DB spectra the subtraction of the background counts under the 0.511 MeV peak constitutes a problem. The apparent background below 0.511 MeV is larger than that above 0.511 MeV partly because of incomplete charge collection and partly because of small-angle scattering in the sample and surrounding materials. Additionally, part of the background is caused by Compton scattering of 1.28 MeV photons in the detector if ^{22}Na is used. Consequently, it is not obvious which part of the background is an inherent feature of the detector resolution and which is the "real" background. The background is normally in the order of 1% of the counts in the top of the peak. In practice, several methods of background subtraction seem to have been used. A reasonable practical method is the following. The average background above, B_a, and below, B_b, the peak are determined. At the i'th spectrum point in the photopeak the subtracted background, $B_i = B_a + (B_b - B_a)A_i/A_t$, where A_i is the area of the peak above the i'th point and A_t the total area of the peak.

With few exceptions, the data analyses of the DB spectra have been very elementary. Various simple line-shape parameters are used. The S-parameter is defined as the counts in an energy interval around the top of the photopeak divided by the total area of the peak, while the W-parameter is the ratio of the sum of the counts in the two symmetrically placed intervals at the wings of the peak and that of the total area [3.3,4]. Normally, the regions used to calculate S and W are not clearly given in papers, and hence, a comparison between data from different laboratories is difficult. In chemistry DB measurements have mainly been used by the Strasbourg group (see e.g., Chap. 6–8), who have used PAACFIT analyses to extract the components of the DB spectra expressed in terms of one gaussian only.

A more detailed analysis comparable to that used on AC spectra cannot, apparently, be performed by use of a program such as PAACFIT (ACARFIT in [3.9]), which can fit the spectra in terms of only symmetric fitting curves. The DB spectra are clearly asymmetric. For example, a resolution spectrum (^{85}Sr) at Risoe was poorly fitted by one gaussian of intensity, fwhm, and shift of 100.0%, fwhm = 1.33 keV, −0.011 keV (var. = 52.5), while the three gaussian fit gave 92.73%, 1.27 keV, 0.023 keV; 5.78%, 1.45 keV, −0.555 keV; and 1.49%, 4.50 keV, −1.286 keV (var. = 1.53) in GAUSSFIT analyses. GAUSSFIT [3.12] was developed for the fitting of spectra by a sum a gaussians plus a constant, the fitting parameters being the intensity, fwhm, and position of the gaussians and the constant background. In the present version up to 20 gaussians could be used, and the possible constraints were almost identical to those of PAACFIT, the main difference being that linear combinations of intensities could be fitted to a constant, too. GAUSSFIT has been tested and found very useful in analyses of some DB spectra, about 250 artificial spectra made by sums of gaussians, and about 150 X-ray fluorescence spectra. A final program ought to include a deconvolution of the DB spectra for a resolution curve given by a sum of shifted gaussians, similar to POSITRONFIT.

3.4 Miscellaneous Measurements

We have until now discussed the three most important techniques used in positron annihilation studies in chemistry. In this section we shall briefly describe various other experimental techniques, namely the 3γ-coincidence, energy-spectrum peak-to-valley ratio, slow positron beam, two dimensional AC, and combined LT + DB techniques.

The measurements of the 3γ-annihilation counting rate are fairly difficult. Furthermore, the interpretation of the results is somewhat problematic partly because the inhibition of Ps formation and Ps quenching cannot be directly separated [1.3]. However, the technique has been of importance in the assignment of LT and AC components to Ps annihilation processes. An absolute determination of the 3γ-annihilation yield is very difficult, indeed. However, the 3γ-rate in a sample can be compared to that of another sample (e.g., Al) in which no Ps is formed.

Dauwe [3.13] discusses, in particular, the technical problems of the triple coincidence technique. The three NaI(Tl) SC detectors were arranged in a plane forming angles of $2\pi/3$ with each other and with the source-sample sandwich in the center. Each detector covered a solid angle of $\Omega/4\pi = 0.05$. The NaI(Tl) SC were chosen instead of plastic SC because of their higher detection efficiencies and much better energy resolution. ^{68}Ge was preferred as a positron source because only 2% of the β^+-decays are accompanied by a nuclear photon (1.08 MeV). Energy windows of 260 keV $< E <$ 440 keV assured that mainly the 3γ photons ($\sim$ 341 keV) were detected. Furthermore, vertical lead shields were inserted between the detectors to limit "crosstalking" between the detectors. The triple coincidence counting rates were about 0.2 s^{-1} with a source strength of 10 mCi. In favorable cases the relative uncertainty in the measured 3γ-rate was about 1%. Additionally, it was concluded that for comparative measurements the source-sample sandwiches should have nearly the same absorption dimensions for Compton scattering, and a correction for partial positron absorption and for Compton effect of the emerging 341 keV photons is necessary. The proper parameter is the 3γ-coincidence rate extrapolated to zero source-sample thickness. The discussion in [3.13] illustrates that great care should be exercised in 3γ-coincidence measurements.

Theoretically, some confusion seems to exist in the literature, where a correct expression for the 3γ-gamma yield seems not to be available. For example, in [3.13], {(1), also misprint} and in [1.3] (25) the 3γ-annihilation during the Ps collisions resulting in pick off annihilation is not taken into account. On the other hand, in [1.1] *p*. 41 the pick-off 3γ-annihilation is, very probably incorrectly, estimated to be insignificant. As discussed in Sect. 2.1, the Pauli-principle will cause the Ps electron to repel electrons with its own spin direction in the molecules. This causes the pick-off 2γ-(3γ-) annihilation of o Ps to be larger (smaller) than that of p Ps. However, these effects seem not to have been realistically estimated, yet. In Sect. 8.1, where the kinetic equations of the Ps reactions are discussed, several of the 3γ-annihilation processes are disregarded!

Approximately, the relative 3γ-yield for a pure liquid is:

$$P_3 = (3/4)P(\lambda_t + \lambda_p/371)/(\lambda_t + \lambda_p) + (1 - P)/371, \qquad (3.2)$$

where P is the total Ps yield, $\lambda_t = (139\ \text{ns})^{-1}$, and λ_p is the pick off rate constant {see (2.11)}. The first, second, and third terms in (3.2) are the o Ps 3γ-, o Ps pick-off 3γ -, and the free-positron 3γ-yields, respectively. If Ps is not formed (e.g., in Al) $P_3 = 0.27\%$, while water (hexane) gives $P_3 = 0.63\%$ (1.4%). In (3.2) we have disregarded the problems associated with the fourth LT component (see Sect. 5.3).

The relative Ps yield, P, can also be determined from the γ-ray energy spectrum, e.g., as measured by use of a semiconductor detector. The following calculation illustrates the principles of the method ([3.14], p. 473). Let us divide the spectrum into two regions: the "total" count rate region T and the "peak" count rate region P which are sensitive particularly to the 3γ- and 2γ-annihilation photons, respectively. The count rates in these regions are denoted T_i and P_i where $i = 0$, 1, and P for zero, total, and P Ps formation, respectively. If all the detected gamma rays are caused by positron annihilation only (e.g., in slow positron beam work) it is evident that $T_P = PT_1 + (1 - P)T_0$ and $P_P = PP_1 + (1 - P)P_0$, where the subscripts 1 and 0 refer to $P = 1$ and $P = 0$, respectively ([3.15], p. 613). On solving for P we get:

$$P = \{1 + (R_1 - R_P)(R_P - R_0)^{-1}(P_1/P_0)\}^{-1}, \tag{3.3}$$

where $R_i = (T_i - P_i)/P_i$. If samples with $P = 1$ and $P = 0$ (or any two known values of P) can be prepared (3.3) can be used to find P. In particular, (3.3) has been used in slow-positron-beam research [1.48,3.14,15]. However, in most Ps chemistry studies ^{22}Na is used as a source and hence, a background due to Compton scattering of 1.28 MeV photons is present in the two regions T and P. These background count rates must be subtracted from T_i and P_i, $i = 0$, 1, and P, before (3.3) can be used. The background counting rates might be determined by use of an isotope emitting only one gamma ray of energy close to 1.28 MeV. Of course, a careful estimation of the absorption of the gamma rays in the samples to be compared must be performed also in this case.

Within the last two decades there has been a strongly increasing interest in using slow positron beams to study the surfaces of solids and positron scattering in gases [1.5,7–9,15,48–52]. However, very few studies of chemistry oriented problems have been performed by use of slow positrons. Some measurements are discussed in Sect. 9.5. In the techniques discussed until now the positron sources are β^+-emitting isotopes, such as ^{22}Na, having energy distributions with a maximum energy of about 0.5 MeV. In a slow-positron-beam experiment essentially monoenergetic positrons of energies from a fraction of an eV up to several thousands of eV are used. This allows a new set of phenomena to be studied. On the other hand, the range of slow positrons is very small in condensed matter. Hence, apart from low pressure gases, mainly surfaces and near-surface structures can be studied. Furthermore, the condensed-matter samples must have very low saturated vapor pressure to permit the positrons to reach the samples.

Let us describe a typical slow positron beam [1.5,7–9,15,48–52]. The fast positrons from a β^+-emitting isotope are thermalized in a moderator, e.g, annealed tungsten, and a fraction of about 10^{-4} is emitted with energies of the order of one eV, with thermal widths. These slow positrons are guided to the samples by use of magnetic and/or electrostatic fields, and they are accelerated and/or decelerated by grids of appropriate electric potentials. The fate of the slow positrons hitting the sample is studied by measurements of secondary particles and photons created as a result of the stopping of the positron. For example, the energy spread of the annihilation photons measured by a semiconductor detector gives information on the trapping of the positrons in defects and on molecules. The number of Ps escaping from the surface can be measured by use of a semiconductor detector, as discussed above. Additionally, the energy of the emitted Ps can be determined by time-of-flight measurements. Also the reemitted positrons, secondary electrons, and fluorescence photons due to electron-ion recombination in the positron track can be studied, in principle. The number of escaped Ps as a function of the slow-positron energy gives information on the diffusion constant of Ps.

The fast positrons can also be obtained from an electron accelerator (e.g., a 100 MeV LINAC). The bremssthrahlung due to the stopping of the fast electrons is used to make the fast positrons by pair creation. In that case the slow positrons from the moderator are pulsed, and special methods have to be used to make a continuous slow-positron beam. The slow-positron flux during the pulses is normally so high (e.g., 10^{11} s^{-1}) that the electronic equipment is overloaded.

Today, a magnetically-guided slow-positron beam of rather poor vacuum (e.g., 10^{-8} torr) using a β^+-emitting isotope is fairly easy to make. About 10^5 slow e^+/s at the sample can be obtained with a source of 25 mCi ^{22}Na. It is to be expected that chemistry-oriented slow-positron studies will be more frequent in the future, although many topics discussed in this book cannot be studied profitably by use of slow positrons.

The two-dimensional angular correlation technique (2D-AC) is expensive and technically demanding [1.5,7–9,3.3]. In a 2D-AC setup the detectors are two-dimensional position-sensitive gamma detector for the 0.511 MeV annihilation photons. They are placed several meters away from the source/sample, which is situated between the two detectors, as in a linear slit AC setup discussed in Sect. 3.2. In particular, in Fermi surface studies the 2D-AC technique has been of great importance. In chemistry oriented research 2D-AC measurements have not been used except in some single crystal (ice, quartz) studies.

Finally, let us mention the combined LT + DB technique, by which one can determine the energy distribution of the 0.511 MeV photons in selected positron lifetime intervals [1.7–9,3.4]. The results of the LT measurement are used to direct the 0.511 MeV DB signal to selected regions of the multi channel analyzer. Each region will then store the DB spectrum corresponding to a certain positron lifetime interval. The LT + DB measurements have not yet been of great importance in positron annihilation. Some results showing that p Ps is not thermalized in SiO_2 aerogel are discussed in Sect. 10.5.

4 Light Particle States in Insulating Matter

In many positron annihilation studies of insulating matter essentially two questions must be answered, namely: A) What are the states of the light particles, the positron, Ps, and the excess electron, in the sample, and B) What are the rates of the transitions between these states. Hence, it is appropriate to start the discussions by giving a description of the normally found states of the light particles in this chapter. In exceptional cases the states change so rapidly that the two questions are not useful guides of the reasoning.

Only three light particles take part in low energy physics and chemistry, namely the electron, positron, and Ps. A light particle is defined as a particle having so low a mass that quantum mechanical phenomena, such as tunnelling, delocalization, and zero-point motion, dominate its behaviour at low energies. The behaviour of the heavier particles, e.g., the muon, muonium, proton, and hydrogen, is only perturbed by quantum mechanical effects.

There are several good reasons for the inclusion of a discussion of the excess electron states and behaviour in this book. First, the excess electron is the mostly studied light particle, and hence, it serves as a standard case of reference in the interpretations of the states and behaviour of the two other light particles. Second, Ps is formed by a reaction of a positron and an excess electron in the positron spur in condensed matter (Chap's. 5,7), a process which depends very much on the behaviour of the two particles. Third, Ps formation can be inhibited by addition of solutes reacting with the excess electron, which means that excess electron reactions influence the Ps yields in many cases.

4.1 General Remarks

By the state of the light particle we usually understand the particle itself and the changes its presence has caused in the surrounding matter. As is found in most chemistry problems, detailed quantum mechanical calculations of the light particle states are normally too difficult. The systems are simply much too complicated. We shall see that the chemistry oriented experiments in positron annihilation can only be explained in terms of qualitative and semiquantitative concepts and models, such as: bubbles states; delocalized/localized states; solvated/nonsolvated states; particle mobilities; diffusion and reaction constants;

wave function spill over, wiggles, and overlap; reaction radii; diffusion and/or kinetics controlled reactions; etc. Hence, we will use the concepts and models of normal reaction kinetics involving light particles, in particular those used in the interpretation of the reactions of excess electrons in radiation physics and chemistry.

We shall mainly be concerned with light particle states in liquids in this chapter. Solids and gases are discussed shortly in Sect 4.5. In liquids of normal viscosity a certain structural configuration of the molecules is typically "forgotten" within about 10 ps. For example, in water it takes about 6–8 ps for a molecule to diffuse a distance equal to the size of the molecule, and a molecule is mainly oriented along an external field within about 5–8 ps. However, the very high electric fields close to charged particles reduce the molecular orientation times markedly in water. Molecular vibrations take typically 0.01–0.1 ps. These times may be usefully compared with the Ps formation times of about 10 ps in hexane, and lifetimes of p Ps of 125 ps, free positrons of about 400 ps, and o Ps of several ns.

4.2 The Excess Electron

The excess electrons have been studied much more than the other two light particles in insulators. They are electrons in the conduction band in insulators. Judged on a microscopic scale they are extra electrons. Excess electrons are normally the products of ionizations of the molecules by charged particle and photo-effect processes. As the emitted electrons slow down over distances of 2–100 nm in most insulators, they are not extra electron on a macroscopic scale, i.e., the positive ions are normally also present in the insulators. Excess electrons can, of course, also be electrons injected into an insulator, in which case they are really extra electrons in the sample under study. Excess electrons in liquids have been studied by mainly radiation chemists. They are probably the most important reactants in radiation chemistry.

We shall mainly discuss the excess electrons in thermal equilibrium in the liquids. Of course, we can only give a very brief review of this topic here. Table 4.1 shows the most important measured properties of excess electrons in selected liquids.

The excess electron mobility, μ_-, varies over almost seven decades. It is mainly determined by the localization of the electron. The more delocalized the electron is, the higher the mobility is. In polar liquids the electron mobility is comparable to those of heavier molecular ions, which are normally found to be within the limits $(7.5 \pm 2.5) \times 10^{-4}$ cm^{2}V^{-1}s^{-1} for a viscosity of 1 cP, and to be inversely proportional with the viscosity [4.1]. In the monoatomic noble gas liquids μ_- is comparable to those of electrons in the conduction band of semiconductors. For example, μ_- is 1900 cm^{2}V^{-1}s^{-1} for silicon and 3000 cm^{2}V^{-1}s^{-1} for germanium at room temperature. For the nonpolar molecular liquids μ_- is mainly determined by the shape of the molecules. The more

Table 4.1. Excess electron properties, at room temperature except where indicated, for selected liquids. μ_-, mobility; V_0, work function; G_{fi}, free ion yield; b, secondary electron range. References [4.1–6]. [a] Uncertain values

Liquid	$\mu_-[cm^2V^{-1}s^{-1}]$	V_0 [eV]	G_{fi} [$(100eV)^{-1}$]	b[nm]
krypton (148 K)	2300	−0.78	–	90
xenon (183 K)	1100	−0.63	–	70
argon (90 K)	400	−0.33	–	140
methane (120 K)	300	–	–	50
tetramethylsilane	100	−0.55	0.74	–
neopentane	70	−0.35	1.09	21.5
isooctane	5.3	−0.24	0.33	9.5
cis-butene-2	2.2	–	–	7.4
cyclohexane	0.23	0.0	0.15	6.0
n-pentane	0.15	0.0	0.15	7.0
benzene	0.13	−0.13	0.081	4.2
n-hexane	0.08	0.05	0.13	7.0
toluene	0.07	−0.22	0.093	4.2
n-decane	0.038	–	0.117	–
n-tetradecane	–	0.21	0.12	–
trans-butene-2	0.029	–	–	5.3
helium (4.2 K)	0.02	1.05		
ammonia (240 K)	0.009	solv.	3.0	2[a]
diethylether	0.005	solv.	0.3	3.5
water (H_2O)	0.002	solv.	2.7	2[a]
water (D_2O)	–	solv.	2.9	2[a]
neon (25 K)	0.0016	0.5		
methanol	0.0006	solv.	1.9	2[a]
ethanol	0.0003	solv.	1.9	2[a]

spherical the molecule is the higher the electron mobility is. In particular, the 870 times higher μ_--value in neopentane than in hexane and the 75 times higher μ_--value in cis-butene-2 than in trans-butene-2 illustrate this fact very well.

The excess electron work function, V_0, is the energy of the excess electron in the delocalized state of lowest energy in the liquid, compared to vacuum. It is experimentally determined as the lowest energy compared to vacuum of the states into which the electrons are injected in an experiment where electrons are photo-injected into the liquid from, e.g., a metal. The energy of the equilibrium state of the electron is somewhat lower than V_0 for linear hydrocarbons, e.g., by about 0.2 eV in hexane, while it is almost identical to V_0 in the high mobility liquids. In polar liquids the electron is strongly attracted to the molecules, and the solvation energy of the equilibrium state is about 1.3–2.0 eV in, e.g., water and alcohols. Hence, the excess electron energy is below vacuum in polar liquids. In very few liquids, namely liquid helium, neon and hydrogen, the excess electron energy is strongly positive, e.g., 1.05 eV for helium and 0.5 eV for neon, i.e., the electrons are repelled by the molecules and by the liquid in these cases.

As is to be expected, the excess electron mobility and work function are strongly correlated. In the polar liquids, where the electron is solvated by more than one eV, its mobility is comparable to, or a little higher than, the heavy ion mobilities. Although the solvated electron is localized, its wave function overlaps several molecules. For example, the "radius" of the hydrated electron is about 0.3 nm, and in reactions with certain ions the reaction radius is about 1 nm [4.2], which shows that it can move by tunnelling through energy barriers. In the three liquids where the electron is strongly repelled by the liquid, i.e., liquid helium, neon, and hydrogen, the electron is localized in a bubble state, and hence, its mobility is fairly low. This state will be discussed for the case of Ps below. Apparently, the highest excess electron mobility for molecular liquids is found for the liquids where the equilibrium state energy, i.e., roughly V_0, is -0.3–-0.5 eV. The electron is attracted to these liquids, and to liquid argon, krypton, and xenon. Of course, the very high mobility in the last mentioned liquids is caused partly by the lack of rotational and vibrational degrees of freedom in these monoatomic liquids. In the linear hydrocarbons the energy of the excess electron is a little below that of the lowest delocalized state. It is, therefore, weakly solvated by about 0.2 eV, and hence, it is somewhat localized, and its mobility is between those of the spherical hydrocarbons and the polar liquids.

The free ion yield, G_{fi}, is the number of pairs of excess electrons and positive ions which escape "geminate" recombination in the spurs and, therefore, recombine by homogeneous interspur reactions. G_{fi} is defined as the number of free ions per 100 ev loss of energy of a particle penetrating the liquid. G_{fi} will be used in the discussion of the Ps yield in Chap. 5 and 7.

The secondary electron range, b, shown in Table 4.1, is the average distance of slowing down to thermal energy of the secondary electrons formed by ionization of the molecules by fast (1–1000 keV) electrons in the liquid. Actually, b is a fitting parameter used to fit theoretical distributions of the electron-ion distances in the ion pairs created by the stopping processes, to the experimentally determined numbers, mainly the free ion yield versus external electric field. Several theoretical distributions, such as a gaussian, exponential, gaussian plus a tail, etc., have been used [4.1,3,5,6]. Hence, the given b-values in Table 4.1 depend somewhat on the assumed function. b is a very important parameter to be used in the spur model of Ps formation, as it is a rough estimate of the positron electron distance on thermalization.

The absorption spectrum of the solvated electron is the most important experimental property of the electron for polar liquids [4.2,7–9]. For those nonpolar liquids, where the excess electron is weakly solvated, an absorption spectrum at long wave lengths can be measured, but it is very difficult to use in the study the electron. For all nonpolar liquids the most important experimental technique is to measure the conductivity due to the electrons and other ions. Pulse radiolysis is probably the most important technique in radiation chemistry [4.2]. In a typical experiment a short pulse of electrons creates the excess electrons and other species, and the absorption spectrum and/or the conductivity of the electron and other ions are measured at various times after the pulse. The best time resolution is about 25 ps. However, in most pulse radiolysis studies pulses of 1–20 ns widths are used. The absorption can be measured with a time resolution of about 0.1 ps in laser flash photolysis experiments [4.10]. In the older steady-state irradiation studies, a solution was irradiated for about 20 minutes, and the products of the irradiation were studied by use of chemical analysis techniques.

About 3000 rate constants of hydrated electron reactions with various solutes have been measured [4.2,11,12]. The fastest reactions are well explained by the Smoluchowski expression (6.13,14), a measured diffusion constant of 4.9×10^{-5} cm^{2}s^{-1}, and a reaction radius of 1 nm. Table 4.2 shows rate constants of excess electron reactions with electron acceptor molecules in six nonpolar liquids [4.13,14]. In the linear hydrocarbons the rates are roughly diffusion controlled, as calculated by use of the Nernst Einstein (6.11) and Smoluchowski (6.14) equations. However, in the liquids of high excess electron mobility the rates are much lower than the diffusion controlled rates. The rate constant of the electron SF_6 reaction in tetramethylsilane, $k = 2.09 \times 10^{14}$ M^{-1}s^{-1}, is probably the highest one measured for a reaction in which one reactants is neutral. The rate constant of the excess-electron positive-ion recombination is diffusion controlled in agreement with (6.18) [4.15]. It is about 5×10^{16} M^{-1}s^{-1} in tetramethylsilane.

It has been difficult to calculate theoretically the properties of the excess electron by use of quantum mechanics. Many attempts to derive the absorption spectrum of the hydrated electrons have been published [4.2,8]. The excess electron mobility in molecular liquids is almost impossible to calculate theoretically [4.16]. Similarly, the secondary electron range in molecular liquids cannot be derived theoretically [4.17].

It must be strongly emphasized that a theoretical approach to solve the main problems in the research of excess electrons in molecular liquids has mainly failed. The numbers of Table 4.1 cannot be calculated at present, except for the noble gas liquids where some theoretical attempts seem to have been useful. Only very rough "interpolation rules" have been used for "calculations" of the mobility and range in molecular liquids.

Table 4.2. Rate constants of excess electron reactions, k, in 10^{12} M^{-1}s^{-1}, with various solutes in nonpolar liquids at room temperature [4,13,14]. See also Fig. 7.4

Liquid	SF_6	C_2HCl_3	CCl_4	N_2O	C_2H_5Br
tetramethylsilane	209	23	41	0.75	0.04
neopentane	198	58	29	2.3	0.34
isooctane	58	35	–	9.6	5.1
n-pentane	–	–	1.9	–	1.5
cyclohexane	4.0	–	2.7	2.4	2.0
n-hexane	1.9	2.6	1.3	1.09	1.46

Clearly, these facts tell us that quantum mechanical calculations of the properties of all the three particles in molecular liquids are not expected to be useful at the present stage of development, since the excess electron is far the best studied particle. This is a very important conclusion in positron annihilation research of liquids.

4.3 The Positron States in Liquids

We shall now briefly discuss the states of the "free" positron, i.e., the positron which has not formed Ps, in insulating liquids. Compared to the excess electron case very little is known on the positron states. Generally speaking, the sparse experimental facts seem to indicate that the positron states in different liquids are qualitatively similar to those of the electron.

In polar liquids the positron is, of course, expected to be strongly solvated. Diffusion controlled rate constants of hydrated positron reactions with some halides could be explained by use of (6.21) and the diffusion constant and reaction radius of the fastest hydrated electron reactions with ions (see Chap. 6). We expect that the positron is hydrated by water molecules oriented such that the negative lone pairs are close to the positron, while the positive protons are oriented towards the electron in its hydrated state. Similar to the electron case the positron wave function is very probably delocalized over several water molecules. In certain hydrocarbons the Ps inhibition results indicate that the positron has a mobility which is much higher than that of the positive ions (see Chap. 7). This applies even in the cyclohexane case, where the positive ion is much faster than the normal heavy ions. Except for the hydrated positron no rate constants have been measured for the positron in liquids.

A few attempts to measure the positron mobility, μ_+, in liquids have been reported [4.18,19]. In [4.18] μ_+ was found by DB measurements to be 6, 8.5, and 11.5 $cm^2V^{-1}s^{-1}$, for n-pentane, n-hexane, and n-decane, respectively, while [4.19] reports $\mu_+ = 100$, 75, and 8.4 $cm^2V^{-1}s^{-1}$ for n-hexane, neohexane, and benzene, respectively, measured by AC technique. The large difference between the values for n-hexane illustrates that such measurements are difficult. However, both groups agree upon that μ_+ is much larger than the mobility of heavy ions. Further measurements should be done!

Absorption by the positron or PPR (Positron Magnetic Resonance) in matter have not been reported yet. As there are normally less than one positron in the sample at any time, these techniques are only expected to give detectable information if the annihilation characteristics of the positron are changed significantly on application of light or microwaves, respectively. As the positron is expected to resolvate rapidly, absorption will probably not influence significantly the annihilation in polar liquids. Experiments in very viscous liquids (slow solvation) might be interesting.

4.4 The Ps States in Liquids

Positronium is repelled by most molecules. Exceptions are molecules of high electron (and/or positron) affinity. It is the electron in Ps which is repelled by the electrons in the molecules because of the Coulomb forces and the exchange effect. The latter is a consequence of the Pauli principle, which states that two or more electrons cannot occupy the same state. Hence, the electron in Ps cannot penetrate into the electron clouds of the molecules except if it occupies states of higher energy than the filled states. This corresponds to a repulsion as the energy of the electron in Ps is increased if it penetrates the other electrons. The positron is only repelled by the molecules if it has penetrated so far into the molecules that it is repelled by the positive atomic cores, i.e., if it has penetrated so far into the molecules that the positive nuclei are not fully shielded by the electrons. Roughly speaking, the positron is kept out of the electron clouds of the molecules partly by its Coulomb attraction to the Ps electron, which is repelled by even the outermost electrons in the molecules. As a result, the pick off lifetime of o Ps, τ_3, in condensed matter is much longer than the lifetime of "free" positrons not forming Ps. In hexane, for example, τ_3 is 3.92 ns, while the "free" positron lifetime is about 0.4 ns, and the lifetime of a positron in an electron gas of a density equal to that of an assumed homogeneous distribution of 38 valence electrons per molecule in liquid hexane is 0.76 ns. Of course, this is very qualitative explanation of the Ps molecule interaction. Only a reasonably correct quantum mechanical treatment can give a precise description. Such calculation has, apparently, not been reported yet.

An excess electron is repelled by the electrons in the molecules in the same way as the electron in Ps. However, the charge of the excess electron polarizes the surrounding molecules. The polarization energy, P, calculated by use of the Born formula (see textbooks on physical chemistry), amounts to 0.5–1 eV depending on the effective radius of the excess electron. P corresponds to an attraction, and it counteracts the repulsion discussed above. Clearly, adding a positron to form Ps removes the polarization energy, and only the repulsion is still present. This simple argument explains well the fact that Ps is strongly repelled by the molecules in normal liquids, while the excess electron is attracted to many liquids.

Because of this repulsion Ps is in a bubble state in all liquids except in high electron (and/or positron) affinity liquids. In a bubble state the inwards pressure due to the surface tension is counterbalanced by the outwards pressure due to the Ps molecule repulsion and the zero point energy of Ps. For the electron the bubble state is only found in liquid helium, neon, and hydrogen, which are polarized very little, repel the electron strongly, and have low surface tension. We shall see, in particularly in Chap's. 5,8, that the bubble state is an important property of Ps in liquids explaining much of its behaviour. Hence, a more detailed description of the Ps bubble is appropriate here.

Various models of the Ps bubble state have been used. They are useful for qualitative and semiquantitative discussions, but they should not be taken too seriously to avoid that the interpretation is carried too far. The mostly used model is to assume that Ps is situated in a square well potential of depth U and radius r in the liquid [4.20–22]. The total energy of the bubble state is

$$E = E_0(U, r) + 4\pi r^2 \gamma + 4/3\pi r^3 p. \tag{4.1}$$

$E_0(U, r)$ is the Ps zero point energy in the potential. The second and third terms are the surface and pressure-times-volume energies, respectively, i.e., they are the work necessary to create the bubble. γ is the surface tension and p the pressure. By standard quantum mechanics $E_0(U, r)$ and $P(U, r)$ can be calculated, where $P(U, r)$ is the probability of finding Ps outside the square well. In equilibrium the energy of the Ps state is in a minimum, and hence, $dE/dr = 0$. This gives one equation for the determination of the two unknown parameters U and r. Normally, a second equation is obtained by use of the following approximation for the pick off lifetime, τ_3

$$\lambda_p = \tau_3^{-1} = \pi r_0^2 c n z_{\text{eff}} P(U, r), \tag{4.2}$$

where r_0, c, n, and z_{eff} are the classical radius of the electron, speed of light, density of molecules, and average number of annihilation electrons per molecule, respectively. Hence, λ_p is assumed to be the nonrelativistic decay rate for a known density of electrons (see Chap. 2) times $P(U, r)$. z_{eff} is normally assumed to be the number of valence electrons in the molecules. For example, for hexane ($z_{\text{eff}} = 38$) we have $\pi r_0^2 c n z_{eff} = 1.31\ \text{ns}^{-1}$ and $\lambda_p = 0.255\ \text{ns}^{-1}$, and hence $P(U, r) = 0.195$. In general, the two equations cannot be solved in terms of known functions, but they can be used to calculate numerically U and r. Typical values for organic liquids are $r = 0.4$–0.6 nm and $U = 0.3$–1.0 eV.

If it is assumed that the square well has infinitely high potential walls and that the pv term is insignificant, as is normally found, we have

$$r = 1.242 \times \gamma^{-1/4}\ \text{nm}, \tag{4.3}$$

$$E = 0.2429 \times \gamma^{1/2}\ \text{eV} \tag{4.4}$$

for γ in dyn cm^{-1}. Hence, for $\gamma = 20$ dyn cm^{-1} we get $r = 0.59$ nm. Furthermore, the surface tension decreases linearly with rising temperature, t, in most liquids [4.23]. For example, in hexane $\gamma = 20.44 - 0.1022 \times t$ dyn cm^{-1} (t in °C), which gives $E = 1.35$, 1.23, 1.10, 1.04, and 0.85 eV at $t = -100$, -50, 0, 20, and 80°C, respectively. Hence, by decreasing the temperature from 80 to -100°C we can increase the energy of the Ps bubble by about 0.5 eV in hexane. If Ps encounters an acceptor which provides a lower energy state for Ps, it will normally react with the acceptor. We shall see in Sect. 8.3 that this effect provides the explanation of the strong temperature dependence of the rate constants of Ps reactions with many solutes in liquids.

In another, very simple model, namely the three dimensional harmonic potential for Ps, we can calculate the probability distribution of a Ps coordinate, z, in terms of known functions. Both this distribution and the angular correlation curve are gaussians, for which

$$\text{fwhm}_z \times \text{fwhm}_\theta = 1.07\ \text{mrad nm}, \tag{4.5}$$

where $fwhm_z$ and $fwhm_\theta$ are the fwhm of the z component of Ps and the angular correlation distribution, respectively. In an anisotropic harmonic potential (4.5) is valid for each component. We measure $fwhm_\theta = 2.3$–2.5 mrad for the p Ps intrinsic annihilation for hydrocarbons (Table 5.2) corresponding to $fwhm_z =$ 0.47–0.43 nm in that approximation. The infinite square well potential gives

$$r \times fwhm_\theta = 1.66 \text{ mrad nm} \tag{4.6}$$

The main difficulty in the application of the models of the Ps state is the connection of the parameters of the model with the observed data. For example, in (4.2) z_{eff} is essentially unknown, and U and r are strongly dependent on the choice of z_{eff}. For an infinitely deep square well potential (4.2) cannot be used as $P(U, r)$ is zero. It has then been assumed that a thin layer of electrons on the surface of the well with an annihilation rate of the positron in Ps of 2 ns^{-1} exists, and the probability of finding the Ps inside this layer has been calculated and used to get the lifetime, similar to (4.2). The thickness of the layer can then be used as an adjustable parameter in the fitting of the model predictions to experimental data [4.24,25]. Of course, such a procedure is somewhat arbitrary. However, a fair agreement with the measured data could be obtained, which shows that the models describe essential properties of the Ps bubble state. In the few cases where the energy terms in (4.1) have been calculated by use of various assumptions the total energy of the Ps bubble versus bubble radius had a rather shallow minimum [4.26]. This indicates that the bubble size will vary much in time because of the temperature induced movements of the molecules. A change in the energy by $\pm$ kT seemed to change the radius by up to 25%. Hence, the surface of the bubble is expected to be rather uncertain.

The measured o Ps lifetime versus cavity volume is shown in Fig. 4.1 [4.24] for various solids and liquids. The empty and filled circles show the lifetimes of o Ps trapped in vacancies in molecular crystals for which the volume is taken to be the monovacancy volume. The filled circles are for crystals known to have a high divacancy concentration. The triangles and squares show the measured o Ps lifetimes in three liquids at various temperatures versus the Ps bubble volume derived by use of a square well potential and a partly empirical correlation between the lifetime and the surface tension and pressure [4.26,27]. ($r' \simeq (\hbar^2/16m_e)^{1/4}(\gamma + \beta p)^{-1/4}$ inserted into (4.8) was used, see [4.27] expressions (5) and (6) and explanations). It must be emphasized that the bubble volumes are not measured but derived for the liquids! The very good continuation of the points in Fig. 4.1 from the solid into the liquid phase indicates that the square well Ps bubble is a fairly good approximation for the liquid state. A measured lifetime of about 9 ns for the α-cages of zeolite 13X of volume 9.05 nm^3 [4.28] agrees with the curve through the hexane and SF_6 points in Fig. 4.1. Hence, this curve might be the best one to use at present.

The o Ps pick off annihilation rate, $\lambda_p = \tau_3^{-1}$, has been correlated to several properties of liquids. In particular, the correlation to the surface tension, γ, has been proposed. However, a plot of λ_p versus surface tension [4.25] shows a

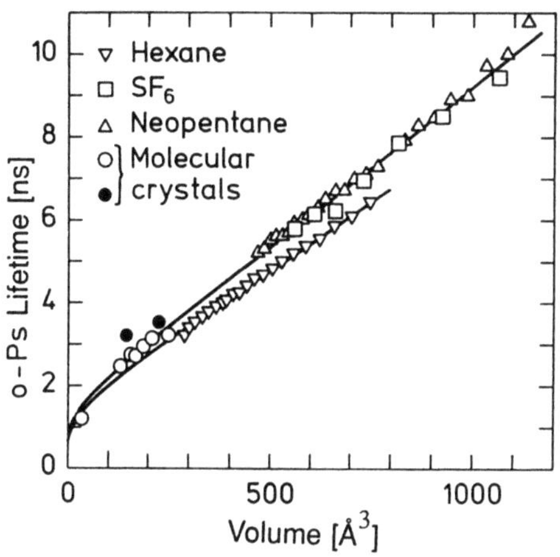

Fig. 4.1. The relationship between the o Ps lifetime in a cavity and the volume of this cavity. The circles show the o Ps lifetimes in various molecular crystals close to the melting point as a function of molecular volume, which is close to the vacancy volume. The open circles in order of increasing volume are for: ice, succinonitrile, norbornylene, cyclohexane, bicyclooctane, adamantane, dl-camphene, and the closed circles: pivalic acid and hexamethylethane which both have high divacancy concentrations. The triangles and squares show o Ps lifetimes in bubbles in liquids as a function of bubble volume. The curves represent model calculations fitted to the data for the liquids

fairly large scatter of the points around curves given by

$$\lambda_p = \kappa \gamma^{\alpha} \tag{4.7}$$

with fitting constants $\kappa = 0.061\ \text{ns}^{-1}$ and $\alpha = 0.5$ for n-alkanes; and $\kappa = 0.046\ \text{ns}^{-1}$ and $\alpha = 0.55$ for compounds having strong hydrogen bonding properties. At higher temperatures close to the critical point the surface tension is small and the saturated vapour pressure high, and hence, the pv term in (4.1) becomes important. *Jacobsen* [4.26,27] proposed, therefore, a semi-empirical expression

$$\lambda_p = A \exp\{-\alpha(\gamma + \beta p)^{-1/4}\} \tag{4.8}$$

where A, α, and β are fitting constants. (4.8) was used in a large temperature intervals to fit the measured λ_p values for SF_6, neopentane, and hexane [4.26,27,29]. The fitting constants varied somewhat from liquid to liquid. For hexane $A = 1.77\ \text{ns}^{-1}$, $\alpha = 4.1\ (\text{dyn cm}^{-1})^{1/4}$, and $\beta = 0.110\ \text{dyn cm}^{-1}\text{atm}^{-1}$ were found for temperatures between −73°C and 160°C.

In [4.30] the correlation between λ_p and the Hildebrandt solubility parameter, δ, was discussed. δ is defined as the square root of the cohesive energy per unit volume. In a good approximation

$$\delta = \{(\Delta_l^g H - RT)/V\}^{1/2} \tag{4.9}$$

where $\Delta_l^g H$, R, T, and V are the molar enthalpy of vaporization, gas constant, temperature in Kelvin, and molar volume, respectively [4.31.32]. *Hansen* [4.33] proposed to divide δ into three components according to

$$\delta^2 = \delta_d^2 + \delta_p^2 + \delta_h^2 \tag{4.10}$$

where δ_d, δ_p, and δ_h are the dispersion, polar, and hydrogen bond components of the solubility parameter, respectively. In [4.30] it was shown that λ_p correlated well with the dispersion component δ_d of δ, but not with δ itself.

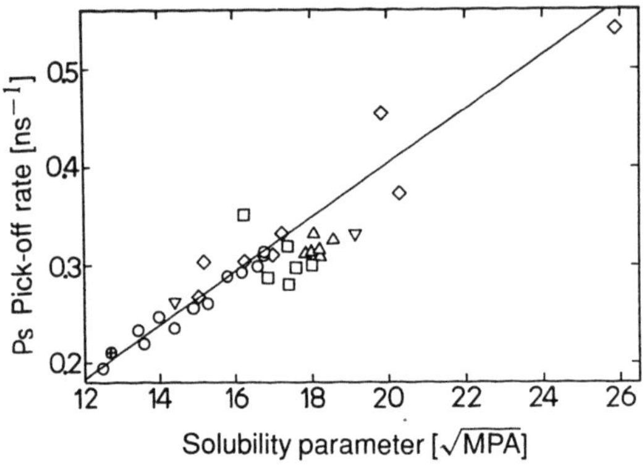

Fig. 4.2. The o Ps pick off annihilation rate versus the combined dispersion and polar components of the solubility parameter $\{(\delta_d^2 + \delta_p^2)^{1/2}\}$ for various liquids: Aliphatic hydrocarbons (O); aromatic hydrocarbons (△); alcohols (□); ethers (▽); tetramethylsilane (⊗); and miscellaneous liquids (◇). The straight line is given by (4.11)

The correlation of λ_p with the solubility parameter, δ, was studied in more detail in [4.34]. In Fig. 4.2 is shown λ_p versus the combined dispersion and polar component of the solubility parameter, $(\delta_d^2 + \delta_p^2)^{1/2}$, for various liquids. The straight line in Fig. 4.2

$$\lambda_p = \tau_3^{-1} = -0.14687 + 0.02755\delta \text{ ns}^{-1}, \quad \delta \text{ in (MPA)}^{1/2} \tag{4.11}$$

where δ is the total solubility parameter, fitted very well the points for aliphatic (linear, branched, and cyclic) hydrocarbons and for hexane at various temperatures between −90.6 and 140°C. These liquids have only a dispersion component. Clearly, the correlation in Fig. 4.2 is good. However, a plot of λ_p versus the total solubility parameter, δ, including all liquids shows a very poor correlation. In particular, the points for the liquids with large hydrogen bond components of the solubility parameter in such plot deviate appreciable from the straight line discovered for the aliphatic hydrocarbons. This concerns, in particular, alcohols, water, acetonitrile, and acetone. For example, the λ_p (ns^{-1}), δ ((MPa)$^{1/2}$) values are: water 0.541,48.0; methanol 0.279,29.7; ethanol 0.286, 26.1; propanol 0.296,24.9; butanol 0.298,23.7; acetonitrile 0.303,24.8; and acetone 0.304,19.7. Smaller shifts to lower solubility parameters on omission of the hydrogen bond component occur for pyridine, dioxane, tetrahydrofuran, cis- and trans-butene-2, and diethylether. In Fig. 4.2 the points for all liquids are close to the straight line.

It is important to realize that several sets of solubility parameters and several ways of determining their components are used [4.31]. We have here used the "Hoy et al." components given in Table 2 in [4.31], while older "Hansen" components were used in [4.30]. The differences between the two sets of components reflects the methods used in their derivation. Apparently, it is not possible to decide which components, dispersion ("Hansen") [4.30] or combined dispersion and polar ("Hoy et al.") [4.34], give the best correlation to λ_p. The components are discussed in [4.31,34]. Furthermore, the scatter of the points in Fig. 4.2 seems to be mainly caused by the uncertainty in the derivation of the solubility parameter components.

The important experimental result, discovered by use of plot of λ_p versus various components of the solubility parameter, is that a good correlation is only found if the hydrogen bond component is neglected. This result is independent of which set of solubility parameter components is used. The only reasonable conclusion is that hydrogen bonds are not broken when the Ps bubble is formed. Hence, almost all the molecules close to the Ps bubble, and at the Ps bubble surface, participate in the same number of hydrogen bonds as in the liquid not disturbed by the presence of the Ps bubble. A similar principle is seen to work for clathrate formation in hydrogen bonding systems, where the hydrogen bonds are also satisfied as a "first priority" in crystal structures. Furthermore, clathrate like structures around neutral solutes in water seem to have been made very plausible by use of molecular dynamics calculations. The clathrate like structure around the Ps bubble is, therefore, in agreement with what is expected from other known structures.

The missing effects of the hydrogen bonds on the Ps bubble state explains well why λ_p versus the macroscopic surface tension, γ, gave a fairly poor correlation, in particular why the hydrogen bonded liquids had to be fitted by use of other parameters than the normal nonpolar liquids. The models of the Ps bubble discussed above can be usefully applied in the discussion of the hydrogen bonded liquids if an effective surface tension is used instead of the macroscopic surface tension. A necessary condition is that the energy of the bubble surface is approximately proportional to the bubble surface area, the effective surface tension being the proportionality constant. Hence, the qualitative and semi-quantitative conclusions reached by use of the Ps bubble models are still valid. The models were only used to obtain a rough understanding of the Ps bubble state, anyway, and they should not be taken too seriously. Of course, a difficulty is that the effective surface tension is unknown and has to be estimated.

The measured o Ps lifetimes for mixtures of hydrogen bonded and nonpolar liquids speak strongly in favour of the lack of effects of hydrogen bonding on the Ps bubble. In the Risoe papers on Ps formation in liquid mixtures the o Ps lifetimes have normally also been reported for the full concentration interval. These results include CS_2, alcohol, acetone, pyridine, C_6F_6, chlorobenzene, CH_3Cl, CH_3Br, and CCl_4 mixtures with one or more hydrocarbons, such as hexane, isooctane, neopentane, cyclohexane, and benzene (tetramethylsilane), and mixtures of hydrocarbons. The references can be found in Chap. 7. In nearly all mixtures the o Ps lifetime varies almost linearly with the mixing ratio (mole-, volume-fraction or mole/liter) with a small tendency of being a little below the straight line joining the values for the pure liquids. In particular, the results for mixtures of alcohols and nonpolar liquids follow that rule, and they are not significantly different from those of other mixtures. At first sight, one might have expected that Ps would make the bubble preferentially in the nonpolar parts of the liquid, and hence that τ_3 versus concentration would be very much above the linear dependence on concentration. This is not found because the energy necessary to create the bubble is not very different in the nonpolar and hydrogen bonded parts of the liquids, as discussed above. As shall be discussed in Sect. 7.6, the Ps yield versus concentration varies very strongly, as the electron and positron are preferentially solvated in the polar parts of the

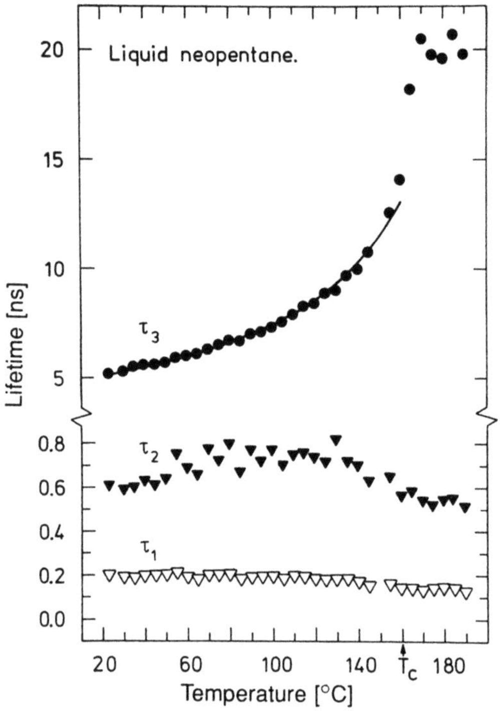

Fig. 4.3. The temperature dependence of the positron lifetimes in liquid neopentane as obtained from lifetime spectra resolved into three exponential components. T_c is the critical temperature

liquids. Hence, the characteristic properties of the light particle states cause strong effects in the measured annihilation results.

The o Ps lifetime versus temperature in neopentane at its saturated vapour pressure is shown in Fig. 4.3 [4.29]. This illustrates the change in the o Ps lifetime at the critical point, T_c. Apparently, neopentane [4.29] and SF_6 [4.27] are the only molecular liquids which have been studied at and above T_c. SF_6 will be discussed in more details in Chap. 5, as it gives two o Ps lifetimes. In both neopentane and SF_6 the o Ps lifetime is constant above T_c, while it increases very strongly for increasing temperatures up to T_c. This is a very dramatic effect! The surface tension decreases with increasing temperature below T_c and becomes zero at, or very close to, T_c. Hence, above T_c the inwards pressure of the Ps bubble is only given by the pv term in (4.1), where p is the saturated vapour pressure of the fluid. As this pressure varies little with the temperature compared with the variation of the surface tension below T_c, the o Ps lifetime is nearly constant above T_c. This is another results which is easily understood in terms of the Ps bubble state.

The discussion of the Ps bubble state in this section is only an introduction to the topic. Several special effects caused by the bubble around Ps will be treated in other chapters. In particular, the "squeezing" of Ps out of the bubble state and into the "fourth positron state" in liquid CS_2 by lowering the temperature will be discussed in Chap. 5. On decreasing the bubble size, the

Fourier transform of the p Ps wave function, and hence, the p Ps AC spectrum, gets broader because of the increase in zero point energy. However, the wave function begins to spill over at still smaller bubble sizes. In the smallest bubbles which still can bind Ps, the Ps wave function contains a large spilled-over component. This results in an AC spectrum of p Ps consisting of a normal narrow component due to the spilled-over, and a fairly broad component due to the strongly localized, components of the wave function. The effects of the Ps bubble on the Ps reactions with solutes is treated in Sect. 8.3, and the change in the Ps bubble state on transition from liquid to amorphous state is described in Sect. 10.1. The related topic of Ps in holes in molecular compounds is discussed in Chap. 9.

4.5 Short Remarks on Light Particle States in Solids and Gases

With the exception of the molecular solids normally treated in chemistry, such as polymers and molecular crystals and glasses, positron annihilation in the solid state, e.g., metals, semiconductors, and ionic crystals (see, e.g., [1.5] and proceedings of the ICPA conferences [1.7–12]), is not a topic of this book. Positron annihilation in gases is not treated either [1.7,8,9,13,15]. However, some remarks on the light particle states in solids and gases in general are appropriate here.

In metals, semiconductors, and other rigid solids the electrons in valence and conduction bands are in delocalized states. The standard delocalized state is a Bloch function, which is a wave function diffracted into itself by the interaction with the atomic cores. In practice the delocalized states are always disturbed somewhat because of the presence of defects and the motion of the atomic cores due to thermal motion and zero point energy. This disturbance is normally described by reference to, e.g., a coherence length over which the electron interferes constructively with itself, a mean free path, or other parameters of semi-quantitative theories. In metals the strong electron-electron interaction complicates the description further. Hence, Bloch functions are idealized states which only describe the real states approximately. Apparently, a reasonably correct and practically useful theory of coherence lengths, mean free paths, etc. is not available, and hence, the models are at best semi-quantitative and should not be taken too seriously. The best investigated "excess electrons" in solids are probably the electrons in the conduction bands in silicon and germanium. Most textbooks on solid state physics discuss these topics in great details. Problems in the description of electrons in solids are discussed in Sects. 5.12,7.1,2 of [4.35].

Very little is known experimentally about the positron in molecular solids. We shall discuss the lifetime and the very few mobility results in Chaps. 9,10. Theoretically, it is expected that the behaviour of the positron is roughly similar to that of the excess electron.

The normally found state of Ps in the non-defected molecular solids where it is formed, is a self-trapped state. The strong repulsion between Ps and the molecules discussed above is the reason for the equilibrium state being a localized Ps in a "cage", i.e., a small hole in the structure which is smaller than the Ps bubble in the liquid. Hence, Ps digs a hole in the solid by pushing away the molecules. Only in the most rigid solids, such as ice and some ionic crystals at lower temperatures, Ps is delocalized in a Bloch function state, which, of course, is somewhat disturbed by imperfections in the lattice. Ps is a good example of the fact that a light particle is only delocalized if it interacts fairly weakly with molecules. Strong attraction or repulsion cause the particle to be localized in local structural deformations created by the presence of the particle, i.e., it is selftrapped.

In principle, all solids contain defects, i.e., deviations from the perfect structure. Ps is trapped in the hole type defects, such as vacancies (missing molecules) and vacancy clusters, if it can reach the defect by random diffusion and transition into the defect within its lifetime. The reason for the lower energy of Ps in the hole is, of course, the repulsion between Ps and the molecules. Ps studies of hole type defects in molecular crystals will be the most important topic of Chap. 9. Positron trapping in hole type defects in metals, semiconductors, and ionic crystals is a very important topic of positron annihilation.

In summary, light particles are "seekers" and "diggers" in many compounds. Only particles interacting weakly with the molecules, or situated in rigid structures where deformation is difficult, are in delocalized states, which in perfect solids are the somewhat disturbed Bloch function states.

In gases of molecules not accepting the particles, the light particles are in free particle states at very low pressures. At somewhat higher pressures, the particles can be described as free particles disturbed by occasional single collisions with the molecules. The free particle states then play the same role in theory as the Bloch function states in solids, i.e., they are the initial wave functions in a perturbation theoretical calculation of the effect of the collisions on the states. At still higher densities, the light particle interacts with several molecules at the same time, and the concepts of single scattering theory, such as cross sections of collisions with single molecules, do not describe well the simultaneously collisions with many molecules. In many papers the cross sections for single collisions are used incorrectly in high density gases, liquids and amorphous solids. In crystalline solids, cross sections of scattering from Bloch function states might be useful concepts in the theories.

The changes in the behaviour of the light particles if the density is increased from gaseous to liquid densities are very interesting [4.36,37]. If the molecules are polar, the charged particles, the positron and the electron, can get trapped in clusters of molecules in gases at densities more than hundred times lower than liquid densities. For example, at increasing density the lifetime of the free positron decreases to that of liquid NH_3 in gaseous NH_3 at 22°C already at a density of about 0.25% of the liquid density. This indicates a positron annihilation from a solvated positron state in a cluster of NH_3 [1.53]. On the other hand, Ps forms a sort of a bubble state in gases at densities far below those of the liquid [1.13]. If the light particle interacts weakly with the molecules, strong changes in the behaviour on increasing the density might first occur at, or fairly close to, the transition into the liquid. For example, the mobility of the excess electron, μ_-, is roughly inversely proportional to the density for densities below a density of 5×10^{20} and 1.5×10^{20} cm^{-3} in methane and ethane, respectively. In methane $n\mu_-$, where n is the density, increases by a factor of 50 at about $n = 5 \times 10^{21}$ cm^{-3}, while $n\mu_-$ decreases by a factor of more than 300 in ethane at around $n = 5 \times 10^{21}$ cm^{-3}, on increasing the density to liquid densities at the vapour/liquid coexistence curves [4.37]. The reasons are that the electron is partly delocalized in liquid methane, but it is partly localized in liquid ethane. Hence, the localization and delocalization might occur close to liquid densities on increasing the density in fluids where the interaction with the light particle is fairly weak. However, localization occurs long before liquid densities are reached if there is a strong interaction between particle and the molecules.

Of course, this is a very short discussion of the light particle behaviour in fluids, as, for instance, the temperature plays a strong role, too. We can only describe very few results here. However, the few discussed cases are very illuminating. Very few results of studies of the positron and Ps in fluids around the critical point have been published compared to what is known about the excess electron. This is partly an open field, which should be investigated much more in the future. No Ps quenching or inhibition by solutes above, or just below, the critical point have been published for any fluid!

4.6 Measurable Properties of Light Particles

From another point of view it is useful to discuss the type of information on light particles which can be obtained in selected cases. In principle, measurements might give various types of information on the light particle state, as illustrated by the remarks 1)–6) given below concerning the wave function of the light particle alone. Of course, a full description of the state must also include the changes in the surrounding structure, and the density matrix of the particle must be referred to. However, as we shall discuss basic principles in a qualitative way and since the measurements normally give information on the light particle itself, the reference to the wave function of the light particle alone is sufficient here.

1) The full information on the light particle alone, namely its wave function, ψ, cannot be determined in a single experiment. This is a general property of quantum mechanics.
2) The probability distribution, $|\psi|^2$, of the light particle alone in real space cannot, apparently, be measured in any detail for any light particle.

3) The square of the numerical value of the Fourier transform of ψ (measurable for p Ps) is similar to the real-space distribution $|\psi|^2$ with respect to the information content of the data.
4) For a bound particle (trapped on a defect or impurity, unpaired electron on a free radical), properties that essentially depend only on its ground-state wave function might be measurable.
5) For a bound particle, other measurable properties (e.g., optical absorption) might depend on both its ground state and excited state wave functions.
6) Properties that can be expressed by single numbers calculated by use of the wave function (e.g., trapping rates, mobilities, etc., for mobile particles) might be measurable for trapped and mobile particles.

The main reasons for the importance of EPR measurements on free radicals and electrons trapped in defects are that the measured electron properties are essentially only dependent on the ground-state (except for the magnetic splitting) wave function (case 4), and that an EPR measurement, being a resonance technique, gives results which are very dependent on details of the structure of the EPR center. More detailed information is available on the structure of defects in silicon, than in any other system, mainly because EPR experiments on the defects (free radicals, trapped electrons) have been systematically carried out. Similarly, very detailed information on the structure of colour centers, etc. in ionic crystals has been obtained by use of EPR and optical absorption measurements. However, optical absorption results for solvated and trapped electrons depend on both the ground and excited states (case 5), and, hence, several theoretical models of the structure of the electron and surrounding molecules can normally fit the measured absorption spectra. Roughly speaking the only measurable properties of partly delocalized excess (conduction band) electrons seem to be single-number properties, such as mobilities, Hall coefficients, and trapping rates (case 6).

For p Ps (trapped or delocalized) the numerical value of the Fourier transform of the center of mass wave function can be determined (case 3). This detailed information on the state is not available for the other light particles in any system. EPR studies are just as informative, though. In particular, much more information seems to be available on Ps Bloch function (case 3), than on the excess electron Bloch function (case 6) in solids. In this context it must be realized that Ps trapping at two different times (the p Ps and the o Ps lifetimes) can be measured (trapping rate determination, case 6). Furthermore, the AC spectrum and the pick off lifetime of o Ps are ground state properties (case 4) given by the overlap of the positron wave function with the wave functions of the electrons in the molecules. Additionally, magnetic quenching of Ps gives information on the density of the electron at the positron (case 6). For free positrons the AC spectrum is determined by the Fourier transform of the positron electron wave-function overlap. As this is an ground-state property (case 4), more detailed information is probably available on the localized positron than on the localized electron in the fairly frequent cases where the EPR measurements on the electrons do not give valuable information. For example, optical absorption studies on solvated electrons do not seem to give as valuable information as the AC studies on the solvated positrons. For the hydrated electron, the main theoretical problem seems to be that many models of its structure can be adjusted to fit the measured absorption spectrum, i.e., the determination of the structure is very difficult. The studies of the positron trapping in defects in metals give information similar in quality to the information obtained in colour center research of ionic crystals, while only scattering properties of the defects can be measured by use for the delocalized electrons in metals. This explains why positron trapping in hole type defects in metals has been one of the best, and in some cases the only available, methods of research of certain defects in metals.

In summary, the structure of a light particle state must be determined by use of the measured numbers, since first-principle quantum-mechanical calculations are normally much too complicated. The best method is to use intelligently chosen, approximate models of the states, and describe the structure in terms of these models. The higher the "quality" of the measured results, the easier this assignment of the structure is. The more averaging, integration, summation over states, etc. one has to perform in the calculation of the measured results, the lower the "quality" of the results is. Because of the characteristics of the annihilation process certain particular properties can be measured for the Ps and positron states which cannot be measured for the other light particle, the electron. Hence, all three light particles must be treated in a discussion of light particle states in matter.

5 The Positron Spur and Ps Formation in Pure Liquids

Positronium formation occurs mainly by a reaction between the positron and an excess electron in the positron spur in liquids, solids, and high density polyatomic gases [1.35]. The positron spur is the terminal spur of the positron track formed when the positron loses the last part ($\sim$ 500 eV) of its kinetic energy. Both particles are almost, or completely, thermalized at the time of the reaction. However, in frequent cases the particles are not fully solvated, i.e., not in thermal equilibrium with the solvent, when they form Ps.

At first we shall discuss the positron spur and compare its properties with those of the spurs of the electron track. The homogeneous, long time irradiation caused by the positron slowing down is then shortly treated. The following section is a discussion of the experimental determination of the Ps yield. The spur reactions in the positron spur in pure liquids are then discussed in details. In the last section we explain the measured Ps yields for pure liquids in terms of the positron spur reactions. In Chap. 7 we shall discuss how the Ps yield is changed if various solutes are added to the liquids. The treatment given in this chapter will be the starting point of the discussions. For short we call excess electrons just electrons whenever convenient.

5.1 The Positron Track and its Terminal Spur

Energetic positrons and electrons penetrating matter lose energy and are deflected from there original course. Changes take place in the matter that is penetrated. In particular, atoms are excited and ionized, molecules are dissociated, and defects are created in solids. These species constitute the particle track. The track species thermalize rapidly, normally within one picosecond or less. Exception are the electrons in the heavier, liquid, noble gases whose thermalization can take 4–6 ns. They then react with the other track species and/or molecules or they diffuse out of the track. In solids parts of the track species might be frozen in and have very long lifetimes.

Charged particle tracks are discussed in many reviews, monographs, and textbooks ([5.1], pp. 1–78 in [5.2], Chap's. 1–6 in [5.3], [3.1]). The early reactions in the tracks are discussed in [5.4–7].

The stopping power is defined as the energy loss suffered by the particle in traversing a unit length in the medium along its track and is synonymously

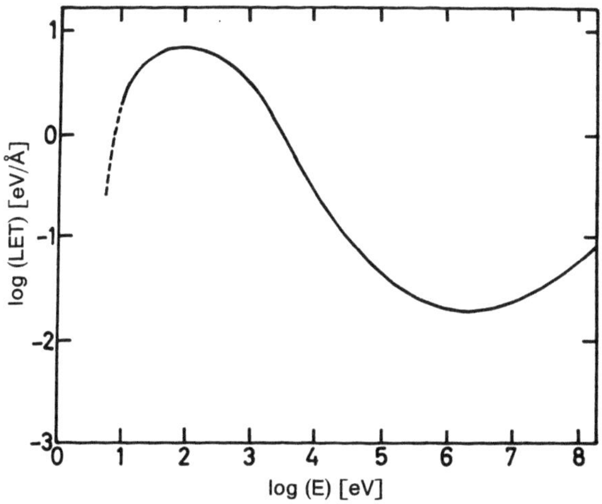

Fig. 5.1. Stopping power, LET, of water for electrons versus electron energy, E, plotted on a log-log scale. LET is in eV/Å($1 Å = 0.1$ nm) and E in eV

known as the linear energy transfer (LET). In Fig. 5.1 is shown log(LET) versus log(E) for an electron in water [5.1]. The positron track is very similar to the electron track except at low energies. For other normal liquids the LET versus E curve for electrons and positrons is fairly similar to that for water. Below an energy of 100 eV the LET-values are uncertain for the electron. Furthermore, very little is known about the track and its LET at subexcitation energies, i.e., below the energy of lowest excited state in the molecules. Still less is known at energies below the vibration and rotation levels of the molecules of the liquid. The stopping power at these very low energies seems to be the processes which mainly determine the range of the secondary electrons, b, shown in Table 4.1.

For fast electrons of energy between 0.03 and 10 MeV the LET is between 0.02 and 0.05 eV/Å. Experiments show that 4–5 electron-positive-ion pairs are formed for a loss of 100 eV by a fast electron. Hence, the average distance between two ionizations is larger than about 40 nm, which is normally large compared to b. To aid the description of this situation the term "spur" has be defined [5.8] as "A cluster of one or more geminate ion-electron pairs. More generally, a spur is one of the clusters of reactive intermediates formed in a dense medium by ionizing radiation. Reaction rates in spurs are described by nonhomogeneous kinetics". Nonhomogeneous kinetics has been defined [5.8] as "Rates and mechanism of reactions between reactants at least one of which is neither uniformly nor randomly distributed in space".

By use of these definitions the electron and positron tracks can be described as strings of spurs. At energies above 0.02 MeV the initially formed spurs are well separated. At lower energies the initially formed spurs might overlap somewhat, and at energies of 100–500 eV they overlap so much that the last part of the track is better described as one large cylindrical distribution of track species, a so called "short track" [5.1].

The most important reactants in the spur are the electron-ion pair. It is important to realize that ionizations and excitations of molecules to higher energy states lead normally to the similar electron-ion pairs after thermalization. Hence, ionizations cannot be well separated from excitations in spurs. Exceptions seem to be the lowest excitations of the larger, rigid molecules. The reasons is that electrons of energies which in vacuum would give ionizations will normally thermalize inside the spur of average size b (Table 4.1). As the interaction between the electron and the molecules is strong, the electron will have a coherence length (mean free path) which is much shorter than the quantum mechanical orbit an electron at that distance to the ion would occupy in vacuum. An electron which in vacuum would be in an excited state will after thermalization be in a similar state. Hence, both electrons are in excess electron states with coherence lengths much shorter than the size the orbits which they would occupy in vacuum at the same distance form the positive ion. Consequently, they can be treated approximately as classical particles diffusing and drifting under the influence of the electric fields of the charged particles. The corresponding diffusion equation will be discussed in the Sect. 6.2, where the reaction kinetics of importance for positron annihilation is discussed in some details.

Tracks are not visible except if the tiny signals of the primary radiation damage are directly amplified into something visible in, for example, cloud chambers, bubble chambers, and nuclear emulsions. By Monte Carlo calculations based on theories of the slowing down processes good "pictures" of tracks and their developments in time can be obtained, as illustrated by Fig's. 2, 3 in [5.9]. Many papers on the calculation of track structures have been published (see, e.g., [5.1,3–5,9,4.12] and Ref's therein).

The positron spur is simply defined as the spur where the positron is thermalized, i.e., the last spur of the positron track [1.35]. The positron track is very similar to the electron track except at very low energies. Actually, our knowledge about the positron spur is limited, except for the information obtained by Ps formation studies. Except for the very low energy positrons, the positron spur is expected to be part of a "short track", i.e., a cylindrical distribution of radiation damage. It is, thus, a high density spur where several electrons, ions and the positron are within each others Coulomb fields. Roughly speaking, a charged reactant is influenced by an electric field due to other charges, if its energy in the field is larger than the thermal energy kT. The Onsager radius, r_c, is the distance from a single charge where the Coulomb energy of an electron is equal to kT (6.19). In nonpolar liquids $r_c \simeq 25$–30 nm, while it is only 0.7 nm in water at room temperature. Consequently, the spur is mainly inside the Onsager radius in nonpolar liquids. The G_{fi}-values of Table 4.1 illustrate this, as the total G-value is 4–5 $(100\ \mathrm{eV})^{-1}$, i.e., only 2–20% of the electron-ion pairs escape recombination in nonpolar liquids. It must be realized that these G_{fi}-values are experimental ones for fast electron spurs, which are normally assumed to be mainly single pair spurs. Actually, the percentage of single pair spurs in a fast electron track seems to be known only with great uncertainty.

This illustrates one of the main problems we encounter in the use of data obtained for fast electron spurs for the interpretation of Ps formation results.

With few exceptions information on spur properties is available only for fast electron, i.e., mainly single pair, spurs, while the positron spur is a high-density, multipair spur. Hence, we cannot expect to be able to explain the Ps yields in terms of the available spur results in details. We shall, thus, mainly use qualitative and semiquantitative correlations to fast electron spur data in our discussion of the Ps yields. In the following chapters it will be shown that the characteristics of the Ps formation processes exclude a detailed comparison of the electron and positron spurs anyway. Ps formation by positrons of such low energy ($E \ll 500$ eV) that a large positron spur is not formed, i.e., in slow positron beam experiments, will be discussed in Chap. 9 on molecular crystals.

In rigid solids the positron spurs will be different from those in typical liquids as the light particles are delocalized with coherence lengths comparable to, or longer than, the sizes of the quantum mechanical orbits involved. The positron spur will then be an "exciton like" spur. The simplest "exciton like", single-pair spur is probably the self-trapped exciton, i.e., an electron in an orbit around a substitutional X_2^- ion, in alkali halides, where it seems to be the most important primary electron-ion pair state created by radiation ([5.10,11] and proceedings of "Defects in Ionic Crystals" conferences). Ps formation in "exciton like" positron spurs might be an interesting topic of future research! In certain inert gas liquids, e.g., xenon, with high excess electron mobility, the spurs may tend to be "exciton like", too. The transition from "classical particle like" to "full quantum mechanical particle" spurs has been discussed very little in the literature.

In gases the transition from "classical" to "quantum mechanical" spurs on increase of the coherence lengths caused by a decrease in the density takes place along with an increase in the secondary particle ranges. Hence, the positron electron distance at thermalization gets longer as the density decreases, and Ps formation by the spur process becomes less probable. The effects of the quantum mechanical behaviour of the particles in the low density positron spurs where the spur process of Ps formation is still important are difficult to estimate. The theoretical problems seems not to have been discussed.

5.2 Homogeneous, Long Time Irradiation by Positron Slowing Down

Until now we have discussed the effects of the species which are close to the positron at thermalization and created by the slowing down of the positron itself. It is appropriate here to discuss briefly the homogeneously distributed, long-time radiation damage from positron slowing down and its effect on positron annihilation. We limit the discussion to insulating liquids and solids. Let us first assume that the species stay where they are created. For a typical lifetime measurement a ^{22}Na source of 20 μCi (1Ci $= 3.7 \times 10^{10}$ Bq) is used, about 0.005 cm^3 is irradiated as the positrons penetrate roughly 0.2 mm, and the average energy per positron is about 0.2 MeV. This results in $n = 3.0 \times 10^{11}$ G species/s cm^3, where G is the yield of species per 100 eV loss of energy, or $c = 5.0 \times 10^{-10}$ G M/s concentration. After one day and for $G = 4$ we get $c = 0.17$ mM. For angular correlation measurement the source is typically 40 mCi ^{22}Na, of which 6 mCi hits a sample surface of about 1 cm radius. This gives roughly a 25 times higher concentration of radiation created species than in the lifetime case. Hence, after one day $c = 4$ mM for $G = 4$ $(100 \text{ eV})^{-1}$. In general, the concentration of the irradiation created species might be reduced at higher concentrations by reactions with the species of the new tracks. Furthermore, in many liquids and solids most electron-positive-ion pairs disappear by recombination.

For liquids the diffusion constant is normally $(2–3) \times 10^{-5}$ cm^2s^{-1}, and hence, the species move about 1 cm in 3 hours by diffusion alone, i.e., apart from motion due to convection. As the sample volume is typically 5–10 ml, the concentration of species created homogeneously by the positron slowing down will be too low to influence the Ps formation and/or lifetime; the only exceptions being the chain reaction case. If chain reactions enhance the irradiation yields very large G-values might be found (e.g., 200–5000 $(100 \text{ eV})^{-1}$), and the Ps parameters can be strongly influenced by the radiation damage. For example, in many alcohol mixtures with halogenated hydrocarbons chain reactions occur. A very strong change in the AC spectrum was found after some days in CBr_4/butanol mixtures due to destruction of CBr_4 and formation of Br^- by chain reactions [5.12]. CBr_4 inhibits Ps formation, while Br^- reacts with the positron. This case will be discussed in Chap. 6.

In insulating solids the species created homogeneously by the positron slowing down processes can be very important in particular cases. The reason is, of course, that the species cannot move and disappear by mutual reactions in many solids, and, furthermore, the trapping rates of the light particles might be large in solids. Some examples illustrate this effect. In ionic crystals the excess electrons created by radiation can get trapped in defects to form colour centers, and electron centers trap positrons and/or react with Ps. Hence, the AC, LT and DB spectra may change strongly at the beginning of a measurement until saturation in electron trapping is reached. As the light particles are partly delocalized concentrations as low as parts per million may trap ten percent or so of the Ps atoms in ionic crystals [5.13]. In ice at temperatures below about 40 K the positron slowing down creates a high concentration of defects, probably OH radicals, which inhibits roughly 40% of the Ps formation after only one hour ($\sim$ 85 krad) in AC measurements [5.14]. Strong external electric fields separate the irradiation created electron-ion pairs in solids, and charging (electret effects) may occur [5.15]. This might strongly influence Ps formation. In particular, a Cu-64 source was used in AC studies of electric field effects on Ps formation in polyethylene [5.16]. The additional electron irradiation of Cu-64 meant that doses of over 200 Mrad per day were applied. One Mrad gives roughly ρG mM concentration of defects for the density, ρ, in g/cm^3, and hence, the polyethylene was probably strongly defected and the Ps yields might be fairly different from those of nonirradiated polyethylene. In conclusion, tests of irradiation effects due to positron slowing down should always be performed for insulating solids. Such test could be a measurement of the time dependence of the positron annihilation characteristics.

5.3 Determination of the Ps Yield

Before we can discuss the Ps yields in various types of compounds it is necessary to describe how the Ps yield is determined experimentally. We will mainly use liquids as examples.

As discussed in Chap. 4, pure Ps states exist only in vacuum. Furthermore, in several systems one can continuously change the positron state from one which can clearly be denoted Ps to another which should not be called Ps. This can be done in Li/NH_3 solutions under variation of the Li concentration [5.17], and in liquid CS_2 by lowering the temperature (see Sect. 5.5). Recently, studies of the magnetic quenching of Ps have demonstrated that strongly distorted Ps states exist in many systems. Hence, in some systems the existence of Ps is only a nomenclature problem, i.e., all states between Ps and some other positron states are found. It is then arbitrary to define a boundary between Ps and non-Ps states. In most compounds, however, the existence of Ps can be decided upon fairly easily.

Traditionally, the Ps yields have been determined by use of lifetime measurements. On assuming that the lifetime spectra consist of three exponentially decaying components, the o Ps yield is simply the relative intensity of the longest lifetime component taking into account small corrections due to three gamma decay. However, an extra lifetime component has recently been discovered by use of high resolution lifetime spectrometry [5.18,19], and the situation has become more difficult. A detailed discussion of the problems is, therefore, appropriate.

Some historical comment might be illuminating here. At the beginning of positron annihilation, i.e., roughly before 1970, the lifetime resolution was poor and the set-ups fairly unstable. A resolution of fwhm = 700–1200 ps was fairly typical, and normally the lifetime spectra were only separated into two components assumed to the long-lived o Ps component and a component composed of the sum of the free positron and p Ps annihilation components. A ruler through the measured points on a semilogaritmic plot of the spectra was used to extract the lifetime and intensity of the long-lived component. Later simple computer programs were used to analyze the spectra. Actually, a carefully performed "ruler method" gave acceptable results, which sometimes were better than those obtained by use of the simple computer programs. In hindsight, we now know that the extracted lifetimes were better than the intensities, and the intensities were normally too small if extracted by use of a computer program which could not take into account the resolution of the set-up.

At about 1970 the "standard" lifetime apparatus with a resolution of 350–450 ps became available in several laboratories. At first they were somewhat unstable, in particular, temperature dependent, and the resolution curve had to be measured frequently. Around 1974 the computer program POSITRONFIT (Chap. 3) began to be used by several Ps chemistry groups, and, hence, the resolution curve could be included in the analysis. The main part of the available lifetime results in the last 10–15 years have been measured with a resolution of about 400 ps and analyzed by use of POSITRONFIT. This is the "standard" procedure today, and most of results of the Risoe and Strasbourg groups belong to that category.

Around 1983 the lifetime set-ups at Risoe were improved to give a resolution of 190–210 ps. It was then decided to do some high quality lifetime measurements on typical liquids to test some of the basic assumptions of Ps chemistry [5.18,19]. This seems to be the only published high-resolution lifetime results for liquids. Apart from the good resolution we used a low source strength in order to reduce the constant background and thereby make the determination of the intensity of the longlived component easier. Furthermore, good statistics were obtained, which made the counting time fairly long. The resolution curve was normally measured just before and after the lifetime measurements. In some cases the lifetime spectra were measured with two different settings of the time-to-amplitude converter; one to get many points at short lifetimes and another to get a good determination of the background and the longlived component. Many careful analyses by use of RESOLUTION and POSITRONFIT were used to obtain the best possible extraction of the lifetimes and intensities.

As a result of these high quality lifetime measurements a new component of the lifetime spectra was discovered. Generally, our studies indicated that the new component is present in the spectrum for all liquids where Ps is formed. The new, fourth component is a distribution of lifetimes of relative intensity about 3–8 %, and with an average lifetime of roughly 0.7–2.0 ns. In the around 40 spectra for various liquids and solutions measured until now this component seems to be present with roughly this intensity and lifetime independent of the type of liquid. It is, of course, most well defined in nonpolar liquids with longer o Ps lifetimes and more difficult to determine in water and CS_2 having short o Ps lifetimes. In the older measurements and analyses this component seems to have been overlooked. Actually, we suspected that the three term analyses of spectra for liquids were somewhat incorrect for many years. In particular, the water lifetime spectrum was notoriously very difficult to analyze, and a fixed free positron lifetime of 395 ps was normally used to stabilize the fitting analyses.

Although we shall mainly use the results of the three term analyses of lifetime spectra in this book, mainly because high quality data is not available, it is useful to realize what experimental facts forced us to accept the fourth lifetime component. In Table 5.1 are given the results of unconstrained three and four term analyses of the same high quality spectra for hexane, 2, 2-dimethylbutane, and diethylether [5.18]. The shown changes in variance on adding the fourth term are typical for most liquids. Hence, three term fits are somewhat wrong. Furthermore, several guesses of the initial parameters were used to test whether the fourth component parameters were not just caused by a local minimum of

Table 5.1. The results of three and four term analyses of lifetime spectra for liquids. *m*. means that the variance is uninteresting, and *f* that the lifetime was fixed. Where two fits are given for a liquid it is for the same spectrum, and the variance is given

Liquid	I_1 [%]	τ_1 [ns]	I_2 [%]	τ_2 [ns]	I_3 [%]	τ_3 [ns]	I_4 [%]	τ_4 [ns]	var.
hexane	20.6	0.155	37.1	0.475	–	–	42.4	4.01	1.238
hexane	15.2	0.127	39.8	0.410	6.0	1.7	39.0	4.18	1.046
2,2-dimethylbutane	21.8	0.149	30.5	0.487	–	–	47.8	4.46	1.716
2,2-dimethylbutane	15.9	0.122	28.8	0.372	8.1	0.8	47.2	4.47	1.116
diethylether	22.8	0.186	44.4	0.460	–	–	32.8	4.04	1.259
diethylether	15.6	0.155	49.4	0.410	8.4	2.1	26.6	4.60	1.033
3-methylpentane	16.8	0.131	39.1	0.429	5.7	2.03	38.4	4.31	*m*.
2-methyloctane	14.9	0.132	41.7	0.407	4.5	1.70	38.9	3.87	*m*.
2,2-dimethylpentane	12.8	0.109	37.3	0.373	7.2	1.08	42.8	4.34	*m*.
2,2-dimethylhexane	13.6	0.122	46.4	0.430	4.2	1.96	35.8	4.14	*m*.
methanol	14.0	0.184	63.1	0.446	6.8	2.3	16.1	4.02	*m*.
carbondisulfide	17.5	0.141	35.1	0.419	46.1	2.107	1.3	3.91	*m*.
heptane	26.7	0.170	32.6	0.508	–	–	40.7	3.81	*m*.
isooctane	22.4	0.139	34.6	0.480	–	–	43.0	4.05	*m*.
neopentane	27.1	0.144	20.4	0.619	–	–	52.5	5.29	*m*.
cyclohexane	26.1	0.214	36.3	0.469	–	–	37.6	3.25	*m*.
benzene	20.5	0.142	36.4	0.440	–	–	43.1	3.18	*m*.
hexafluorobenzene	30.6	0.189	18.2	0.725	–	–	51.2	3.87	*m*.
tetramethylsilane	22.7	0.139	22.2	0.514	–	–	55.1	4.75	*m*.
methanol	17.6	0.150	60.8	0.464	–	–	21.6	3.53	*m*.
propanol	16.8	0.191	60.3	0.442	–	–	22.9	3.38	*m*.
water	25.6	0.211	48.4	0.454	–	–	26.1	1.89	1.13
water	11.6	0.137	61.0	0.395f	–	–	27.5	1.85	1.17
acetonitrile	12.9	0.196	71.7	0.492	–	–	15.4	3.30	*m*.
anilin	38.4	0.204	28.6	0.477	–	–	32.9	2.37	*m*.
carbontetrachloride	9.8	0.157	89.6	0.458	–	–	0.6	1.24	*m*.

the variance of the fit in the parameter space. A fourth component, such as the ones shown in Table 5.1, always was the result. Consequently, the fourth component is a well defined analysis result for most liquids. However, if several lifetime spectra for the same liquid were measured and analyzed a large scatter in the lifetime and intensity of the fourth component was found. This indicates that it is not just one exponentially decaying component, but a distribution of exponentials. The very small change in the variance of the fit for the same water spectrum on fixing the second lifetime to 395 ps instead of the 454 ps found in the unconstrained fit (Table 5.1) illustrates that the minimum of the variance in the parameter space is very shallow because the model, namely three terms, is wrong as the fourth component is missing.

Actually, the most reasonable explanation of the fourth component was proposed before its discovery [5.20]. Also in this case, experimental facts forced us to propose the new explanation. In Table 5.2 the intensities and fwhm's of three gaussian fits to the angular correlation spectra for many liquids are

Table 5.2. Angular correlation and lifetime results. I_i (fwhm$_i$) is the relative intensity (fwhm) of the i'th gaussian in a fit of a sum of three gaussians to the spectra. H is the number of counts in the top of the curves divided by the area between ±13 mrad as calculated from the fitting curves. I_3' (τ_3) is the intensity (lifetime) of the long-lived component in a three term fit to the lifetime spectra. a) I_3 (fwhm$_3$) for sidepeaks at ±4.2 mrad. b) fwhm$_2$ was fixed. *m.* not available

Compound	I_1 [%]	fwhm$_1$ [mrad]	I_2 [%]	fwhm$_2$ [mrad]	I_3 [%]	fwhm$_3$ [mrad]	H [mrad^{-1}]	I_3' [%]	τ_3 [ns]
n-pentane	17.91	2.53	79.88	8.59	2.21	22.45	0.1553	41	4.25
n-hexane	16.66	2.54	80.88	8.57	2.45	20.34	0.1519	42.4	3.92
n-heptane	16.97	2.60	81.16	8.64	1.87	24.19	0.1510	41	3.81
n-octane	15.32	2.73	82.22	8.71	2.46	18.52	0.1431	39.6	3.49
n-dodecane	14.58	2.69	83.96	8.70	1.47	24.60	0.1427	38	3.43
n-tetradecane	15.65	2.77	82.08	8.76	2.27	19.49	0.1426	36	3.35
2-methylbutane	17.45	2.52	80.07	8.54	2.48	21.26	0.1548	*m.*	*m.*
2.2-dimethylbutane	17.37	2.60	79.42	8.50	3.21	18.79	0.1528	48	4.46
3-methylpentane	16.64	2.57	81.15	8.58	2.21	22.98	0.1512	*m.*	*m.*
isooctane	17.47	2.68	79.64	8.54	2.89	17.83	0.1509	43.1	4.05
neopentane	19.98	2.50	74.95	8.30	5.06	15.88	0.1634	52.5	5.15
cyclohexane	14.73	2.71	83.35	8.74	1.92	21.21	0.1420	37.9	3.24
methylcyclohexane	15.73	2.75	82.62	8.71	1.65	21.44	0.1440	41	3.50
benzene	13.70	2.80	81.19	8.40	5.11	15.64	0.1401	43	3.15
toluene	13.74	2.82	82.80	8.53	3.46	17.62	0.1392	42.4	3.24
ethylbenzene	10.83	2.70	85.05	8.69	4.12	17.86	0.1322	36	3.02
o-xylene	13.58	2.79	83.28	8.57	3.14	18.15	0.1391	44	3.08
m-xylene	14.19	2.75	82.87	8.54	2.95	18.71	0.1416	45	3.20
p-xylene	14.33	2.75	81.38	8.47	4.29	16.40	0.1421	45	3.21
mesitylene	14.27	2.75	82.84	8.58	2.88	18.94	0.1413	46	3.21
1,2,3,4-tetramethylbz.	13.59	2.77	83.25	8.59	3.15	17.94	0.1392	46	3.02
perfluorohexane	17.57	2.44	63.47	10.39	18.95	17.15	0.1378	*m.*	*m.*
hexafluorobenzene	17.42	2.66	67.76	10.21	14.83	17.54	0.1337	51	3.78
tetramethylsilane	21.17	2.53	72.89	8.11	5.94	14.93	0.1671	55	4.75
diethylether	16.81	2.57	77.66	8.86	5.53	17.16	0.1475	32	3.82
dioxan	20.61	3.09	71.50	9.11	7.89	15.50	0.1418	52	3.02
15-crown-5	9.58	3.19	84.77	9.93	5.65	19.45	0.1121	23	2.66
methanol	8.80	2.57	85.27	9.59	5.92	19.82	0.1195	21	3.58
ethanol (99%)	8.22	2.62	85.20	9.45	6.59	17.86	0.1184	22	3.50
propanol	9.00	2.67	85.24	9.37	5.78	18.09	0.1208	22.9	3.38
butanol	8.70	2.76	86.60	9.40	4.70	19.13	0.1192	23.4	3.36
octanol	8.75	2.82	86.98	9.35	4.26	19.20	0.1193	24.5	3.13
1,2-etanediol	7.17	3.17	87.56	9.84	5.27	19.99	0.1082	*m.*	*m.*
water (H_2O)	7.85	3.26	87.55	10.18	4.60	39.75	0.1069	28	1.85
water (D_2O)	5.67	3.10	89.95	10.08	4.38	23.48	0.1038	23	1.95
acetone	7.76	2.71	80.64	9.34	11.61	16.26	0.1156	17	3.29
acetonitrile	5.14	2.70	87.86	8.89	7.01	17.39	0.1152	15.4	3.30
anilin	9.88	2.98	87.51	8.76	2.61	21.52	0.1266	33	2.37
cyclohexylamine	10.79	3.03	84.15	9.07	5.06	17.14	0.1239	*m.*	*m.*
pyridine	3.60	2.75	90.34	9.08	6.06	17.90	0.1096	14	2.68
carbontetrachloride	−93.18	7.14	189.25	8.00[b)]	3.93	23.89	0.1020	0.56	1.24
bromoform	−53.33	5.69	146.61	7.29	6.72	22.70	0.1050	0.0	
2,2,2-trichloroethanol	−55.19	6.65	149.71	8.09	5.47	18.88	0.0992	*m.*	*m.*
dimethylsulfoxide	4.19	2.92	90.41	9.65	5.41	21.71	0.1049	14	2.29
carbondisulfide	10.55	2.62	162.52	10.58	−73.08	8.14[a)]	0.1422	44.5	2.20
nitromethane	0.85	2.43	92.28	10.26	6.87	21.37	0.0920	*m.*	
nitrobenzene	−6.52	5.57	95.18	9.57	11.34	17.54	0.0895	*m.*	

given [5.20,21]. Corrections for the angular resolution of about 1.15 mrad have not been performed. Additionally, the intensity and lifetime of the longlived component of the three term fits to the lifetime spectra (resolution about 400 ps) are shown in Table 5.2. Many three terms fits are shown fully in Table 5.1. If one reasonably assumes that the intensity of the narrowest gaussian, I_1, is caused by the p Ps intrinsic annihilation, and the long lifetime intensity, I'_3, is caused by the o Ps pick off annihilation, a ratio of $I'_3/I_1 = 3$, except for small corrections due to three gamma decay, is expected. Clearly this is in disagreement with experimental facts, as shown in Table 5.2. For most liquids I'_3/I_1 is smaller than three, which will be explained for the CS_2 case in Sect. 5.5. Detailed discussions are found in [5.20].

It was realized that the most reasonable explanation of the fact that I'_3/I_1 was smaller than three was that it was caused by reactions of Ps with the other spur species of the positron spur [5.20]. As o Ps lives much longer than p Ps it has a much larger chance of reacting with the spur species than p Ps. Of course, such o Ps reactions will show up as a reduction in the intensity of the longlived component and a distribution of shorter o Ps lifetime components, i.e., the fourth component actually found in the lifetime spectra. Most o Ps will be formed fairly far away from other spur species and give the longlived component. Only the o Ps formed close to spur species will react with them. Probably, the most important reaction is Ps oxidation by the positive ions in the spur. However, excess electrons are expected to form bound states with Ps, which will give a narrow component of the angular correlation spectrum. At present we do not know much about the details of these secondary Ps reactions in the positron spur.

In the three term fits of the lifetime spectra the fourth component will influence the fitting parameters in the following way. The longest lifetimes of the distribution will be included in the longlived component, while the shortest lifetimes will be included in the second lifetime component. In particular, the second lifetime component will be changed as most of the distribution is fairly shortlived compared to the o Ps lifetime in most liquids. In general, the fitting parameters will be correlated and hence, all parameters will be influenced by the presence of the fourth component. The three term fits in Table 5.1 show that τ_1 is longer than the expected p Ps lifetime of about 120 ps, its intensity is much larger than the expected $I'_3/3$, and the second lifetime is about 450–500 ps.

Certain older lifetime results could be explained in terms of the fourth lifetime component and its interpretation. We expect that the fourth component can be inhibited and/or quenched if strong electron acceptors are added to the liquid. Ps inhibition is discussed in details in Chap. 7 and Ps quenching in Chap. 8. If the second lifetime, τ_2, is too long because it is influenced by the fourth component, it is expected to shorten if the fourth component is removed by inhibition and quenching. This has clearly been found in several cases. For example, for pure hexane, benzene, and diethylether τ_2 in three term fits was found to be 470 ± 20 ps, 460 ± 20 ps, and 470 ± 20 ps, respectively as results of many measurements [5.21]. Adding CCl_4, a Ps inhibitor, to benzene reduces τ_2 to 416 ps at 0.2 M and 400 ps at 1 M CCl_4. In hexane 2 M CCl_4 gives $\tau_2 = 416$ ps, while $\tau_2 = 409$ ps in 1 M CCl_4/ diethylether. Nitrobenzene in benzene, an inhibitor and quencher gives $\tau_2 = 390$ ps at 1 M. In pure nitrobenzene the only lifetime is 390±5 ps. The CCl_4 lifetimes (mainly 458 ps) are shown in Table 5.1. This effect on τ_2 is found in all the cases where it is expected, in the available older Risoe lifetime spectra. High quality lifetime spectra measured with a resolution of fwhm = 350–400 ps for several molecular crystals seem also to have the fourth component [5.21].

In conclusion, the fourth component of the lifetime spectra seems to be well established now. However, it must be realized that it was discovered very recently, and that, as yet, fairly few high quality lifetime spectra have been measured. Careful tests of the explanation are necessary in future experiments. As mentioned above, the main part of the discussions in this book will be based on the three term fits. The conclusions will only depend little on the fourth component, as the discussions are only qualitative and semiquantitative anyhow.

At Risoe we have attempted to separate the AC curve into its components for three liquids: hexane, benzene, and water (H_2O) (see Table 5.3). This seems to be the only detailed separations attempted. The benzene curves have been used in the interpretation of AC spectra for nitrobenzene/benzene solutions, in which nitrobenzene inhibits Ps formation and reacts which Ps. The results are discussed in detail in Sect. 8.2. The water components marked by 1 in Table 5.3 have been used extensively in studies of hydrated positron reactions and Ps inhibition by halides and pseudohalides in water, (see Chap. 6), where also the assumptions used in the derivation of the components are mentioned. The components for hexane, benzene, and water (1) were derived before the fourth component of the lifetime spectra was discovered. The differences between the water components marked by 1 and 2 in Table 5.3 are examples of the, assumption dependent, changes in the components on introducing the effects of

Table 5.3. Angular correlation components for hexane, benzene and water *m.* not available

Fitting curve for	I_1 [%]	$fwhm_1$ [mrad]	I_2 [%]	$fwhm_2$ [mrad]	I_3 [%]	$fwhm_3$ [mrad]	H [$mrad^{-1}$]
hexane, e^+	−22.36	8.00	57.84	8.95	4.23	16.74	*m.*
hexane, Ps pick off	0.36	2.60	180.20	8.90	−137.61	9.29	*m.*
hexane, Ps intrinsic	17.34	2.60	–	–	–	–	*m.*
benzene, e^+	−5.08	5.36	43.18	8.61	4.57	15.88	*m.*
benzene, Ps pick off	0.92	2.95	29.96	7.48	12.68	9.24	*m.*
benzene, Ps intrinsic	13.79	2.86	–	–	–	–	*m.*
1, H_2O, e^+_{aq} (PPS)	−11.05	6.26	64.20	9.32	10.72	15.24	0.0865
1, H_2O, Ps pick off	2.10	3.82	24.11	10.13	1.47	25.41	0.1029
1, H_2O, Ps intrinsic	2.34	2.68	6.11	3.82	–	–	0.2748
2, H_2O, e^+_{aq} (PPS)	−10.69	6.26	62.10	9.32	10.37	15.24	*m.*
2, H_2O, Ps pick off	−3.40	5.36	25.05	9.11	5.40	15.47	*m.*
2, H_2O, Ps intrinsic	11.17	3.42	–	–	–	–	*m.*

the fourth component (see Chap. 6). The fourth component makes the separation of the AC curve into its components more difficult. However, as long as only the total Ps component is used in the analyses the resulting yields are only weakly influenced by the effect due to the fourth component. Detailed interpretations of the AC spectra, e.g., by use of its components, are the only way of sorting out the type of the reactions between Ps and many solutes. Hence, even if the resulting shapes of the AC components are somewhat uncertain important new information can be obtained by use of the components.

5.4 Formation of Ps in the Positron Spur

The best explanation of the measured Ps yields takes into account all the available data, including, in particular, the Ps yields for solutions (see Chap. 7). Hence, one cannot, in principle, understand the Ps yields for pure liquids fully consistently before parts of the other chapters have been read, and vice versa. This is a typical situation in science. However, we have to start somewhere, and I have chosen to discuss the pure liquids first, although the Ps yields in pure liquids are fairly difficult to understand compared with the situation for non-polar liquid mixtures. Accordingly, some references to the three next chapters must be used in the discussion. The best way of understanding the Ps yields is probably to read the relevant chapters several times; the first reading being only a brief skim for the basic ideas and results.

In Sect. 5.1 we discussed how the positron spur is formed. The spur formation is followed by intraspur reactions resulting in the formation of new reactants or solvent molecules. Some of the spur species might also escape the spur reactions and eventually react with each other or the solvent molecules in homogeneous interspur reactions.

The most important nonhomogeneous positron spur reactions in pure liquids are:

$$e^+ + M \rightsquigarrow M^+ + e^- + e^+ \quad \text{Creation of ions} \quad (5.1)$$

$$e^+ + M \rightsquigarrow R + e^+ + R' \quad \text{Creation of free radicals} \quad (5.2)$$

$$e^+ + M \rightsquigarrow M^* + e^+ \quad \text{Creation of excited states} \quad (5.3)$$

$$M^+ + e^- \rightarrow M \quad \text{Recombination} \quad (5.4)$$

$$e^+ + e^- \rightarrow \mathrm{Ps} \quad \text{Ps formation} \quad (5.5)$$

$$e^+ + M^* \rightarrow \mathrm{Ps} + M^+ \quad \text{Ps formation} \quad (5.6)$$

$$\mathrm{Ps} + M^+ \rightarrow M + e^+ \quad \text{Ps oxidation} \quad (5.7)$$

$$\mathrm{Ps} + e^- \rightarrow \mathrm{Ps}^- \quad \mathrm{Ps}^- \text{ formation} \quad (5.8)$$

and outdiffusion of the spur species.

A schematic drawing of the last part of the positron track and its terminal spur is shown in Fig. 5.2. Only some of the reactants are shown. As mentioned

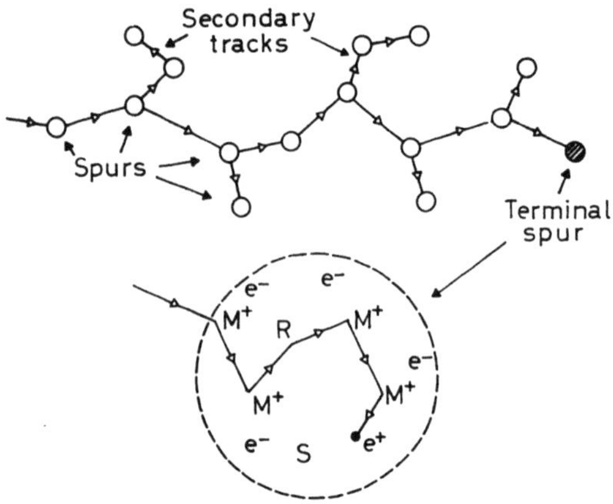

Fig. 5.2. Schematic diagram of the last part of the positron track and its terminal spur, which is called the positron spur. M^+, e^-, e^+, and R denote positive ions, excess electrons, the positron, and a free radical, respectively. S is an electron, positron, or positive ion scavenger. Only the most important spur reactants are shown

in Sect. 5.1 the details of the positron spur, e.g., the number of electron-ion pairs, are unknown. It is normally assumed that the electrons are situated in the outer parts of the spur and the positive ions in the core of the spur. S indicates that scavengers of the electrons, positron and/or positive ions can be added to the liquid. A scavenger reacts with certain spur species.

The Ps yield is mainly the result of the competition between (5.5) and (5.4). (5.6) has been shown in order to indicate that some excited states might be different from just an electron ion pair, as the electron might be situated mainly on the parent molecule in the excited states, e.g., in larger aromatic hydrocarbons. However, (5.6) is probably unimportant in many liquids.

It is important to realize that it is very difficult, if not impossible, to calculate the Ps yields theoretically. This follows from a discussion of the calculations normally performed on the simplest spur, namely the low-LET electron spurs. In Sect. 6.2 a general discussion of reaction kinetics in positron annihilation research will be given. In particular, some of the basic assumptions of the mostly used spur diffusion theories will be treated. We will apply the resulting conclusions here. Only single pair spurs are treated in the normal spur diffusion calculations. It is assumed that the excess electron and the positive ion diffuse like classical particles under the influence of their mutual Coulomb field, i.e., (6.7) is used, and that they react if the distance between them is smaller than the reaction radius, R. As discussed in Sect. 5.1 this is probably a reasonably good approximation for most liquids. In liquids with very high electron mobility quantum mechanical effects might be important and the used approximations might be incorrect.

A diffusion equation (6.7) for the relative motion of the two reactants is solved with the initial condition that the distribution of mutual distances is a certain function, e.g., an exponential or gaussian, characterized by parameters. The diffusion constant and the mobility of the two reactants are assumed to be identical to those found for homogeneous reactions outside the spurs. Furthermore, a dielectric constant equal to that measured macroscopically at weak fields is used. The probability of recombination as a function of time is derived. It includes the probability that the pair escapes the recombination. The recombination probability depends on the parameters which describe the initial distribution of mutual distances. These, essentially unknown, parameters are used to fit the theoretical results to the experimentally determined quantities. The average secondary electron range, b, in Table 4.1 is just such fitting parameter. This theory can also be formulated for the case where an external electric field is present. b has mainly been determined by fitting the theoretical free electron yields, G_{fi}, to the measured ones as a function of an applied external electric field. Table 4.1 shows some G_{fi} values for zero external field. This is the simplest version of the spur diffusion theories applied to the simplest spur. The approximations used will be discussed below.

However, let us discuss how this simple model must be extended to treat the Ps formation case. The simplest positron spur consists of one positron, one electron, and one positive ion. Hence, the diffusion of three species in the total electric field must be treated. Furthermore, the initial distribution functions of the relative distances must be known or parametrisized, and used as initial conditions for the solution of the diffusion equation for all three reactants. It seems to be unlikely that a solution of this complex problem can be expressed in terms of known functions. This is probably also the case if only a limited solution, e.g., the total Ps formation probability, is attempted. Hence, very probably numerical and/or Monte Carlo calculations must be performed. As the initial distributions are unknown, except for rough ideas about the initial distances extracted from electron spur results, and the mobility and diffusion constant of the positron are mainly unknown, as yet, the resulting Ps yield will be very dependent on several parameters, and any measured Ps yield can probably be fitted by a variation of these parameters. Consequently, a theoretical study might not give useful results even for the simplest positron spur.

This estimate of the theoretical difficulties will be even more pessimistic if we take into account the serious approximations of the simple model used in the electron spur calculations. Let us list the most important of these approximations.

The simple model treats only single pair spurs, while a large percentage of the spurs are multipair spurs.

A dielectric constant, ε, independent of the very large electric field in the inner part of the spurs is used. For example, in hexane with $\varepsilon = 1.89$ the electric field is 1.9, 0.31, 0.16, and 0.076 MV/cm at distances of 2, 5, 7, and 10 nm, respectively, from a unit charge. It seems to be unlikely that the effective ε-value can be 1.89 at fields of about 1.0 MV/cm. It will probably be smaller because of saturation, electrostriction, and time dependent effects.

The mobility values measured at low fields (Table 4.1) may well increase by a factor of ten if fields of about 1.0 MV/cm are introduced. In [5.22] the mobility increased by a factor of two on introduction of an external field of only 0.2 MV/cm in liquid ethane and propane, and a similar increase in the trapping rate of excess electrons of SF_6 was also measured. The field is about

0.2 MV/cm at a distance of 6 nm, while the secondary electron range is roughly 7 nm, in hexane. Liquid propane and hexane have very similar excess electron mobilities.

The excess electron is assumed to be a classical particle. In liquids, where the mobility is high, i.e., above about 1.0 cm^2/Vs, this might be a poor approximation. The mobility is above 1.0 cm^2/Vs in liquid spherical hydrocarbons and in liquid noble gases except for neon and helium, and probably also in the presolvated states in other liquids.

These comments on the errors involved in the approximations of the simple spur diffusion model, as used in the single pair electron spurs, apply just as well for the application of the simple model on the simple positron spur. Some of the approximations are even poorer when used on the positron spur. For example, quantum mechanical effects in the behaviour of two light particles in the positron spur might be important. Furthermore, the slowing down distance, mobility, and diffusion constant are not known for the positron except for rough estimates. Moreover, the positron spur is part of a "short track", and hence, it is more likely that multipair spur effects play a role for the positron than for the low LET electron spur.

Hence, we must conclude that it is unlikely that a diffusion model calculation of the positron spur reactions will be successful. Such calculation has not been attempted, as yet. We must use qualitative and semiquantitative models to explain the Ps yields. This is a very useful conclusion, indeed.

5.5 The Measured Ps Yields for Pure Liquids

In the previous sections we had to conclude that a detailed quantitative theory of Ps formation is unavailable at present. We shall, therefore, discuss the Ps yields in pure liquids by use of simple qualitative explanations. More detailed discussions on the Ps yield will be postponed to Chap. 7 on Ps formation in mixtures. The reason is that changes in the Ps yield on addition of solutes are easier to explain than the absolute Ps yields.

The main differences in the Ps formation from one liquid (solution) to another can be qualitatively explained by reference to three properties of the liquids in most cases.

First, the liquids (solutions) can be classified according to the reactivity of the molecules towards the electron and/or positron. Extreme cases are the liquids (solutions) where the products of the reactions of the light particles are not available for Ps formation. In other liquids (solutions) the light particles bound in the product can still form Ps, but normally with a lower probability than that of the free (solvated), unreacted, light particle.

Second, the ability to solvate the positron and/or electron influences strongly the Ps yields in polar liquids (solutions). Solvation influences the Ps formation in two ways. The mobility of the light particles is strongly reduced on solvation, and hence, the positron might annihilate ($\sim$ 400 ps lifetime) before it can react with an excess electron to form Ps. Furthermore, the Coulomb attraction between the electron and positron is shielded for the solvated particles, as expressed by the large dielectric constant, ε, of polar liquids. Hence, at thermalization the Coulomb attraction might be so low for an electron-positron pair that they do not form Ps.

Third, because of inefficient slowing down of the light particles the positron excess electron distances at thermalization can be much larger than the range of the Coulomb force, i.e., roughly the Onsager radius (6.19), resulting in low Ps yields. In particular, this effect explains why the Ps yield is strongly dependent on the density in high pressure molecular gases.

It is important to realize that the three mentioned effects are not independent. In some liquids all three are present. Furthermore, other, less frequent, effects influence the Ps yield. The formed Ps might react with molecules (solutes) of the liquids, and hence, Ps can have a very short lifetime and cannot be detected. The Ps yields can also be influenced by changes in the properties of the spur species which react with spur electrons and/or positron. In particular, the Ps yields might increase if the positive ions react with molecules (solutes) to form products with a reduced electron scavenging property compared with that of the positive ion itself. Other special conditions for Ps formation will be discussed below.

Roughly speaking, liquids of normal viscosities have the following intensities of the longest lifetime, I_3', in three term fits (see Table 5.2). I_3' is roughly equal to the o Ps yield, i.e., I_3' is approximately $0.75 \times P$, where P is the total Ps yield.

A) The nonpolar liquids, unreactive towards electrons and/or positron, such as linear and cyclic, aliphatic hydrocarbons; the smaller aromatic hydrocarbons; and tetramethylsilane, have $36\% < I_3' < 54\%$. The higher the electron mobility, μ_-, is the larger is the I_3'-value.
B) The strongly polar liquids, such as alcohols and water, have $20\% < I_3' < 28\%$, while $28\% < I_3' < 36\%$ for the less polar liquids such as diethylether ($I_3' = 32$).
C) Liquids (solutions) whose molecules (solutes) react with the electron and/or positron give I_3'-values in the interval $0\% < I_3' < 52\%$. If the electron (and/or positron) are bound in shallow traps in the product of the reaction, from where it can be picked off, before the positron annihilates, by a fast moving positron (and/or electron), the I_3'-value is high, i.e., $30\% < I_3' < 50\%$. On the other hand, if the reacted particle is unavailable for Ps formation $0\% < I_3' < 25\%$. For example, CS_2 and C_6F_6 belong to the first type, while CCl_4, $CHCl_3$, C_6H_5I, and other strong electron scavengers belong to the second type.

In the rest of this section we shall discuss in details some selected liquids from each of the three categories as examples, and finish by a subsection on liquids where special effects are influencing the Ps formation.

At first, let us discuss in detail the nonpolar liquids. We use mainly hexane as an example. The excess electron, positron, and Ps properties were described in Chap. 4 (see Table 4.1,2). The secondary electron range $b = 7$ nm compared to an Onsager radius of $r_c = 30.2$ nm at $T = 293$ K and for a dielectric constant of $\varepsilon = 1.89$ (6.19). The Onsager radius is the distance at which the Coulomb energy is equal to the thermal energy kT for unit charges. In the simplest theory [5.23] the probability of a pair of separated unit charges of escaping recombination by Brownian motion is $\exp(-r_c/r)$, where r is the distance between the charges. As $b \ll r_c$ very few electron ion pairs escape recombination in low LET electron spurs. Hence, $G_{fi} = 0.13\ (100\ \text{eV})^{-1}$ compared with the total yield of pairs of $G = 4 - 5\ (100\ \text{eV})^{-1}$, i.e., only about 3% of the pairs escape recombination. The electron mobility, $\mu_- = 0.08$ cm^2/Vs is fairly low for hexane.

The time of recombination for an electron positive ion pair separated by r is very roughly the time it takes the electron to drift the distance r in the field of the ion, namely

$$t = 4\pi\varepsilon_o\varepsilon r^3(3\mu_- e)^{-1} \tag{5.8}$$

where e is the unit charge and $\varepsilon_o = 8.854 \times 10^{-12}$ As/Vm. In neopentane and hexane we get $t = 0.5$ ps and 21 ps, respectively at $T = 293$ K for $r = b$. In [4.1] a geminate recombination time of 15 ps, 2.5 ps, and 0.7 ps for hexane, cyclohexane, and isooctane, respectively is estimated from experimental results of scavenging the electrons undergoing geminate spur reactions. Our estimate for hexane is probably too long because we have neglected the field dependence of the mobility and dielectric constant. Furthermore, as shown by (5.8) the time of recombination is strongly dependent on r. It is estimated that several percent of the geminate recombination has still not taken place at 400 ps for hexane and cyclohexane [4.1]. A direct measurement of the half-life against recombination for hexane gave about 9 ps at 293 K and 70 ps at 214 K [5.24].

We have above discussed the low-LET, mainly single-pair spur for a typical linear hydrocarbon, namely hexane. The positron spur is a high-LET, multipair spur. Compared to single-pair spurs fewer electrons escape recombination in multipair spurs [5.25]. It is important to realize that almost all charged reactants in the positron spur are inside the range of the Coulomb force of the other charged reactants, i.e., they are normally inside the Onsager radius of several charged reactants, at thermalization in nonpolar liquids and solids. Hence, a model in which an independent electron positron pair reacts to form Ps is unrealistic. Furthermore, simple, homogeneous reaction kinetics is, of course, inappropriate; nonhomogeneous kinetics must be used. This applies as well for polar liquids. Furthermore, the effective reaction radius (Sect. 6.2) of the positron spur reactions might be large, e.g., 1–2 nm, for reactions involving light particles. Consequently, initial encounter pairs, i.e., reactants initially within the reaction radius, might be frequent in the positron spur.

It is normally assumed that the first recombinations of electron ion pairs occur very fast in multipair electron spurs because of the strong Coulomb attractions if several positive ions are situated in the core of the spur. The last electron ion recombination is then fairly similar to that of a single pair spur. In the positron spur the first "recombinations" might include the Ps formation, and hence, we cannot reasonably assume that the electron positron reaction occurs preferentially in the simplest positron spurs where the only extra reactant is the positive ion. Hence, the Ps might form also in the crowded parts of the positron spur. However, in nonpolar liquids the rapidly moving electrons disappear fast by recombination leaving Ps and the remaining positive ion in the spur. If they are close together they might react resulting in an oxidation of Ps. It is difficult to estimate theoretically the importance of such reaction.

The four term fits of the high quality lifetime spectra indicate that secondary o Ps reactions take place in hexane (see Sect. 5.3). Because of the fast electron ion recombinations, one might, at first sight, assume that mainly o Ps oxidation takes place in the spur in nonpolar liquids. However, if the fast electron reacts with Ps both p Ps and o Ps might be transferred into Ps^-, as the electron reactions are fast compared to even the p Ps decay. Hence, it is probably most reasonable to assume that both Ps oxidation and Ps^- formation take place. Ps^- gives a narrow AC component.

For hexane the long lifetime intensity, I_3', in three term fits increases linearly with temperature from 35% at −90°C to 43.5% at 50°C, while the narrowest gaussian intensity in three gaussian fits increases linearly with temperature from 11% at −90°C to 19% at 50°C [5.21].

Experimentally a clear, fairly universal correlation between the electron mobility and the Ps yield has been found both on change of the liquid and on change of the temperature for a given nonpolar liquid [4.26,5.20] (see Tables 4.1 and 5.2). The larger the mobility is the larger is the Ps yield other things being equal. The reason is unclear, as discussed in details in [4.27]. In agreement with that the long lifetime intensity, I_3', is large in neopentane and tetramethylsilane (52–54%), and in between these values and the hexane value for the other spherical hydrocarbons (Table 5.2). Furthermore, a small minimum in μ_- is correlated to a minimum in the I_3'-value at the critical point for neopentane. Apparently, the Ps yield has been studied at and above the critical point only for neopentane and SF_6 [4.26,27].

The only liquids for which a fast moving positive hole has been found seem to be cyclohexane, methylcyclohexane and cis- and trans-decalin [4.1]. In cyclohexane the fast hole probably leads to a lower I_3' value of 37.9% compared to 42.4% for hexane, although the μ_--value is higher for cyclohexane. The reasons are fairly uncertain [5.26].

The second class of liquids to discuss are the polar liquids, exemplified by water and alcohols. Because the light particles are solvated in polar liquids the spur reaction times are very much longer than in nonpolar liquids. For example, in water the spur reactions of the hydrated electron take several ns at 293 K [5.27,4.8]. As the positron lifetime is about 400 ps the reaction of a fully solvated positron and excess electron to form Ps is normally too slow. The very important conclusion is that Ps is almost exclusively formed by reactions between a positron and electron, of which at least one is not yet fully solvated, or by the initial close encounter effect. In the latter case the positron and electron are situated within the reaction radius for fully solvated particles after thermalization and solvation. In radiation chemistry many nonsolvated electron states have been used in theoretical discussions, such as "dry", "infrared", "damp", "presolvated", etc. states, and the nomenclature is certainly not agreed upon yet. Experimental evidence for two nonsolvated states seems to have been found for alcohols [4.9]. As the secondary electron range, b, is small in polar liquids ($b = 2$ nm, Table 4.1), and since the positron spur is very probably a multipair spur, there is a high probability of initial close encounters of the positron and electron. For nonsolvated particles the reaction radius might be 1–2 nm, and the largest reaction radii for hydrated electrons have been measured to be 1 nm.

The high dielectric constant in polar liquids means that the range of the Coulomb forces is short. The Onsager radius is only 0.71 nm for solvated reactants in water. Hence, more than 55% of the electron ion pairs escape recombination for low LET electron spurs, as described by the very high $G_{\mathrm{fi}} = 2.7\ (100\ \mathrm{eV})^{-1}$.

As shown in Table 5.2 the Ps yields are lower by roughly a factor of two in water and alcohols than in nonpolar liquids. The reason is probably mainly that the fully solvated positron and electrons cannot form Ps in time before the positron decays. However, the reduced range of the Coulomb forces might contribute somewhat to this effect, too. In amorphous solid water and alcohols the Ps yields are in between those of the crystalline solid and the liquid, normally closest to the yields of the liquid. This is in agreement with our interpretation, as many positron electron pairs are trapped far from each other in the amorphous solids and thereby prevented from making Ps, i.e., an effect similar to the solvation effect, while Ps formation by pretrapped particles and by particles trapped very close to each other (tunnelling, close encounter effect) is possible, as in the liquid. Ps formation in polar liquids is a more complicated process than in nonpolar liquids.

Reactions of Ps with spur species occur very probably also in polar liquids. Opposite to the nonpolar liquid case the electron ion recombination takes long time (nanoseconds) in polar liquids. Hence, Ps reactions with the electrons and ions can take place over *ns* compared to over 10–20 ps in a typical nonpolar liquid such as hexane. In normal, nonpolar liquids only the remaining positive ion has a long lifetime. The measured lifetime spectra for polar liquids seem to indicate that the fourth component is present with an intensity comparable to that of nonpolar liquids even if the normal longlived component intensity is only half of that of nonpolar liquids. However, only very few high-quality lifetime spectra for polar liquids have been reported.

The third class of liquids to discuss is the liquids where the molecules react with the electrons and/or the positron. For these liquids the Ps yield depends mainly on the properties of the products of the reactions. Let us consider CCl_4 as an example. The following reactions occur in liquid CCl_4:

$$e^- + CCl_4 \rightarrow (CCl_4^-)^* \quad \text{Electron reaction} \tag{5.9}$$

$$(CCl_4^-)^* \rightarrow CCl_3^{\cdot} + Cl^- \quad \text{Dehalogenation} \tag{5.10}$$

$$e^+ + Cl^- \rightarrow [Cl^-, e^+] \quad \text{Bound state formation} \tag{5.11}$$

$$e^+ + CCl_4 \rightarrow [CCl_4, e^+] \quad \text{Bound state formation} \tag{5.12}$$

apart from the reactions (5.1–5). The evidence for these reactions will be discussed in Chap's. 6, 7, and 8. $(CCl_4^-)^*$ is very shortlived and hence, the pick off of the trapped electron by the positron is insignificant. Reactions (5.9–11) make the electron unavailable for Ps formation. The unbound electrons can probably pick off the positron from $[CCl_4,e^+]$, but the electrons are trapped very fast on CCl_4 and have no time for the pick off. Ps does not react with CCl_4 and other halogenated hydrocarbons in nonpolar liquids. Apparently, the lifetime spectrum for CCl_4 contains more than just a single lifetime component (Table 5.1), which might somehow be a result of (5.9–12).

CCl_4 is a typical example of a liquid, where the product of a reaction of the electron with the molecules is unavailable for Ps formation. Many chloronated, bromonated, and iodonated (but not fluoronated, see below) hydrocarbons belong to that category, for which, however, the electron reaction and following dehalogenation are normally not fast enough to totally exclude Ps formation [1.34]. A slow dehalogenation of the primary anion will permit that the electron on the anion is picked off by the positron to form Ps. For example, I_3' is 9% for chlorobenzene, 3.4% for bromobenzene, 1.6% for iodobenzene, and 0% for $CHCl_3$ and CCl_4 [1.34].

In liquids where Ps reacts strongly with the molecules Ps is, of course, not found experimentally as its lifetime is very short. In general, such molecules have a high electron affinity, and the Ps formation is inhibited by electron scavenging anyhow. For example, no Ps is formed in nitrobenzene, which reacts with Ps in hydrocarbons. The only exceptions seem to be SO_2 and O_2, which react with Ps but show little if any Ps inhibition in hydrocarbons (see Sect. 7.4).

In liquids where the molecules react with the electron and/or positron to form products, in which the light particle is weakly bound, Ps is formed by a pick off process, where the bound particle is picked off by the oppositely charged particle. In these liquids the Ps yield might be high, when certain condition are fulfilled. Examples are many fluoronated hydrocarbons, such as hexafluorobenzene ($I_3' = 51\%$) and perfluorohexane (see Sect 7.4). The reason is that the carbonfluorine bond is much stronger than the other carbon-halogen bonds, and hence, the formed anion radicals do not defluoronate but are stable in nonpolar liquids.

Concerning Ps formation, CS_2 is a unique liquid, as described in [5.12,5.28]. A discussion of Ps formation in CS_2 is very instructive, indeed. The long lifetimes (intensities) in three term fits were found to be 2.65 ns (46%), 2.20 ns (44.5%), 1.73 ns (43.8%), 1.58 ns (40.8%), and 0.80 ns (5.8%) at 105°C, 20°C, −36°C, −51°C, and −111°C, respectively. The lifetime and intensity varied linearly between 105°C and −36°C and between −51°C and −111°C. Above about −40°C the Ps bubble state is the state of lowest energy. As the surface tension is very high (32 dyne/cm [5.29]) and not associated with hydrogen bonding, the o Ps lifetime is short. The narrowest gaussian in a three gaussian fit of the

AC spectrum (Table 5.2) is of fwhm = 2.62 mrad (10.55%) at 20°C. This is a very low intensity compared to that expected from the long lifetime intensity. The reason is probably that the p Ps center-of-mass wave function consists of a strongly varying part at the bubble giving a middle-broad component of fwhm $\simeq$ 4.5 mrad included in the two broader gaussians, and "spilled-over", broader part giving the narrow gaussian (see Fig. 5 in [5.30]). Hence, experiments indicate clearly that Ps is not far from being "squeezed" out of the bubble at 20°C.

Because of the increasing "effective" surface tension for decreasing temperature, the Ps bubble is not the lowest energy state at −111°C, i.e., Ps is mainly "squeezed" out of the bubble at −111°C. The transition from Ps bubble state to the mainly non-Ps state takes place between −40°C and −111°C. However, at −111°C the positron-electron pairs which would have formed Ps above −40°C, are still bound by their mutual strong Coulomb attraction, as the dielectric constant is low (2.64 at 20°C and 3.00 at −110°C [5.29]). They cannot make the final transition to the Ps bubble state because of the rather large energy needed to form the bubble at −111°C and because they are both fairly strongly bound to the CS_2 molecules. CS_2 has an electron affinity of 0.51 eV in vacuum, and hence, about 1.5 eV in the liquid because of the polarization energy. Furthermore, Ps yields for CS_2 hydrocarbon mixtures indicate clearly that the positron is bound to CS_2 in these liquids and hence, very probably also in pure CS_2. The detailed structure of the positron-electron state at −111°C is unknown. Such state has been called the "fourth positron state"; the three other states being the free positron, p Ps, and o Ps states. It is an electron and positron attached to the same or different molecules at a short distance from each other, being unable to form Ps and to escape their mutual Coulomb attraction. Somewhat similar states, except for the bubble effects, for two oppositely charged ions in nonpolar liquids are fairly well known.

A transition from a Ps state to a non-Ps state in a liquid by decreasing the temperature, and thereby increasing the surface tension and "squeezing" Ps out of the bubble, has only been found for CS_2, as yet. Liquid SF_6 under its vapour pressure gives unusual lifetime results, too [4.27]. Two o Ps long lifetimes were found, one of about 2 ns and the other, normal, o Ps lifetime increasing from about 5.7 ns at −45°C to about 19 ns at the critical point of 45.7°C. At −45°C the two intensities are roughly equal (16% each), while at 46.7°C the short o Ps lifetime has an intensity of 5% and the normal lifetime about 46%. The electron affinity of SF_6 is 1.05 eV (Table 8.1), and Ps yield results indicate that the positron is bound to SF_6 in hydrocarbons. Compared to CS_2, however, the surface tension in liquid SF_6 is low. The strong effects of an external pressure [1.45,46] and temperature on the rates of Ps reactions with various electron acceptors are discussed in Sect. 8.3.

Transitions from Ps to non-Ps states on decrease of the temperature occur also in some aromatic solids, such as naphthalene, biphenyl, para-terphenyl, and quarterphenyl (see Chap. 9), where the non-Ps state is probably the "fourth positron state". Probably Ps is not formed because the molecules have positive electron, and maybe positron, affinities and because Ps is strongly "squeezed" in the solid, i.e., the situation is fairly similar to that in liquid CS_2. Ps disappears also on transition from liquid to crystalline solid anthracene for roughly the same reasons. The molecular solids are discussed in details in Chap. 9.

Some of the liquid noble gases have abnormal Ps formation properties. However, we shall not in this book discuss these liquids except for few comments. Liquid argon has a very large secondary electron range, $b = 140$ nm, at 90 K (Table 4.1). Consequently, many electron positron pairs are separated by more than the Onsager radius at thermalization, and the o Ps yield is only about 6%. Addition of small amounts of propane increases the Ps yield strongly because the b-value is decreased drastically on addition of propane [5.31]. Liquid xenon has a very long slowing down time for electrons (6.5 ns at 163 K, [5.32]) and probably positrons. It is much longer than the positron lifetime. Hence, the Ps yield is low [5.33]. The Ps yield in liquid xenon is still partly unexplained. Long slowing down times for electrons are also found for liquid krypton [5.32]. Ps yield results for some other liquids are discussed in [5.34,28,1.35].

The discussion of the Ps formation will be continued in Chap. 7 on the effects of addition of solutes to liquids, in Chap. 9 on molecular crystals and solid-liquid transitions, in Chap. 10 on polymers, glasses, and miscellaneous systems, and occasionally in the other chapters, too.

6 Solvated Positron Reactions. Models of Reaction Kinetics

In this and the next three chapters we shall discuss in detail the main topics of this book, namely the reactions of the positron and Ps with various reactants added to liquids and solids. We shall in this chapter discuss the positron reactions with heavy reactants, i.e., the reactions with the excess electron are not included. The main reason for starting the discussions with the solvated positron reactions is simply that these reactions are the easiest ones to discuss at present. Only very few positron reactions with heavy reactants have been studied. We shall compare the positron with the excess electron reactions whenever appropriate. As usual we call the excess electron just electron in most cases.

A second purpose of this chapter is to use the positron and electron reactions as examples in the discussion of the models of reaction kinetics, which are of great importance in positron and Ps chemistry in general. In particular, the role of diffusion, time dependence of the rate constant, non-homogeneity, initial close encounters, and high concentration of reactants in reaction kinetics will be illustrated in the discussion. The results of the discussion will be used in the rest of the book.

Third, the discussion of positron reactions will illustrate in great detail how the AC spectra can be analyzed and used to get information on the fate of the positron.

The first rate constant of a solvated positron reaction, namely that of the hydrated positron reaction with Cl^-, was measured at Risoe in 1974 [1.37].

The first direct studies of the hydrated electron rates began in 1960–1962 [4.2], while the rapidly moving electrons in liquid hydrocarbons were first detected in 1968–1970 [4.1]. However, the solvated electron in liquid ammonia has been studied for many decades.

Rate constants of the reactions of the third light particle, Ps, in liquids were first measured around 1954–1956 [1.4].

However, for all three light particles reactions were first studied in gases (probably with the exception of electrons in liquid ammonia). Furthermore, the excess electron in ionic crystals, for example, is an old topic of solid state physics.

As suggested by P.E. Cade we use the notation $[A, e^+]$ for the bound state of the positron and the reactant A [6.1,2].

6.1 Hydrated Positron Reactions with Halides

We shall illustrate the main principles of the solvated positron reactions by a detailed discussion of how the hydrated positron (e^+_{aq}) reactions with halides (X^-) in water have been studied. This is an obvious choice as $e^+_{aq} + X^-$ reactions are the best investigated cases.

At first it must be realized that mainly the AC measurements have given useful information on the positron reactions with heavy reactants in liquids. It is a surprising fact that the differences between the solvated positron lifetimes of roughly 400 ps and the until now measured lifetimes of the positron bound states with heavy reactants are too small to be measurable in practice. Hence, LT measurements have only given information on solvated positron reactions through the fairly weak influence of the reactions on the Ps formation. Of course, DB studies give similar information as the AC measurements. However, quantitative studies of the solvated positron reactions are fairly difficult even if the AC technique is used, and hence, because of its poorer resolution the DB techniques is mainly useful in qualitative investigations.

The main ideas of the positron reaction studies are A) to determine whether a bound state is formed, B) if affirmative then to derive the shape of the AC (DB) curve of the formed bound states, C) to measure the intensity of this curve in the AC spectra as a function of the concentration of the reactant, D) to use the obtained results to make conclusions about the kinetics of the positron bound state formation, and E) to explain the kinetics in terms of the physics of the solutions of the reactants.

In order to do this it is necessary to know the shapes of the components of AC spectrum of the pure solvent in a reasonable approximation. If the Ps formation is mainly inhibited by a high concentration of a Ps inhibitor the shapes of the Ps components need only to be known in a very rough approximation though (see below). Hence, we shall, at first, discuss the determination of the AC components for the pure solvent using water as an example.

In principle, the AC spectrum for the hydrated positron is determined as the spectrum for water in which no Ps is formed. The Ps component can be reduced to zero by the addition of a strong electron acceptor which inhibits the Ps yield and reacts with Ps by oxidation (see Chaps. 7,8). A weaker electron acceptor inhibits only the Ps yield to a certain low value, and the remaining Ps components can be subtracted. In water 1.74 M 1-pyridiniumpropanesulfonate-3′ (PPS) [6.3,4] or 3 M NO_3^- [1.37] were used as Ps inhibitors. As discussed in Sect. 7.3, the Ps yield versus Ps inhibitor concentration is well fitted by (7.30). In water the fitting constants are $\alpha = 5.85\ M^{-1}$ for PPS, $\alpha = 3.4\ M^{-1}$ for NO_3^-, and $\beta = 1.0$ for both. Hence, in both cases the spectra contained 3.23% of remaining Ps components, which had to be subtracted. The shape of the Ps component cannot, in principle, be determined independently from the determination of the shape of the e^+_{aq} component, but must be derived by an

Table 6.1. Fitting parameters for some AC spectra. The last seven curves are for aqueous solutions. I_i (fwhm$_i$) is the relative intensity (full width at half maximum) of the i'th gaussian in a fit of three gaussians to the spectra. H is the counts in the top of the fitting curves divided by the area between ±13 mrad

Fitting curve for	I_1 [%]	fwhm$_1$ [mrad]	I_2 [%]	fwhm$_2$ [mrad]	I_3 [%]	fwhm$_3$ [mrad]	H [mrad^{-1}]
[Cl^-,e^+]	−14.72	5.15	108.11	8.66	6.61	17.89	0.0944
[Br^-,e^+]	−26.02	5.20	118.44	7.71	7.57	18.62	0.1020
[I^-,e^+]	−37.60	4.70	129.73	6.62	7.87	19.03	0.1138
[SCN^-,e^+]	−38.32	5.40	136.64	7.50	1.68	40.00	-
[CN^-,e^+]	−0.04	0.64	93.88	9.22	6.16	19.20	-
[S^{--},e^+]	−18.07	5.05	115.93	7.53	2.14	25.74	-
1.74 M PPS	−4.40	5.10	98.97	10.04	5.42	22.19	0.0878
0.4 M $KMnO_4$	−20.64	6.58	105.28	9.34	15.36	16.25	0.0861
0.5 M $K_2Cr_2O_7$	−15.89	6.29	100.27	9.40	15.62	16.18	0.0865
3 M $NaNO_3$	4.01	5.08	87.88	10.23	8.12	18.88	0.0866
1.3 M KF + 1.55 M PPS	−11.47	6.09	102.63	9.70	8.84	18.56	0.0871
3 M $MnCl_2$	20.08	3.34	75.30	8.92	4.63	18.97	0.1389
1 M TMOPO	−50.01	8.00	140.82	9.22	9.18	18.73	0.0903

iterative procedure [6.4]. The shapes marked 1, H_2O in Table 5.3 are the results of the latest iteration step. The fitting curves for the AC spectra for 1.74 M PPS and 3 M $NO_3{}^-$ are given in Table 6.1.

After the correction for the Ps components the two curves do not have exactly the same shapes ([6.4] and Fig. 6.1). There are strong reasons to believe that the corrected 1.74 M PPS curve gives the correct shape of the e^+_{aq} AC component, as discussed in detail in [6.4]. Its shape (normalized to 63.87%) is the curve marked 1,H_2O e^+_{aq} (PPS) in Table 5.3. The most important reasons are that the e^+_{aq} (PPS) curve is equal to the measured curves for 0.5 M $K_2Cr_2O_7$ and 0.4 M $KMnO_4$ (see Table 6.1) within the experimental uncertainty [5.21]. $Cr_2O_7{}^{--}$ and $MnO_4{}^-$ inhibit the Ps yield and oxidize the non-inhibited Ps, so that practically no Ps component is expected for these spectra. In summary, we conclude that the e^+_{aq} curve is shaped as the corrected 1.74 M PPS curve, and that the addition of 1.74 M PPS, 0.5 M $K_2Cr_2O_7$, or 0.4 M $KMnO_4$ to water does not change the shape of the AC spectrum appreciably away from the e^+_{aq} curve. The 1 M TMOPO spectrum in Table 6.1 [5.21] shows that a shape different from that of the e^+_{aq} curve can also be obtained if Ps is almost totally inhibited and quenched. TMOPO is a free radical reacting with Ps probably to form some bound state of the positron in water, and inhibiting the Ps formation, as discussed in Chap. 8. The 3 M $MnCl_2$ curve in Table 6.1 [5.21] illustrates that another free radical, Mn^{++}, converts o Ps into p Ps resulting in a very high p Ps narrow peak intensity. These examples show clearly that the selection of the shape of the e^+_{aq} curve must be done with great care.

It is very important to realize that the shape of the "free" (i.e., solvated, not bound in Ps) positron AC component for a liquid can only, at present, be

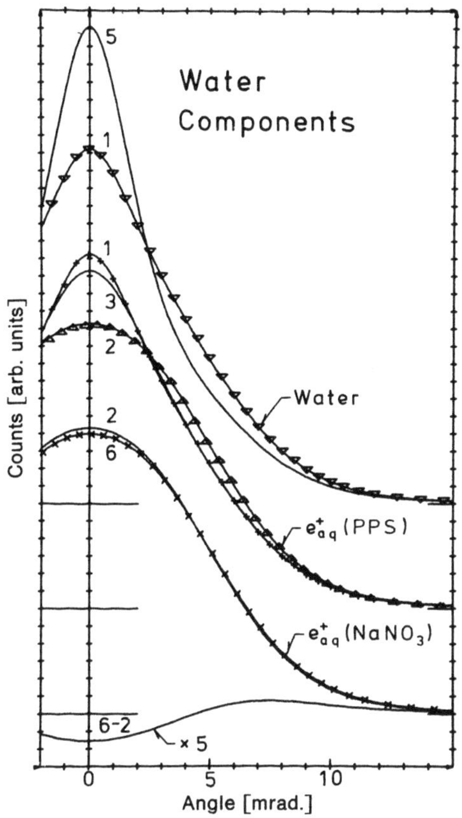

Fig. 6.1. The AC curves for pure H_2O (1) and some of its components: (2) the e^+_{aq}, (3) the pick-off, and (5) the total Ps curves. The lower curve shows the difference between the e^+_{aq} curves in PPS (2) and $NaNO_3$ (6) solutions. All curves are normalized to equal area

determined by a procedure such as that briefly described here, and described in more details in [6.4]. It is very useful to test such assignments in more details in future experiments.

The derivation of reasonably good approximations for the Ps AC components under the assumption that the shape of the solvated positron AC component is known, is fairly difficult. Further approximations must be applied. In the main part of the Risoe work the following assumptions were used [6.4]. The fourth LT component due to secondary spur reactions of o Ps (see Sect. 5.3) had not been discovered, and hence, it was assumed that the third long lifetime was caused by pick off annihilation of all the o Ps. Our best values of the intensity and lifetime of this component for water were $I_3 = 27.1\%$ and $\tau_3 = 1.82$ ns, respectively. Furthermore, we assumed that the ratio of o Ps to p Ps yield was three, and that the p Ps pick off rate was equal to that of o Ps. Hence, the narrow AC component intensity was $I_N = I_3\lambda_0/3(\lambda_0 + \lambda_3) = 8.45\%$ and the total pick off intensity $I_{po} = I_3\{1 + \lambda_3/3(\lambda_0 + \lambda_3)\} = 27.68\%$, which gives the e^+_{aq} intensity $I_+ = 1 - I_N - I_{po} = 63.87\%$. $\lambda_3 = \tau_3^{-1}$, $\lambda_0 = 8 \times 10^9$ s^{-1} the intrinsic p Ps decay rate, and 3γ-annihilations were neglected. Thus we subtracted

63.87% of the e^+_{aq} (PPS) curve from the pure water spectrum. The "rest-curve" was well fitted by a sum of four gaussians. We then used the approximation that the intrinsic annihilation p Ps component is composed of the narrowest gaussian plus enough of the second narrowest gaussian in that sum to give 8.45%. The rest of the four gaussian curve was assumed to be the pick off component (mainly o Ps) of intensity $I_{po} = 27.68\%$. In that way the shapes of the pure water components marked 1 in Table 5.3 were derived. The selected e^+_{aq}, pick off, total Ps, and the pure water curves are shown in Fig. 6.1. Clearly, the pick off curve is more narrow than the e^+_{aq} curve and of a shape fairly similar to that of pure water.

Very recently the AC components of water marked by 2 in Table 5.3 were derived taking into account the fourth LT component [5.21]. The following assumptions were used. The shape of the e^+_{aq} component was assumed to be equal to that of the corrected 1.74 M PPS curve as used above. The total o Ps intensity was chosen to be 30%, of which 25% was assumed to be the normal pick off component and 5% the component associated with the secondary Ps spur reactions. This gives a total p Ps component of 10%, of which 9.40% is narrow and 0.60% pick off assuming $\tau_3 = 1.95$ ns. Very roughly the 5% o Ps reacting in the spur was assumed to give 1.45% of pick off shape due to pick off annihilations, 1.77% due to o Ps spur reactions with electrons and shaped as the narrow p Ps component, and 1.78% of the e^+_{aq} shape caused by secondary spur reactions of o Ps with the positive ions and/or the reaction products of the positive ions. The secondary spur reactions of o Ps are discussed in Sect. 5.3. The e^+_{aq} shaped component of 61.78% was subtracted from the pure water curve, and the "rest-curve" was well fitted by a sum of three gaussians constraining the narrowest gaussian to have 11.17% intensity relative to the pure water curve. The resulting AC components of pure water are marked by 2 in Table 5.3. These components are probably better approximations than the previously used ones marked by 1.

However, the shapes of the total Ps component are not very different in the two cases. Hence, in the analyses where only the total Ps components are used the results obtained by use of the two different shapes are not expected to differ very much. In all the hydrated positron reaction studies only the total Ps components were used, and in most cases the Ps formation was even mainly inhibited by added PPS, and hence, the obtained results would probably change little if the components marked 2 were used in stead of those marked 1. This situation might change completely if the intensity of both Ps components can vary independently in the analyses of AC spectra. As discussed in Sect. 8.5, this was the case in analyses of AC spectra for aqueous solutions of MnO_4^-, $Cr_2O_7^{--}$, and NO_3^-. These analyses in terms of the three components of water marked 1 in Table 5.3 gave essentially useless results. It has not been possible for me to find time for a new analyses of all the previously measured AC spectra for aqueous solutions (stored on diskettes) in terms of the new water components. As the new pick off component is fairly different from the pure water curve, three component analyses in terms of the new components might give well defined results. However, we are still left with the problem of deciding which shapes are actually the best approximations to the real shapes. Apparently, the only way out of this situation is to go on deriving better shapes of the components and testing them in detailed analyses of many measured spectra. The above described detailed analyses of AC spectra for liquids seem only to have been performed at Risoe. The Strasbourg group has used similar analyses of DB spectra, but their derived components were only given by one gaussian each.

The addition of I^-, Br^-, and Cl^- to the 1.74 M PPS aqueous solution changed the measured AC spectrum [6.4], which in 1.74 M PPS is that of the e^+_{aq} AC curve except for the small correction due to remaining Ps components. This effect was independent of the positive ion (normally Na^+), which does not react with e^+_{aq}. The reason is that the positron is trapped on the X^- ion at higher X^- concentration and annihilates in the trapped state. No significant detrapping of the positrons from the ions has been detected.

The practical criterion for a positron bound state formation with a solute is that addition of the solute to the liquid changes the solvated positron AC component. If a bound state is formed giving an AC spectrum very close to that of the solvated positron we cannot, in practice, detect the bound state formation. A rough indication of a positron bound state formation can sometimes be detected from changes in the Ps yields determined from LT spectra.

It is fairly difficult to determine the shape of the AC component for the positron bound state, $[X^-,e^+]$, because both the shape and the intensity must be derived in the same analyses of the measured AC spectra versus solute concentration. The following analysis methods were used [6.4]. All measured and corrected AC spectra for 1.74 M PPS + c_s M X^- containing a bound state component of fairly high intensity were analyzed in terms of the e^+_{aq} component plus a sum of enough gaussians, normally three, to give acceptable variances of the fits. Just enough extra gaussians were used as too many extra gaussians can fit almost any spectra, and hence, would give very uncertain shapes of the $[X^-,e^+]$ AC component. The rest curves given by the sum of extra gaussians were normalized to equal areas, plotted, and compared visually.

In the case of I^- the shape of the rest curve was fairly independent of the I^- concentration. Visually one of these rest curves having a shape which was close to be an average of the shapes of all the rest curves was chosen as the best approximation of the $[I^-,e^+]$ curve. This curve was well defined.

The same procedure was used for the determination of the shape of the $[Br^-,e^+]$ curve, which was more uncertain than the $[I^-,e^+]$ curve.

Similarly, the shape of the $[Cl^-,e^+]$ curve was determined. However, the uncertainty in the shape of the $[Cl^-,e^+]$ was rather large, because the addition of Cl^- to the 1.74 M PPS solution resulted in rather small changes in the corrected AC spectrum away from that of the e^+_{aq} curve.

The $[X^-,e^+]$ curves are shown in Fig. 6.2 and compared with the probably best available theoretical curves [6.2]. Clearly, there is a rather good agreement between the experimental and theoretical curves at lower angles. At angles above about 8 mrad the theoretical curves are above the experimental ones,

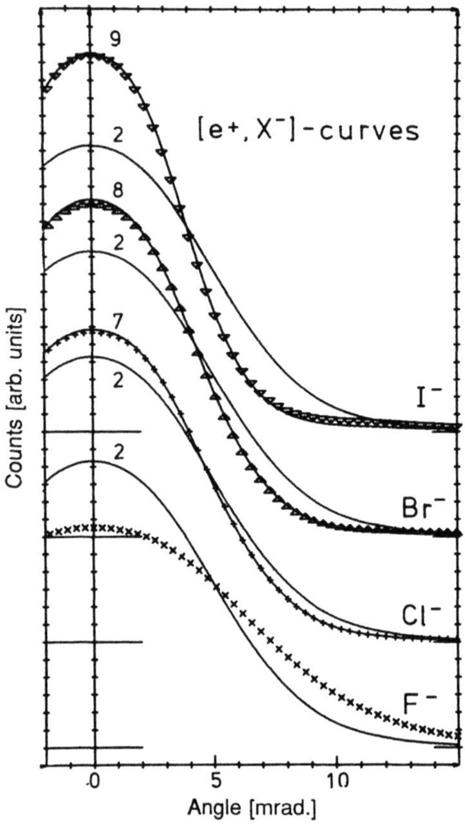

Fig. 6.2. The AC curves for the $[X^-,e^+]$ states. The points are the theoretical points [6.2], while the lines through the points are the measured curves, Table 6.1. No $[F^-,e^+]$ state was found in water. At all curves the e^+_{aq} curve (2) is shown. All curves are normalized to equal area

very probably because the theoretical curves were derived by use of the same "enhancement factor" for valence and core electrons. The "enhancement factor" accounts for the fact that theoretical derivations of the annihilation rate normally give much too low values, probably because of the difficulties of calculating the "correlation" effect on the density of electrons at the positron correctly. As the core electrons are less polarizable than the valence electrons the density of electrons at the positron is enhanced less for the core electrons than for the valence electrons. This is not taken into account in the calculations, and hence, the calculated AC curves at higher angles caused mainly by core electron annihilations are too high compared with the valence electron contribution. The e^+_{aq} curve is also given in Fig. 6.2. The $[X^-,e^+]$ curves given by a sum of three gaussians are shown in Table 6.1. The positron annihilates mainly with the outermost electron in X^- in the bound state [6.1,2]. Hence, the widths of the bound state curves reflect mainly the fact that this electron in the ions is moving slowly in I^- and much faster in Cl^-, Br^-being in between. The theoretical curve for the $[F^-,e^+]$ state shows the effect still clearer. The positron does not form a bound state with F^- in water, see below.

The e^+_{aq} reactions with the pseudohalide ions have also studied [6.5]. Bound states are formed with SCN^-, CN^-, and S^{--}, but not with OCN^-. The reaction of e^+_{aq} with SH^- seems to give $[S^{--},e^+]$. The determination of the bound state AC curves, shown by three gaussian fits in Table 6.1 was very difficult for the pseudohalides. The obtained pseudohalide curves are shaped somewhat like triangles, while the halide curves are shaped like parabolas, in the top.

Competitive e^+_{aq} reactions with halides and pseudohalides have also been studied [6.6,7,5]. For example, 1 M I^- + 1 M Br^- + 1.74 M PPS in water gave the $[I^-,e^+]$ spectrum except for the usual small correction for the Ps components. This shows that the positron is stronger bound to I^- than to Br^-. In that way it was found that the binding energy of the $[X^-,e^+]$ bound state increases in the series Cl^-, SCN^-, Br^-, S^{--}, SH^-, and I^-. Theoretically, the binding energy increases in the series I^-, Br^-, Cl^-, and F^- for the states in vacuum [6.1,2]. This difference is probably caused by the hydration energy of the ions which is roughly 3.05, 3.48, 3.76, and 5.24 eV for I^-, Br^-, Cl^-, and F^-, respectively. The high hydration energy of F^- due to its ability to take part in hydrogen binding with the water molecules, is very probably the reason for the inability of F^- to bind the positron in water.

F^-, OH^-, $SO_4{}^{--}$, $MnO_4{}^-$, $Cr_2O_7{}^{--}$, acetate ions, and oxalate ions do not form bound states with the positron in water, i.e., they do not change the spectra away from the e^+_{aq} spectrum when added to a 1.74 M PPS aqueous solution [6.3–5].

The exchange of the normally used Na^+ positive ions with H^+ or Cs^+ ions seems not to influence appreciably the strength of positron bound state formation.

Bound states of the positron are also formed in alcohols and acetonitrile with roughly the same ions as in water [6.3,4]. We shall discuss positron bound state formation with various molecules in nonpolar liquids in Chap. 7.

Lifetime measurements for D_2O give $I_3 = 23\%$ and $\tau_3 = 1.95$ ns (Table 5.2). On taking this into account and by using the same procedure as for H_2O, the AC components of D_2O were determined [6.3]. Surprisingly, the shapes of the AC components for D_2O were the same as for H_2O; only the intensities were different in D_2O. The positron bound state curves were assumed to have the same shapes in H_2O and D_2O.

As mentioned above the only other attempt to derive the shapes of the $[X^-,e^+]$ curves has been done by the Strasbourg group. As they used DB measurements their shapes are expected to be of much poorer quality than the shapes obtained from AC spectra. Furthermore, the Strasbourg $[X^-,e^+]$ curves are approximated by only one gaussian. Some of the Strasbourg and Risoe $[X^-,e^+]$ curves are compared in Fig. 1 of [6.8]. For example, the Strasbourg curves for S^{--}, SCN^-, CN^-, Cl^-, Br^-, and I^- are above the Risoe curves in the top by 2–12%.

The derived shapes of the pure water and positron bound state AC components have been used in the analyses of many measured AC spectra. The AC spectra were measured for various concentrations of NaCl, NaBr, and NaI in pure H_2O and D_2O and in 1.74 M PPS solutions of H_2O and D_2O [6.3]. The spectra for the pure water solutions were fitted by a sum of the three curves with the shapes of the e^+_{aq}, total Ps, and $[X^-,e^+]$ components, respectively. The curves marked 1, H_2O in Table 5.3 for the H_2O components, and curves with practically the same shapes for the D_2O components, were used. The spectra for the $c_s MX^- + 1.74$ M PPS solutions were first corrected for the 3.23% Ps component and then fitted by a sum of two curves with the shapes of the e^+_{aq} and $[X^-,e^+]$ curves, respectively.

Two results are obtained by such analyses. First, the fits were found to be acceptable in all the cases. Hence, the experimental results were in good agreement with our basic model of positron bound state formation and the assumptions used in the derivation of the shapes of all the components. This is a very important result. Second, the percentage of the positrons forming bound states at various X^- concentrations was determined by the fitting analyses. We shall here show some of the results for iodide solutions. The results for the solutions of Br^- and Cl^- were fairly similar [6.3].

In Fig. 6.3 the intensity in % of the positrons not forming Ps which are trapped before the annihilation, i.e., the intensity of the $[I^-,e^+]$ component divided by the sum of the intensities of the $[I^-,e^+]$ and the e^+_{aq} components in %, is shown. Furthermore, Fig. 6.4 shows the intensity in % of the total Ps component for the pure H_2O and D_2O solutions. For comparison the total Ps intensity in % as derived from the intensity of the o Ps long lifetime component in the LT spectra in three term fits is shown in Fig. 6.5 [6.3]. The concentration (c_s) axis in Fig. 6.3 can be divided into intervals. At $c_s < 0.001$ M I^- the percentage of trapped positrons is very low, and hence, very uncertain. Furthermore, at $c_s > 0.03$ M I^- the concentration of I^- is so high that the usually used dilute concentration models of reaction kinetics do not apply. However, in the interval 0.001 M $< c_s <$ 0.03 M we can expect to be able to explain the

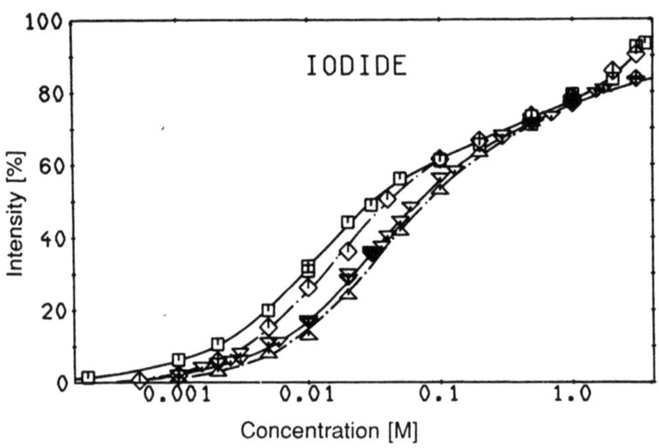

Fig. 6.3. Percentage of "free" positrons bound to I^- at annihilation versus I^-concentration in H_2O (□), D_2O (◇), 1.74 M PPS/H_2O (▽), and 1.74 M PPS/D_2O (△). Solid lines are visual fits

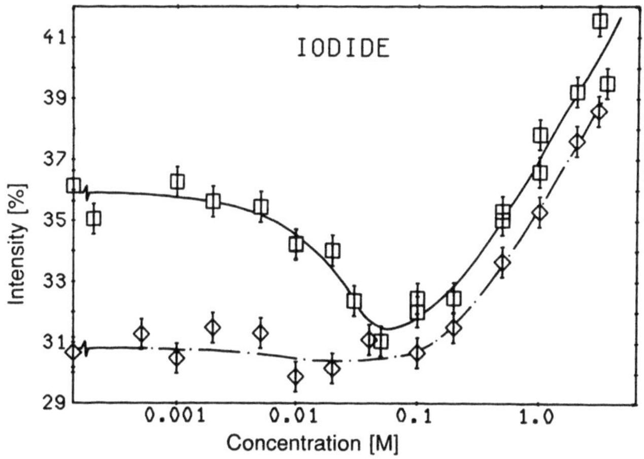

Fig. 6.4. Total Ps yield from AC spectra versus I^- concentration in H_2O (□) and D_2O (◇) (see [6.3]). The lines are visual fits

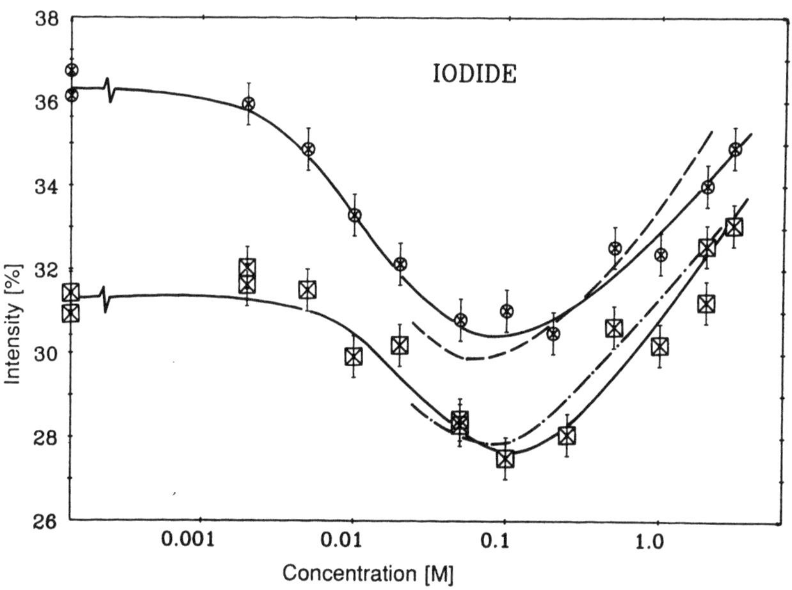

Fig. 6.5. Total Ps yield from LT spectra versus I^- concentration in H_2O (⊗) and D_2O (⊠) (see [6.3]). The dashed and dot-dashed lines show the yields obtained in Strasbourg. Solid lines are visual fits

results by use of the "normal" theories. The results will be discussed in detail in Sect. 6.3.

There are several reason to state that the main number of reactions take place after the positron is hydrated in water for $c_s < 0.03$ M. This is discussed in detail in [6.3,8]. Three reasons will be mentioned here. First the average distance

Table 6.2. Percentage of positrons reacting in initial close encounters for an effective radius $R' = 1.4$ nm in (6.27) ($R = 1$ nm) and the average distance between halide ions, versus halide concentration in water

Conc. [M]	Close Enc. [%]	Dist. [nm]
0.000464	0.319	15.3
0.001	0.685	11.8
0.00215	1.47	9.2
0.00464	3.14	7.1
0.01	6.65	5.5
0.0215	13.38	4.3
0.0464	27.32	3.3
0.1	49.72	2.6
0.215	77.27	2.0
0.464	95.89	1.5
1.0	99.90	1.2
2.15	100.0	0.9

between the iodides is 4.3 nm at 0.02 M I^- (Table 6.2), which is fairly large compared to the reaction radius of 1 nm or smaller expected for e^+_{aq}. Second, only about 3% of the excess electrons seem to react before hydration with Cd^{++} at 0.01 M Cd^{++} in water [6.9], while 31% of the hydrated positrons have formed $[I^-,e^+]$ at 0.01 M I^-in water. Cd^{++} is one of the ions, which shows the highest reaction rates with non-hydrated excess electrons in water. Third, the ionic strength effect on the positron reaction with I^- in water has been found to agree fairly well with the normal theory for a hydrated positron [6.10].

6.2 Reaction Kinetics in Positron Annihilation

The discussions of the e^+_{aq} reactions must be based on selected parts of reaction kinetics [6.11-14]. In addition, reaction kinetics is used in the discussions of Ps formation, Ps reactions, and Ps trapping in defects in solids in the following chapters. It is, therefore, appropriate to discuss certain aspects of reaction kinetics in some detail here. We have postponed such discussion to this chapter because we can then use the experimental results of the e^+_{aq} reaction studies to illustrate directly that it is necessary to use fairly complicated parts of reaction kinetics to explain the results. Standard textbooks on physical chemistry (see, e.g., [6.15,16]) do not treat reaction kinetics at the level needed here. Of course, we cannot derive the expressions used in this book; only a guide to the literature where they are derived and a short discussion of the basic principles, can be given.

At first sight, the positron reactions are second order, bimolecular reactions, which, by use of simple kinetics to describe the reaction between reactants A

and B [6.11,15], are written

$$A + B \longrightarrow \text{product}. \tag{6.1}$$

The rate expression is assumed to be

$$d[A]/dt = -k[A][B], \tag{6.2}$$

where the rate constant, k, is assumed to be independent of the concentration of A, $[A]$, and B, $[B]$, but dependent on temperature and pressure. t is the time.

In the cases of the positron (and Ps) reactions the concentrations of the reactant B (e.g., the halide concentration) are large, and the loss of reactant B during the measurement can be neglected. Hence, our positron reactions are pseudo-first order reactions [6.11,15], i.e., the rate expression is

$$dc_+(t)/dt = -kc_s c_+(t), \tag{6.3}$$

where $c_+(t)$ is the positron concentration, which is time dependent, and c_s is the reactant concentration independent of time. Accordingly, the normally used kinetic equations for the reactions of positrons are

$$dc_+(t)/dt = -\lambda_f c_+(t) - kc_s c_+(t), \tag{6.4}$$

$$dc_b(t)/dt = -\lambda_b c_b(t) + kc_s c_+(t), \tag{6.5}$$

where c_+, c_s, and c_b are the concentrations of positrons, positron scavengers, and bound positrons, respectively, λ_f and λ_b are the annihilation rates for the "free" (i.e., not bound or in Ps) and bound positrons, respectively, and k is the rate constant. These equations are called the "trapping model" in studies of positron trapping in defects in solids [1.5].

The initial conditions ($t = 0$) are assumed to be $c_+(0) = 1$ and $c_b(0) = 0$, i.e., all positrons are assumed to be "free" initially. The kinetic equations are then solved for the physical properties that can be measured, i.e., the LT and AC (DB) spectra. The solution for the LT spectra is a sum of two exponentials. Hence, we can characterize the solution by τ_1 (I_1) and τ_2 (I_2), the resulting lifetimes and their intensities. In some cases, where the lifetimes are difficult to separate, e.g., because of limited experimental resolution, an average lifetime is derived and compared with the measured average lifetime. In AC studies in particular the percentage of positrons which annihilates from the bound state can be measured, because the AC spectrum is a "fingerprint" of the state of the electron-positron pair at the time of the annihilation. The calculated probability of annihilation in the bound state can be compared with that determined experimentally.

Two extreme cases may come out of the comparison of a typical theory and experimental facts. In some cases the theoretical model cannot explain the measured results at all, even if adjustable parameters are varied between reasonable limits. Of course, this disproves the theory, and a new, usually more complicated, theory must be looked for. On the other hand, the theory might explain the measured data quite well with reasonable values of the adjustable parameters. This is a *necessary* condition for the statement that the model is the correct model. However, it is not a *sufficient* condition for a correct model. The agreement between theory and experiment might be fortuitous, or much more likely, the agreement is obtained by variation of adjustable parameters, whose values are unknown within intervals limited by widely separated numbers. Of course, in some comparisons of theory and experiment no adjustable parameters are involved for that particular experiment. Furthermore, some models make use of so many adjustable parameters (or even functions) that almost any experimentally obtained results might be fitted well by variation of the parameters.

These short remarks on the estimation of the validity of theoretical models will serve as a reminder in this and the following chapters, in which we shall discuss the applicability of many models for the explanation of the experimental results.

We can use (6.4,5) to derive the fraction of positrons, x, annihilating in the $[X^-,e^+]$ bound state, if k is assumed to be independent of the time, t, [6.4]

$$x = kc_s/(\lambda_f + kc_s). \tag{6.6}$$

We find $x = 20\%$ at $c_s = 0.005$ M I^-in pure H_2O, and further $\lambda_f = \tau_f^{-1} = 2.53 \times 10^9$ s^{-1}. Hence, $k = 12.7 \times 10^{10}$ M^{-1}s^{-1}. This is roughly twice the value expected for the simple, time-independent, diffusion-controlled, theoretical rate constant (see below). Above $c_s = 0.03$ M the rate constant derived from (6.6) decreases for increasing I^- concentration. At 3 M I^- it is roughly ten times smaller than at 0.005 M.

Essentially, the formulas, which are commonly used in chemistry, and from which the basic ideas of (6.2–5) are taken, have been proposed on a mainly empirical basis many years ago. However, the basic rate expression (6.2), and hence, the corresponding terms in (6.4,5), can be derived theoretically by use of approximations and by assuming that certain models of the physical processes which control the behaviour of the reactants are correct.

We shall shortly discuss the appropriate theories of the reaction rates in order to get an understanding of the basic ideas and the limits of applicability of the theories. The rate theories are the most important, general, mathematical tools in the interpretation of the main parts of the experimental data in greater parts of positron annihilation, including the chemistry oriented parts. In most experiments the electron, positron, and/or Ps react with molecules, ions, atoms, defects, clusters of defects, etc., and the rate expressions are used to interpret the reaction, while the quantum mechanical theories, as described in Chap. 2, are used in the understanding of the products of the reaction.

In a typical kinetic equation, as exemplified by (6.4,5), the left hand side is the time derivative of the concentration of a certain reactant, or of the equivalent probability of observing the reactant, while the right hand side consists of one or more terms describing the processes by which that reactant is created or destroyed. In principle, each of these terms can be derived theoretically, and the derivation of one term is normally done independently of that of another term. Some of the terms can be determined in other experiments.

For example, the first terms on the right hand side of (6.4,5) can be measured in LT experiments, which for halides in water give $\lambda_f = \lambda_b = 2.53 \times 10^9\ s^{-1}$. Furthermore, the quantum electrodynamic (QED) theory discussed in Chap. 2 gives as result, that each state is expected to give a certain lifetime. A distribution of fairly similar states, calculated by use of statistical mechanics, is expected to be well described in terms of an average lifetime. Hence, theory tells us that a certain decay rate is to be expected in a good approximation, and experiments confirm that result. Similarly, QED tells us that each state of the positron gives a certain AC spectrum, and this is confirmed by experiments, too.

The second terms in (6.4,5) describe elementary, bimolecular reactions. Such reactions have been studied theoretically in very many papers and by use of many approximations, which are frequently not controlled, i.e., the errors caused by the used approximations are not estimated. Some approximations are sometimes not even stated explicitly. Because of lack of space we cannot, of course, give a reasonably complete discussion of these theories in this book. Instead we shall give a short guide to the theories and, in particular, discuss some of the approximations which are of importance in light particle reactions.

Let us discuss (6.1) and how (6.2) is derived. Apparently, the calculation of the rate constant has not been done in a fully complete way, i.e., the calculations in the literature seem only to be correct if certain uncontrolled approximations are valid. The rate constant for reaction (6.1) can be derived by use of the equation [6.17]

$$\partial n/\partial t = \nabla D\{\nabla n + [zn/(|z|k_B T)]\nabla U\}. \qquad (6.7)$$

n is the concentration of, or the probability of observing, reactant B, t is the time, D the diffusion constant of B, k_B is the Boltzmann constant, T the temperature in K, and U the potential energy of reactant B. z is the charge of reactant B (e.g., $z = -1$ for the electron). An identical equation is used for reactant A, and the resulting rate constant is obtained by adding the constants obtained for B diffusing towards A and A diffusing towards B. This seems to be a somewhat uncertain way of starting the calculation of the rate constant. The model assumes a continuous distribution of the reactants. Apparently, the more correct, stochastic theories give very similar results.

Expression (6.7) tells us that a reactant moves by two physical processes, namely by random diffusion as given by the first term, and by drift in a potential given by the second term, on the right hand side of (6.7). The first term is based on Fick's law [6.15,16]

$$\Phi = -D\nabla n, \qquad (6.8)$$

where Φ is the flux of B.

The second term is based on the normal assumption that the velocity of a reactant in an electric potential, U, is [6.15,16]

$$\mathbf{v} = \mu \mathbf{E} = -\mu \nabla U, \qquad (6.9)$$

where μ is the mobility and E the electric field strength. The flux due to this drift is

$$\Phi = \mathbf{v}n = -(z/|z|)\mu n \nabla U. \tag{6.10}$$

Furthermore, by use of the Nernst-Einstein equation [6.15,16]

$$D = (k_B T/|z|e)\mu, \tag{6.11}$$

where e is the elementary charge, the second term in (6.7) is obtained. For an isolated pair of ions

$$U = z_1 z_2 e^2/4\pi\varepsilon_0\varepsilon r, \tag{6.12}$$

where $z_1 e$ and $z_2 e$ are the charges of the ions, ε_0 and ε are the vacuum and relative dielectric constants, and r is the distance between the ions.

Expression (6.7) is a partial differential equation. It is normally solved only partly to give (6.2) for a given initial distribution of the reactants in space, i.e., a normal differential equation in time only is the result of the mathematical derivations. During the derivations the density, $n(x, y, z, t)$, in (6.7) can be replaced by a space-independent, average density (or concentration) for dilute solutions. This explains how (6.2) is obtained.

The following rate constants (6.13–23) are obtained by use of the initial condition ($t = 0$) that the reactants are randomly distributed in space, i.e., for homogeneous reactions [6.11–14,17]. It is also assumed that the solution is so dilute that the number of initially formed encounter pairs, i.e., pairs with a center-to-center distance smaller than the reaction radius, R, is negligible. Furthermore, a weak extra dependence on time, t, for $t < 10$ ps in normal liquids is disregarded [6.13,17].

Let us first give the rate constants for reactions where the potential, U, is zero, i.e., where the second term in (6.7) is missing. Let us further assume that the reactants react immediately, if the distance between them during the diffusion happens to become smaller than the reaction radius, R. This gives a rate constant, in units $M^{-1}s^{-1}$

$$k = 4\pi RD'\{1 + R/(\pi D't)^{1/2}\}N_0/1000, \tag{6.13}$$

where $D' = D_A + D_B$ (cm^2/s) is the sum of the diffusion constants for A and B, $t(s)$ is the time, R (cm) is the reaction radius, and N_0 is the Avogadro number ($N_0 = 6.04 \times 10^{23}$ $mole^{-1}$). At longer times, $t \gg R^2/\pi D'$, where the second term in (6.13) is negligible, we get

$$k = 4\pi RD'N_0/1000. \tag{6.14}$$

Expressions (6.13,14) are the rate constants for diffusion controlled reactions, first derived by Schmoluchowski. They give the rate constants if A reacts immediately with B upon encountering in the process of random diffusion. Of course, this means in practice only that they react very fast. Nonradiative processes which normally remove the energy of a reaction always take finite times (e.g., some picoseconds).

A less restrictive condition is the case in which A and B react with a finite rate constant, k_R ($M^{-1}s^{-1}$), when in encounter pair position. It results in a rate constant [6.13]

$$k = k_D/(1 + k_D/k_R)[1 + R/(\{1 + k_D/k_R\}\{\pi D't\}^{1/2})], \tag{6.15}$$

where $k_D = 4\pi RD'N_0/1000$. At $k_D \ll k_R$ (6.15) reduces to (6.13). At $k_D \gg k_R$ we get

$$k = k_R[1 + k_R R/\{k_D(\pi D't)^{1/2}\}] \simeq k_R \tag{6.16}$$

At longer times where the second term in (6.15) is very small, i.e., for $t \gg R^2/\pi D'$, (6.15) reduces to

$$k = k_D k_R/(k_D + k_R). \tag{6.17}$$

Next, we give the rate constants for reactions between two charged reactants interacting through the Coulomb potential [6.11–14,17]. Hence, the second term in (6.7) accounts for the drift of a charged reactant in the Coulomb potential of the other charged reactant.

At longer times, $t \gg R^2/\pi D'$, and for $k_D \ll k_R$, Debye showed that

$$k = 4\pi r_c D'N_0/[1000\{\exp(r_c/R) - 1\}], \tag{6.18}$$

where the potential energy is given by the Coulomb energy (6.12) and the Onsager radius, r_c (cm), is given by [5.23]

$$r_c = z_1 z_2 e^2/4\pi\varepsilon_0\varepsilon k_B T. \tag{6.19}$$

For $r_c \gg R$ (6.18) gives

$$k = 4\pi r_c D'N_0/1000, \tag{6.20}$$

i.e., r_c is the effective reaction radius.

Rice et al. [6.17] have derived the following expression which takes into account all three effects, namely 1) the Coulomb field between the reactants, 2) the time dependent effect, and 3) the finite rate of reaction in the encounter pair states. For times $t > 10$ ps for normal liquids they derived the following expression for k

$$k = 4\pi R_{\mathrm{eff}} D'(N_0/1000)[1 + R_{\mathrm{eff}}/(\pi D't)^{1/2} \exp(r_c/R)\{1 + 4\pi r_c D'/k_R\}], \tag{6.21}$$

where

$$R_{\mathrm{eff}} = r_c/[\{1 + 4\pi r_c D'/k_R\}\exp(r_c/R) - 1]. \tag{6.22}$$

Expression (6.21) reduces to (6.15) for $r_c = 0$, and to

$$k = 4\pi R_{\mathrm{eff}} D'(N_0/1000)[1 + R_{\mathrm{eff}}\exp\{r_c/R\}/(\pi D't)^{1/2}], \tag{6.23}$$

where $R_{\mathrm{eff}} = r_c/\{\exp(r_c/R) - 1\}$ for $k_D \ll k_R$. At longer times, $t \gg R^2/\pi D'$, (6.23) reduces to (6.18), the Debye rate constant for diffusion controlled reaction between ions.

Rice et al. [6.17] discuss also the rates if the initial distribution of one reactant around the other reactant is the Boltzmann distribution

$$n(r, t, 0) \propto \exp(r_c/r), \tag{6.24}$$

where r is the distance between the centers of the ions. They get

$$k = 4\pi R_{\text{eff}} D'(N_0/1000)\{1 + R_{\text{eff}}/(\pi D't)^{1/2}\} \tag{6.25}$$

with R_{eff} given by (6.22).

Apparently, (6.21,25) are the best general, theoretical rate constants available for the interpretation of many positron annihilation experiments, including the e^+_{aq} reactions with halides. However, two important effects are not taken into account in the derivation of (6.21,25), namely the effects of the initially formed encounter pairs [6.18] and the ionic strength [6.11,15,16].

The initially formed encounter pairs, i.e., the reactants which are within the reaction radius of the other reactant at time $t = 0$, can be very important at high concentration of reactants, such as in the hydrated positron studies at I^- concentrations above roughly 0.03 M. The above given general equations for k are for dilute solutions only! *Czapski and Peled* [6.18] calculate the probability that n reactants, S, each of an "encounter volume" $v = (4/3)\pi R^3$ do not cover a random spot in the solution of volume V. If it is assumed that $R \gg r_S$, where r_S is the physical radius of S, the probability is given by

$$P = \{1 - (v/V)\}^n \simeq \exp(-c_s v'), \tag{6.26}$$

where c_S is the concentration of S in unit M, and v' is the partial molar volume of S in M^{-1}. Hence, the probability of initial encounter formation is

$$P' = 1 - \exp\{-(4/3)\pi R^3 c_S N_0/1000\}. \tag{6.27}$$

As the concentration of species reacting with positrons, Ps, or electrons in positron annihilation research is large in many studies, (6.27) is very useful. It seems only to be approximately correct. At high concentration, c_S, and if the reactant is excluded from an appreciable volume (e.g., e^+ is excluded from a volume occupied by Na^+ ions), significant errors might result if it is used.

The effect of ionic strength on the rate of reactions of charged reactants is given by the classical Broendsted-Bjerrum theory [6.11,15,16]

$$\log(k) = \log(k_0) + c_0 z_1 z_2 I^{1/2}/(1 + \alpha I^{1/2}), \tag{6.28}$$

where $k(k_0)$ is the rate constant at ionic strength I (0), z_1 and z_2 are the charge numbers of the ions (e.g., $z_1 = -1$ for e^-_{aq}) and $\alpha = (r_1 + r_2)/3$, where r_1 and r_2 are the radii of the two reactants, in unit 10^{-10} m. The ionic strength is

$$I = (1/2)\sum_{i=1}^{s} c_i z_i^2,$$

where c_i is the concentration (M) of i'th ion and z_i their charge number. The sum is over all types of ions present in the solution. The constant $c_0 = 1.02$ in water solutions, if the ionic sphere around both ions is formed, while $c_0 = 0.51$ in water solutions if it is not formed around one of the ions (e.g., in dry electron reactions) [6.19,20].

Expression (6.28) can be used on diffusion controlled as well as slower reactions. However, it is only valid if the ions have enough time to form an ionic sphere around the reactants for $c_0 = 1.02$. It should not be used for $I > 0.05$ M, and it is not valid if ion-pair complexes are formed [6.11].

It has not been possible to find theoretical treatments in which all five effects, namely time-dependent rate, finite rate at R, electric potential, initial encounter pairs, and ionic strength effects, are combined in one rate constant expression. This was found only for the three first effects, which are combined in (6.21,25).

In the preceding discussion we have assumed a homogeneous distribution of both reactants, except for the discussion of (6.25). However, nonhomogeneous reactions are very common in positron annihilation research. Nonhomogeneous kinetics has been defined as [5.8]: "Rates and mechanisms of reactions between reactants at least one of which is neither uniformly nor randomly distributed in space". The reactions in the positron spurs, and in the tracks of charged particles in general, are good examples of reactions which must be interpreted by use of nonhomogeneous kinetics in most cases. We give more detailed discussions of the positron spur in Chaps. 5,7, where some of the assumptions used in the spur diffusion models, based essentially on (6.7), are discussed, too.

Finally, our short guide to reaction kinetics must include some comments on the effects of quantum mechanics on the reactions of the light particles. The above given theory concerns, in principle, classical particles characterized by a center, a charge, and reaction radii in reactions with various other reactants. Quantum mechanical effects are very important in the behaviour of the three light particles. However, "semi-classical" theories are normally used to explain the transport phenomena of electrons, even in semiconductors and metals where the electrons are delocalized [6.21]. Typically, a classical diffusion term in the transport equation and a density of electrons in space are amongst the concepts normally used to describe electron motion in the bulk. Furthermore, kinetic expressions, similar to (6.4,5), are usually applied to describe the trapping of electrons in, e.g., semiconductors ([6.22], Chap. 14). *Ziman* ([6.21], Sects. 5.12, 7.1, and 7.2) discusses the problems of using the "semi-classical" models on electrons in solids.

Actually, this seems to be a partly unsolved problem, even for electrons in Si and Ge, for which the band structure is well known, and the physics of phonons etc., which might absorb the energy released as the electron is trapped, is well studied. Apparently, the "semi-classical" theories of the behaviour of electrons work well even for very delocalized electrons in pure solids. The problems of using such model on the behaviour of the three light particles in liquids are, of course, much smaller than for electrons in very perfect solids. Furthermore, and very importantly, other, more quantum mechanical, theories seem not to be available, i.e., we have essentially no other choice than to use the "semi-classical" theory at present. Of course, if it is possible the parameters describing the behaviour of the light particles in the "semi-classical" theory can be profitably calculated by use of quantum mechanics. In that way tunnelling can be included in an effective reaction radius, and detailed trapping mechanisms can be used to calculate rates of trapping, for example. Trapping of Ps from Bloch function states will be discussed in Chap. 9.

6.3 Discussion of e^+_{aq} Reactions. Low Concentrations

We shall now continue the interpretation of the e^+_{aq} reactions with iodide. It is expected that the theories discussed in the preceding section can be applied in the interpretation of the measured fraction of positrons annihilating in the bound state with I^- for I^- concentrations $0.001 < c_S < 0.03$ M (see Fig. 6.3).

On referring to the hydrated electron reactions, we cannot expect that the values of D_+, the positron diffusion constant, and R are very different from $D_- = 5 \times 10^{-5}$ cm^2/s [4.2] and $R = 1$ nm measured for e^-_{aq} reactions with strongly oxidizing ions such as $IO_4{}^-$ [6.23]. Lifetime measurements give $\lambda_f = \lambda_b = 2.53 \times 10^9$ s^{-1}. Furthermore, $D \simeq 2 \times 10^{-5}$ cm^2/s for halides, and $r_c = -0.71$ nm for water. Hence, if we choose $D' = 7 \times 10^{-5}$ cm^2/s and $R = 1$ nm, we get $k = 5.3 \times 10^{10}$ M^{-1}s^{-1} and $k = 7.4 \times 10^{10}$ M^{-1}s^{-1} from the time independent expressions (6.14) and (6.18), respectively. We measured $k = 12.7 \times 10^{10}$ M^{-1}s^{-1} in pure H_2O on the assumption of a time independent k.

However, the time dependence of k is important for the e^+_{aq} reactions with I^-. The second term in (6.13), $R/(\pi D't)^{1/2} = 1.51$, 0.67, 0.34, and 0.21 at $t = 20$ ps, 100 ps, 400 ps, and 1000 ps, respectively. As the average lifetime of e^+_{aq} in water is 395 ps, this term cannot be disregarded.

In [6.3] we used (6.21) to calculate k, as this seems to be the best expression available. Furthermore, we corrected for the initial encounter pair formation by use of (6.27). We substituted (6.21) for k in (6.4,5) and solved these equations for $c_+(t)$ and $c_b(t)$ using the initial conditions mentioned above. We then calculated the fraction of positrons annihilating in the bound state $[I^-,e^+]$, x. We found

$$x = c_1[-k'c_S/G + \{(\pi)^{1/2}\lambda_f B/(2G(-G)^{1/2})\}\exp(-B^2/4G) \times (1 - \mathrm{erf}\{-B/(2(-G)^{1/2})\})], \quad (6.29)$$

for $G = -\lambda_f - k'c_S$, $B = -2k'c_S R/\{a(\pi D')^{1/2}\}$, and c_1 is the initial positron concentration. We expressed (6.21) as

$$k = k'[1 + R_{\mathrm{eff}}/\{a(\pi D't)^{1/2}\}], \quad (6.30)$$

and assumed a diffusion controlled reaction, i.e., $k_R \gg 4\pi r_c D'$. Hence,

$$k' = 4\pi R_{\mathrm{eff}} D' N_0/1000, \quad (6.31)$$

and

$$a \simeq \{\exp(r_c/R)\}^{-1}. \tag{6.32}$$

Of course, (6.29) reduces to (6.6) if the second term in (6.30) is missing.

We assumed that a fraction of the positrons, given by P' of (6.27), formed initial encounter pairs. Hence, the fraction $1 - P$' reacted according to (6.4) and (6.5), i.e., $c_1 = 1 - P$'.

In this treatment the reactions are divided into three cases:

A) The initially formed encounter pairs.
B) The rapid reactions close to the halide but outside R_{eff} are taken into account by the second term in (6.30). It must be emphasized that the time dependence in (6.30) is really a "space effect", namely an effect due to the faster than "steady state" reaction of reactants close to but outside the reaction sphere at shorter times, causing a gradient in the distribution to be formed around the sphere.
C) The "steady state" reactions at longer times, where the second term in (6.30) is very small.

This division into three cases is very important for the electron and positron reactions in the positron spur and, therefore, for the understanding of Ps formation.

The percentage of positrons reacting as initial encounter pairs for an effective reaction radius of $R_{eff} = 1.4$ nm, and the average distance between halide ions, versus halide concentration are shown in Table 6.2. Clearly, initial encounter pair formation is not important for $c_S < 0.02$ M.

The most important result of the given interpretation of the e^+_{aq} reactions in terms of (6.29,27) with the chosen parameters of the fastest e^-_{aq} reactions is that the percentage of trapped positrons is fairly well explained for $c_S < 0.01$ M I^- (see Fig. 8 in [6.3]). It must be noted that no adjustable parameters are used in that comparison of theory and experiment. As the results for Br^- and Cl^- are very close to those of I^- for $c_S < 0.01$ M, also the results for the other two halides are well described by the theory. This speaks in favour of the assumption of a diffusion controlled reaction with the hydrated positron for all three halides. It speaks against a hypothesis that the positron reacts mainly before hydration, since this implies that, at least, for one of the halides a rate constant different from that expected for the hydrated positron would be anticipated.

An assumption of D' being inversely proportional to the viscosity, which is obtained from (6.18) using Stokes expression for the viscosity [6.15,16], explained only roughly half the variation between the four solutions of I^- (Fig. 6.3). We note that the "steady state" rate constant $k' = 7.4 \times 10^{10}$ $M^{-1}s^{-1}$ in pure H_2O. Hence, the time dependent term in (6.30) doubles roughly the "effective" rate constant. Very few experimental results in chemical kinetics demand for their interpretation the time dependent term in the rate constant. For e^-_{aq} reactions these problems are discussed in [6.23,24,18].

It must be emphasized that the good agreement with the theory with the chosen parameters from e^-_{aq} reactions speaks only in favour of the theory. It is not sufficient to prove that the positron actually has the same D and R values as the electron in water. Other values can explain the data. No other solvated positron reactions are available for a similar detailed interpretation at present. We have thus shown that the advanced theory of diffusion controlled reactions explains well our results on the fraction of trapped positrons versus I^- concentration for $c_s < 0.01$ M. We discussed this in detail because we need the theories in the rest of the book. However, we have to refer to more experimental results, and several complicated effects, in order to finish the discussion of the positron reactions. Because of lack of space we must limit this treatment to the most important facts, and refer to details in the original papers, some of which are not yet published. Let us first discuss some experimental results which support the given interpretation.

The AC spectrum for 1 M CBr_4 in methanol, measured at Risoe, is practically identical to the $[Br^-,e^+]$ curve, extracted as described above. In alcohol mixtures with CCl_4 and CBr_4 the radiation damage caused by charged particle penetration is strongly enhanced because of chain reactions, which destroy CCl_4 and CBr_4 and create Cl^- and Br^-, respectively. The normally found yield of radiation created species per 100 eV of 1–5 can be enhanced to 100–1000, if chain reactions occur. Apparently, only CCl_4/alcohol mixtures have been studied in radiation chemistry [6.25]. However, we expect that chain reactions are even more important in CBr_4/alcohol mixtures. We tested this effect in CBr_4/butanol mixtures by AC measurements and Br^- concentration determinations. We found that the Br^- ions created by the homogeneous radiation damage due to the slowing down of the positrons and the chain reactions explained well the changes in the AC spectra caused by the reduction in the Ps inhibition by CBr_4 and the increase in the positron reaction with Br^-.

In the CBr_4/methanol mixture all Ps formation was inhibited by the remaining CBr_4, and the Br^- concentration was high enough to trap all positrons. The new, important result was that the $[Br^-,e^+]$ curve was measured i.e., no alcohol component was present. We always measured that at least 10% of the spectra for aqueous solutions of Br^- was the e^+_{aq} curve. Apparently, a similar alcohol component is not found in methanol. This result speaks strongly in favour of the methods used to extract the $[X^-,e^+]$ curves. Of course, more experiments are very important on that point.

As mentioned above, we also studied the ionic strength effect on the $e^+_{aq}+I^-$ reaction. The I^- concentration had to be rather high because of the short e^+_{aq} lifetime, and hence, the added inert ion concentration had to be so high that it was very close to or above the concentration which is the upper limit for the applicability of the "classical" theory (6.28). In spite of these problems we found that log(k) decreased linearly with $1.02 \times I^{1/2}/(1+I^{1/2})$, as given in (6.28) for $c_0 = 1.02$, at I^- concentrations of 0.005 M and 0.01 M, if the inert ions were Na^+ and $ClO_4{}^-$. The slope indicates an e^+_{aq} reaction, and not a positron reaction before hydration, with I^-. This is the first and only ionic strength study of a solvated positron reaction. The results support our interpretation of the e^+_{aq} + X^- reaction, in general.

At halide concentration $c_S < 0.03$ M roughly the same fraction of trapped positrons was measured for Cl^-, Br^-, and I^-. This is expected for a hydrated positron reaction which is diffusion controlled. On the other hand, it would be a strange coincidence for a nonhydrated positron reaction, if all three halides gave the same rates, which also coincide with the largest rates of e_{aq}^- taking into account the time dependent effect. Again our general assumptions are supported.

At Risoe we have also studied aqueous solutions (H_2O) of two halides plus 1.74 M PPS in great detail. The results were analyzed by use of the simplest possible model of competition between three processes, namely annihilation as e_{aq}^+ and trapping on the two halides. Furthermore, the trapped positron on the shallower trap (e.g., Br^-) was allowed to transfer to the deeper trap (e.g., I^-). The measured spectra were analyzed by use of PAACFIT into its components, as discussed above. In positron studies of defects in solids this model is called the two trap trapping model with transfer from shallow to deep trap. The annihilation rates in water and on the traps are known, as discussed above.

The rate constants for positron trapping were chosen as the apparent rate constants at the halide concentration equal to the total halide concentration, as calculated from the halide plus 1.74 M PPS curve (Fig. 6.3 for I^-). The only adjustable parameter was the positron transfer rate constant, k_t. For Cl^- and Br^- added to 0.001 M $I^- + 1.74$ M PPS aqueous solution up to 0.1 M, the measured results were well represented by the model for $k_t = (2.0 \pm 2.0) \times 10^{10}$ $M^{-1}s^{-1}$, i.e., k_t was very uncertain. For Br^- added to a 0.03 M I^- + 1.74 M PPS solution up to 0.3 M, the results were well represented by the model for $k_t = (2.0 \pm 0.5) \times 10^{10}$ $M^{-1}s^{-1}$. For Cl^- the same k_t value gave the best fit, but the measured e_{aq}^+ intensity was higher than that obtained from the model, a disagreement which was found to be still more pronounced at higher halide concentrations.

At halide concentration of approximately 1 M the agreement between model and the measured intensities was very bad. However, a model where all positrons annihilate from a bound state with the deepest trap explained well the results at roughly 1 M halide concentrations for all ions. At such high concentrations encounter pair effects dominate the reactions and the positrons need not to diffuse to reach the deeper trap.

This is the first and only study of positron transfer reactions between heavy reactants. Electron transfer reactions are very important in chemistry, while Ps transfer reactions seem not to have been clearly identified, as yet.

The measured k_t value is approximately the expected rate constant. Compared to the "steady state" value of $k' = 7.4 \times 10^{10}$ $M^{-1}s^{-1}$ obtained for the $e_{aq}^+ + X^-$ reactions in pure water (i.e., without PPS), we expected a decrease in the rate constant caused by an increase in the viscosity due to the added PPS, and furthermore, a decrease in the diffusion controlled rate for a heavy halide compared to that of the lighter e_{aq}^+. Additionally, the reaction is between a neutral and a charged reactant, not between two charged reactants as in the $e_{aq}^+ + I^-$ reaction. The agreement between our model of simple competition and the experimental facts supports our interpretation of the e_{aq}^+ reactions, in general.

The LT and AC measurements show that the Ps formation is inhibited partly if Cl^-, Br^-, and I^- (and some pseudohalides) (but not F^-) are added to pure water, as shown in the iodide case in Fig's. 6.4,5. A similar Ps inhibition in parts only is found if ions which trap the electron in shallow traps (e.g., Cd^{++}) are added to pure water. According to the calculations in [6.1,2] the positron is only weakly bound on the three halides in vacuum. In water the effective binding energy is modified by the hydration of the halides, as mentioned above. The Ps inhibition in parts is interpreted as caused by a further reaction of the trapped positron (electron) and a spur electron (the positron) to give Ps. We shall discuss these secondary processes in more detail, using Cd^{++} which traps electrons, as a standard example, in Sect. 7.5. Here we can conclude that the Ps formation data for the solutions of the three halides, and for some pseudohalides too, support our general interpretation.

As expected, positron reactions with the three halides and Ps inhibition in parts occur also in other polar liquids. In particular, LT and DB studies in Strasbourg have shown that [6.26,27].

In nonpolar liquids the solubility of ions is very low, and hence, positron reactions with halides cannot be studied, except under very special conditions [6.28]. However, clear cases of positron reactions with various neutral molecules, such as crown ethers, perfluorohexane, CCl_4, CBr_4, and some other halogenated hydrocarbons, have been found. We shall postpone a discussion of these cases to Chap. 7.

6.4 Discussion of e^+_{aq} Reactions. High Concentrations

Above an I^-concentration of 0.03 M the fraction of positrons annihilating from a bound state, $[I^-,e^+]$, is much smaller than the theoretical value, as shown in Fig. 6.3. At 1 M I^- the effective rate constant is decreased by a factor of ten compared with the value at intermediate concentrations. This effect decreases in the series I^-, Br^-, and Cl^-. It is difficult to interpret these results, which are discussed in great detail in [6.3,4]. We shall here give a short repetition of parts of this discussion with special reference to the reaction kinetics aspects.

It is important to note that the above given detailed discussion of the theories of the reaction rate constants, except for the encounter pair effect, is applicable only for dilute concentrations. Actually, it is difficult to find clearly stated limits of applicability in the literature. Clearly, Table 6.2 shows that the initial encounter pairs become important above 0.05 M, and they dominate the reactions at 1 M, i.e., no diffusion is necessary for $R = 1.4$ nm at concentrations of 1 M where the average distance between halide ions is only 1.2 nm. Additionally, the positrons are excluded from a volume around the positive ions, Na^+.

Strictly speaking, the positron reactions with halides are partly nonhomogeneous reactions. The positrons are expected to be mainly randomly distributed with respect to the halide, when they react. However, some positrons are thermalized in the positron spur and are not randomly distributed with respect to the other spur reactants. In particular, the spur electrons and the positive ions are expected to influence the behaviour of the positrons because of their charge. In PPS solutions most electrons are trapped on PPS, from where the positron cannot pick them off. Nevertheless, PPS^- will influence the positrons somewhat.

Clearly, it is not the same positrons which react with the halides in pure water and in the PPS solutions. In the latter the main number of positrons which would form Ps in pure water, will react with the halides, and as they are expected to be close to the negative charge on PPS^-, their reaction behaviour might be somewhat different from that of the other positrons. On the other hand, the decrease in the effective rate constant is roughly similar in pure water and in the PPS solutions, and hence, it is unlikely that it is caused by the different state of the most important negative reactant.

In [6.3,4] viscosity, salt, and spur effects are discussed together with positron tunnelling and delocalization on X^-, early processes, and ion association. Intuitively, we feel that the ion association between the positive Na^+ ions and the halide might be the best way of explaining the high concentration effect in the rates of positron reactions. The stronger decrease in the rate for I^- could then be caused by a stronger tendency for iodide to form complexes with Na^+ ions, because of the higher polarizability of iodide compared with that of the other halides, Cl^- and Br^-. It must be emphasized that this is a preliminary interpretation. Further experiments are needed!

Perhaps the way the bound state curves was extracted is somewhat in error. The bound state curves might contain some fraction shaped as the e^+_{aq} curve, which our method cannot detect. In [6.1,2] the bound state wave functions for the ions in vacuum were calculated. It was found that a large part of the positron wave function was outside the halides. This has lead to a proposal of a "bubble state" of the positron halide bound state in water [6.4].

As mentioned above, only one detailed study of the solvated positron reactions exists, namely the one discussed in this chapter. The solvated electron has been studied in much greater detail. Because the hydrated electron has a much longer lifetime than the hydrated positron, and also because of experimental difficulties at shorter times, the hydrated electron reactions have normally been studied at very low electron scavenger concentrations, typically at concentrations of 1 μM–1 mM [4.2]. This makes a study of the ionic strength effect much easier in the electron case.

In summary, the discussion of the very few results of studies of solvated positron reactions in this chapter showed that the solvated positron is a genuine chemical reactant along with the solvated electron. "Classical effects", such as the ionic strength effect and positron transfer reactions from one heavy reactant to another, have just been demonstrated. As the solvated positron reactions have been studied very little the field is wide open for further work.

7 Positronium Formation in Solutions

In particular, studies of Ps formation in solutions have given valuable information on the processes of Ps formation. In a typical experiment, a solute which has a positive electron affinity is added to a solvent, and the intensity of the longlived o Ps lifetime is measured as a function of the solute concentration. In many cases the Ps intensity decreases with the concentration and becomes low or zero at a concentration of roughly 1 M. This is called Ps inhibition. The addition of other electron acceptors enhances the Ps formation. Enhancement of the Ps yield can also be obtained on the addition of certain positive ion scavengers to a solvent. In this chapter we shall discuss the processes influencing the Ps formation in solutions in detail.

At first, we discuss qualitatively some typical Ps yields for solutions of nonpolar liquids. Furthermore, a short discussion of the theory, and simple expressions for the Ps yield versus solute concentration, are given. This is followed by a more detailed description and a theoretical discussion of the Ps formation in nonpolar mixtures, in general. In polar liquids the Ps yield as a function of the concentration of ions has been studied very much. In particular, aqueous solutions have been investigated in great detail. We shall discuss how the electron acceptor properties of the ions determine their Ps inhibition ability. It is necessary to correlate the Ps inhibition strength of an ion to a "special" redox potential of that ion. Furthermore, Ps formation in mixtures of polar and nonpolar liquids, and the effects of positive ion scavenging, will be discussed. At last the older models of Ps formation are shortly described.

7.1 Special Cases of Ps Formation in Nonpolar Mixtures

Figure 3.2 shows the measured lifetime spectra for mixtures of CCl_4 in benzene [3.7]. This is a clear case of strong Ps inhibition. Each mark in the figure is the average of two points, and a constant background has been subtracted from each spectrum. The average lifetime of o Ps, τ_3, in the mixtures is the reciprocal of the slope of the long-lived component, while its relative intensity, I_3, is the area under the long-lived component divided by the total area of the spectrum. We see that the o Ps lifetime does not change on addition of CCl_4, i. e., Ps does not react with CCl_4.

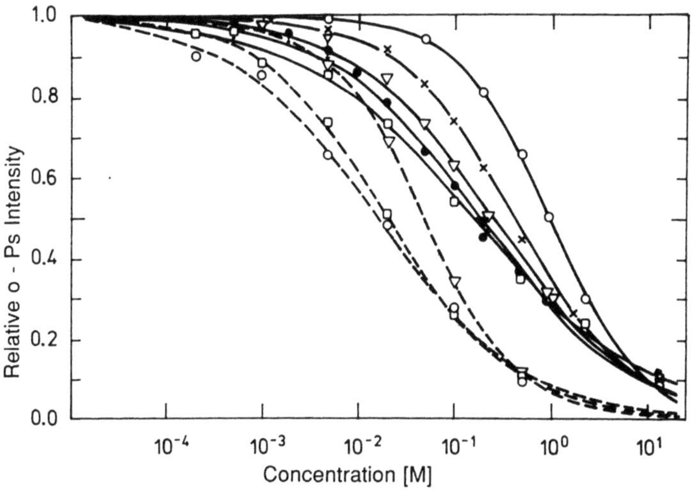

Fig. 7.1. Inhibition of Ps formation. Relative o Ps intensities versus inhibitor concentrations in various solvents: (▽) tetradecane, (●) hexane, (□) isooctane, (X) neopentane, (O) tetramethylsilane (TMS), and a (▼) TMS/tetradecane mixture. Solid curves for C_2H_5Br and dashed curves for CCl_4 are given by (7.30) with inhibition parameters of Table 7.1, except for TMS/hexane where the solid curve is omitted [7.1]

In Fig. 7.1 the relative o Ps intensity, i.e., roughly the relative Ps yield, versus the concentration of CCl_4 and C_2H_5Br in various nonpolar liquids is shown [7.1]. Clearly, CCl_4 is a stronger Ps inhibitor than C_2H_5Br . The Ps yield in pure CCl_4 is zero, while 5.2% of the positrons form o Ps in pure C_2H_5Br.

Figure 7.2 shows the o Ps yield versus the concentration of C_6F_6 [7.2] and C_6F_{14} [7.3] in hexane. Although these solutes are also electron acceptors, they influence the Ps formation very differently from CCl_4. Furthermore, the addition of C_6F_6 (and CS_2) to a 0.05 M CCl_4/hexane solution results in a strong increase of the Ps yield [7.2], as shown in Fig. 7.3.

The following nonhomogeneous reactions in the positron spur explain the measured Ps yields. Only the most important reactions are shown. The reactants and products are shown within square brackets if they are situated mainly inside the positron spur.

$$e^+ + M \rightarrow [M^+ + e^- + e^+] \quad \text{spur formation} \quad (7.1)$$

$$[e^- + M^+] \rightarrow [M] \quad \text{recombination} \quad (7.2)$$

$$[e^+ + e^-] \rightarrow [Ps] \quad \text{Ps formation} \quad (7.3)$$

$$[e^+] \rightarrow e^+ \quad e^+ \text{ outdiffusion} \quad (7.4)$$

$$[e^-] \rightarrow e^- \quad e^- \text{ outdiffusion} \quad (7.5)$$

$$[e^- + CCl_4] \rightarrow [(CCl_4{}^-)^*] \quad e^- \text{ scavenging} \quad (7.6)$$

$$[(CCl_4{}^-)^*] \rightarrow [CCl_3 \cdot + Cl^-] \quad \text{dechloronation} \quad (7.7)$$

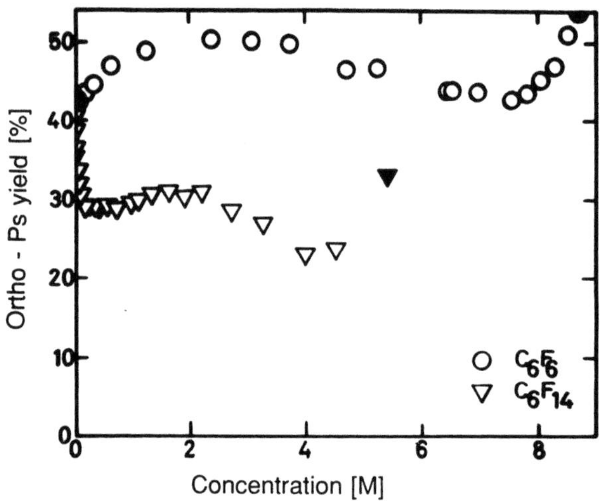

Fig. 7.2. The o Ps yield versus concentrations of C_6F_6 [7.2] and C_6F_{14} [7.3] in hexane. Filled marks are for pure liquids

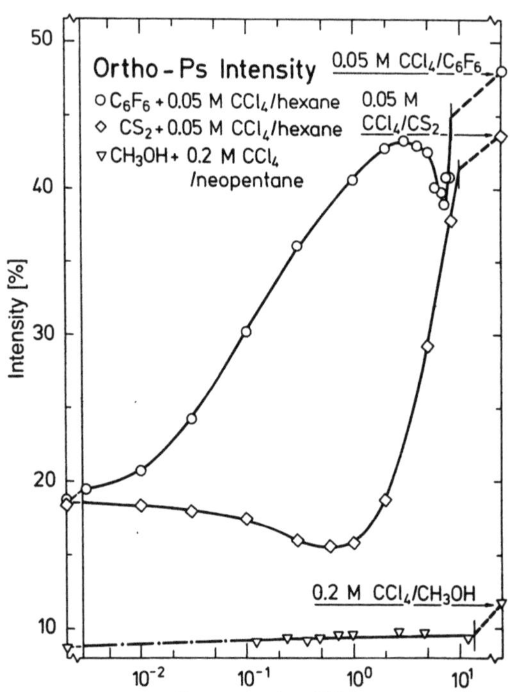

Fig. 7.3. The o Ps yield versus concentrations of C_6F_6 and CS_2 in a 0.05 M CCl_4/hexane solution, and of CH_3OH in a 0.2 M CCL_4/neopentane solution. The curves are best fits to experimental points [7.2]

$$[e^+ + (CCl_4^-)^*] \rightarrow [Ps + CCl_4] \qquad e^- \text{ pick off} \qquad (7.8)$$

$$[e^+ + Cl^-] \rightarrow [[Cl^-, e^+]] \qquad \text{bound state formation} \qquad (7.9)$$

$$[e^- + [Cl^-, e^+]] \rightarrow [Ps + Cl^-] \qquad e^+ \text{ pick off} \qquad (7.10)$$

$$[e^+ + CCl_4] \rightarrow [[CCl_4, e^+]] \qquad e^+ \text{ scavenging} \qquad (7.11)$$

$$[e^- + [CCl_4, e^+]] \rightarrow [Ps + CCl_4] \qquad e^+ \text{ pick off} \qquad (7.12)$$

If C_6F_6 is added to the CCl_4/hexane solution the Ps formation is mainly influenced by the following additional nonhomogeneous reactions.

$$[e^- + C_6 F_6] \rightarrow [C_6 F_6{}^-] \qquad e^- \text{ scavenging} \qquad (7.13)$$

$$[e^+ + C_6 F_6{}^-] \rightarrow [Ps + C_6 F_6] \qquad e^- \text{ pick off} \qquad (7.14)$$

$$[M^+ + C_6 F_6{}^-] \rightarrow [M + C_6 F_6] \qquad \text{recombination} \qquad (7.15)$$

$$[CCl_4 + C_6 F_6{}^-] \rightarrow [CCl_4{}^- + C_6 F_6] \qquad \text{electron transfer} \qquad (7.16)$$

Let us discuss these reactions in some details.

Reactions (7.1)–(7.5) illustrates Ps formation in the positron spur in pure hexane, which was discussed in Chap. 5. We recall that only about 4% of the electrons in the spurs in hexane, formed by high energy electron irradiation, escape recombination with the positive ions in hexane.

The inhibition of Ps formation by CCl_4 is described by the reactions (7.6)–(7.12). Electrons are trapped on CCl_4, and the dechloronation time is so short that the amount of Ps formation by reaction (7.8) is expected to be insignificant. We know that the positron reacts with Cl^-, Br^-, and I^- in water, and hence, it reacts with Cl^- in hexane, too, as e^+ and Cl^- are much less solvated in hexane than in water. Consequently, the electron trapped on CCl_4 is normally lost for Ps formation. Angular correlation results indicate that the positron is trapped on CCl_4 in hexane. However, we do not know the significance of this reaction and that of the positron pick off reaction (7.12). Actually, the evidence for a strong Ps inhibition by CCl_4 in hexane indicates that reactions (7.6), (7.7), and (7.9) are very probably the most important reactions.

Addition of C_6F_6 to pure hexane at concentrations below 3 M results in an enhancement in the Ps formation by the so called "antirecombination" effect [7.2]. The spur electrons are trapped on C_6F_6 to form stable anions from which the positron can pick off the electrons to form Ps, as illustrated by the reactions (7.13) and (7.14). The positrons have high mobilities, while the positive ions are slowly moving in hexane. Thus, reaction (7.15) is slow compared to reaction (7.14). Some of the electrons which in pure hexane recombine with the positive ions will therefore in C_6F_6/hexane form Ps, i.e., Ps formation is enhanced, as seen in Fig. 7.2.

Similarly, reaction (7.13) competes with reaction (7.6) if C_6F_6 is added to a CCl_4/hexane solution, and Ps formation is strongly enhanced, as shown in Fig. 7.3. This is called "antiinhibition". The minima in the Ps yields at high C_6F_6 concentration in hexane, shown in Figs. 7.2 and 7.3, will be discussed below.

On the other hand, the addition of C_6F_{14} to pure hexane results in a decrease in the Ps yield, although C_6F_{14} has an electron affinity of roughly the same value as C_6F_6. The reason is that C_6F_{14} traps both the positron and the electron. Hence, both particles become slowly moving, and the Ps formation process takes a time comparable to the positron lifetime in the bound state on C_6F_{14} ($\sim$ 400 ps). Positronium formation thus decreases for increasing C_6F_{14} concentration up to 0.4 M, as shown in Fig. 7.2. The higher concentrations will be discussed below.

7.2 Theory of Ps Inhibition and Antiinhibition

In Chap. 5 we discussed the spur reactions and Ps formation in the positron spur for pure liquids. Section 6.2 contained a short discussion of reaction kinetics in positron research. We shall now attempt to use these discussions to understand how electron and positron scavenging in the positron spur influence the Ps formation. The conclusion will be that the possibilities for a detailed theoretical treatment are very limited, and we have to accept that only simple, semiempirical models are useful at present.

Obviously, quantum mechanical calculations of the spur reactions are much too complicated for a successful description of Ps formation. Quantum mechanics might maybe be used to derive properties of the states of the light particles in a liquid in the future, but the transitions of the particles between states, or the mobilities of the particles, cannot be treated quantum mechanically at present. The results are much too uncertain (see [4.16,6.22]).

The theory of particle diffusion in the spur constitutes a simpler description. It treats the spur processes by assuming that the spur reactants diffuse "classically" in the spur under the influence of the electric fields of the charged particles, and by assuming a certain probability of reaction if the particle distances are within a reaction radius characterizing the reaction. Apparently, this is the only theory which has given useful results in the description of electron scavenging in spurs, mainly one electron-ion pair spurs, in radiation and photo chemistry.

The theories of electron scavenging in the one electron-ion pair spur nearly always use (6.7). The scavenging is assumed to be described by a term, $-kcn$, added to the right hand side of (6.7). Here k is the rate constant, c the concentration of the scavenger, and n the probability distribution of the electron. n is a function of space and time. In some calculations the rate constant is assumed to be the diffusion controlled constant (6.13). However, in most cases k is taken to be independent of time to make the calculations simple. Apparently, the initial encounter pair effect (6.27) is always neglected, except in one calculation [6.18]. As (6.3), in which the rate constant is defined in the diffusion theory for homogeneous initial distribution of reactants, is obtained by solving the diffusion equation (6.7), it is, of course, incorrect to add the term, $-kcn$, to the same equation (6.7). This is an assumption which seems not to be controlled, i.e., the errors introduced by use of this assumption have apparently not been estimated. The same assumption is always used in the theories of positron trapping on defects during its diffusion to surfaces, etc. in slow positron research. The rate constant, k, is normally assumed to be that measured for the electrons escaped from the spurs. D is obtained from the measured electron mobility by use of the Nernst-Einstein equation (6.11). The initial distribution of electron-ion distances is normally taken to be a simple function, e.g., a gaussian or exponential, characterized by one or several constants, which are used as fitting constants in the comparison of theory and experiment. The experimental numbers to be compared to theory can be the yield of the product of the electron reaction with the scavenger. Normally, the theoretical results can describe the experimental results fairly well for reasonable values of the fitting parameters.

In chapter 5 we discussed some of the important approximations used in the single pair spur diffusion calculations. We concluded that the following ones might constitute severe limitations on the validity of the theory. A) Single pair spurs are assumed, although maybe 40% of the spurs are multipair spurs even for low LET electron spurs. B) A constant dielectricity constant is used, although it is strongly dependent on the electric field in the main part of the spur. C) A constant electron mobility is used. In particular, the mobility is expected (and measured) to be strongly electric field dependent. D) Similarly, the rate constant of the electron reactions is expected (and measured) to be strongly dependent on the electric field. E) Quantum mechanical effects are normally neglected.

A calculation of the particle reactions in the simplest positron spur consisting of an electron, a positron, and a positive ion is much more difficult than that of a single electron ion pair spur. This was concluded for the cases of pure liquids in Chap. 5. Apart from the difficulties discussed above the positron spur must be described by the distributions of the mutual distances of the three spur reactants, the positron spur is expected to be part of a "short track", i.e., a multipair spur, and two quantum mechanical reactants are present in the positron spur. Hence, we can conclude that a spur diffusion calculation of the particle scavenging in the positron spur is expected to very difficult, indeed. Apparently, such a calculation has not been attempted, yet. On the other hand, it is useful to estimate the difficulties involved, as done above. Spur diffusion calculations of the positron spur have been proposed, and we might see some results in the future.

A very simple model of the particle scavenging in the positron spur has been used in some cases (see e.g. [7.4]). It was assumed that the reactants are placed in a box or sphere, and "rate constants" of their mutual reactions and of the outdiffusion are assumed or estimated. Although such a model can be solved without too much trouble, its usefulness is very questionable. For example, for nonpolar liquids the charged particles are thermalized inside the practical reaction radii, i.e., the Onsager radii, in most cases. However, the concept of a rate constant is only justified in the diffusion theory used to obtain the expressions given in Chap. 6 if the reactants are randomly distributed initially. Hence, the use of the "rate constants" in that model is theoretically without basis. The spur reactions are nonhomogeneous reactions!

7.3 Empirical Expression for the Ps Yield

The main conclusion of the preceding section was that the application of a detailed spur diffusion theory to the positron spur problem is not promising at present. However, an empirical expression for the Ps yield can be obtained by correlation to a similar expression used in the description of the scavenging of reactants in the electron spurs in radiation and photo chemistry. The Ps yield expression will only be valid if certain conditions are fulfilled!

In a typical steady state scavenging experiment a solution of, for example, a nonpolar liquid and an electron or a positive ion scavenger is irradiated during times which are long compared to the lifetimes of intermediate reactants created by the irradiation, and the yield of stable products is determined after the irradiation. For instance, *Warman*, *Asmus*, and *Schuler* (WAS) [7.5] irradiated cyclohexane solutions of CH_3Cl, CH_3Br, and C_2H_5Br at concentrations from 10^{-4} M to 0.5 M with ^{60}Co gamma rays for 4.8 min at a rate of 5×10^{-18} eVcm^{-3}h^{-1}, and determined the yield of alkyl radicals, $R\cdot$, after the irradiation. The radicals were products of the dissociative electron capture reaction

$$e^- + RX \rightarrow R\cdot + X^- \tag{7.17}$$

analogous to (7.6) and (7.7). The yields of the products, $G(p)$, in units of number of reactants per 100 eV deposited energy, were well fitted by the empirical expression

$$G(p) = G_{\text{fi}} + G_{\text{gi}} F(c), \tag{7.18}$$

for

$$F(c) = (\alpha_s c)^{1/2} / (1 + (\alpha_s c)^{1/2}) \tag{7.19}$$

Here G_{fi} denotes the yield of "free ion pairs," while G_{gi} represents the yield of "geminate ion pairs". $F(c)$ is the scavenging function, c the scavenger concentration, and α_s is a fitting parameter. G_{fi} and $G_{\text{gi}} = 0.12$ and 3.8 (100 eV)$^{-1}$, and $\alpha_s = 16.2$, 7.82, and 5.41 M^{-1} for CH_3Br, C_2H_5Br, and CH_3Cl, respectively, gave good fits of the experimental data.

If two scavengers were simultaneously present at concentrations c_1 and c_2, the total yield, $G(p_1 + p_2)$, could be well represented by the empirical expression [7.5]

$$G(p_1 + p_2) = G_{\text{fi}} + G_{\text{gi}} F(c_1, c_2), \tag{7.20}$$

for

$$F(c_1, c_2) = (\alpha_1 c_1 + \alpha_2 c_2)^{1/2} / (1 + (\alpha_1 c_1 + \alpha_2 c_2)^{1/2}), \tag{7.21}$$

which is a simple extension of (7.18) and (7.19). α_1 and α_2 are the fitting constants for the individual scavengers. The yield of the product, p_1, in the presence of the second scavenger could be well described by

$$G(p_1) = G(p_1 + p_2) \alpha_1 c_1 / (\alpha_1 c_1 + \alpha_2 c_2) \tag{7.22}$$

if no secondary reaction took place. An analogous expression described the yield $G(p_2)$. For example, (7.20) to (7.22) described well the total yields and the individual yields of methyl and ethyl radicals if the electrons in cyclohexane were scavenged by CH_3Br and C_2H_5Br simultaneously present in the solution.

However, if the two solutes CH_3Cl and C_2H_5Br were present in cyclohexane only the total yield was well represented by (7.20) and (7.21). The yield of methyl radicals was too low and that of the ethyl radical too high compared to that given by (7.22). The results were explained by the assumption that the reactions

$$CH_3Cl^- + C_2H_5Br \rightarrow CH_3Cl + C_2H_5Br^- \qquad (7.23)$$

$$C_2H_5Br^- \rightarrow C_2H_5\cdot + Br^- \qquad (7.24)$$

took place in the solution. For this to be true the lifetime of CH_3Cl^- must be fairly long while the dehalogenation of CH_3Br^- and $C_2H_5Cl^-$ must be a rapid process. Electron transfer from CH_3Cl^- to SF_6 and N_2O was also detected in the same way [7.6]. By use of a more elaborate theory the lifetime of CH_3Cl^- in pure cyclohexane could be estimated to be roughly 30 ns [7.6].Recent discussions of the steady state scavenging by CH_3Cl and CH_3Br in nonpolar liquids, and in general, are found in [4.1], [7.7], [7.8], and refs. therein. Apparently, the lifetimes of the anions CH_3Cl^- and CH_3Br^- have not been studied in fast pulse radiolysis or laser flash photolysis experiments.

The Ps yield may be correlated to the scavenging of electrons in the electron spurs by use of the reasonable hypothesis that the relative positron-electron "recombination" in the positron spur, i.e., the Ps yield $P(c)$, is equal to the relative electron positive-ion recombination in the electron spurs [7.9,10,1]. Hence, we may assume that

$$P(c) = I_3(c)/I_3(0) = (G_{fi} + G_{gi} - G(p))/G_{gi}. \qquad (7.25)$$

Consequently, we get

$$I_3(c) = I_3(0)/(1 + (\alpha c)^{1/2}) \qquad (7.26)$$

by use of (7.19), if one scavenger is present. In addition,

$$I_3(c) = I_3(0)/(1 + (\alpha_1 c_1 + \alpha_2 c_2)^{1/2}) \qquad (7.27)$$

from (7.21), with two scavengers in the liquid [7.11].

If a simple characteristic lifetime could be ascribed to the geminate recombination process we could expect that [7.12,13]

$$F(c) = \alpha c/(1 + \alpha c), \qquad (7.28)$$

and hence

$$I_3(c) = I_3(0)/(1 + \alpha c) \qquad (7.29)$$

The expressions (7.28) and (7.29) correspond to the well known Stern-Volmer expression for the inhibition of fluorescence of excited molecules in photochemistry (see, e.g., [6.16,7.14–16]). Equation (7.29) was the first expression used in the description of inhibition of Ps formation [1.4].

The analyses of measured Ps yield results show that fairly good fits could be obtained by use of (7.26) for most strong Ps inhibitors in many nonpolar liquids, and (7.29) for Ps inhibition by many solutes in polar liquids and by some weak inhibitors in typical nonpolar liquids. To improve the goodness of the fits, and to join the two expressions (7.26) and (7.29) into one, a slightly modified empirical expression using a second adjustable parameter, β, in the exponent

$$I_3(c) = I_3(0)/(1 + (\alpha c)^{\beta}) \tag{7.30}$$

was proposed [7.1]. Expression 7.30 corresponds to the scavenging function

$$F(c) = (\alpha c)^{\beta}/(1 + (\alpha c)^{\beta}), \tag{7.31}$$

which seems not to have been used in radiation chemistry. Several groups have used (7.30) with fair success. In some cases a constant has been added to the right hand side of (7.30) to improve the fit [6.3]. In Sect. 7.5 we shall discuss the expressions used by the Strasbourg group for polar solvents [7.17,18]. They used, e.g., a sum of two terms like (7.29) with $\beta = 1$, and with the relative intensity as a third fitting constant.

It must be emphasized that the empirical expressions are only expected to give good fits if mainly a primary scavenging of one of the particles influences the Ps yield. In most cases several scavenging reactions are of importance, and secondary, and even tertiary, reactions influence the Ps formation, too. Specific cases will be discussed in details in the next sections.

7.4 Positronium Formation in Mixtures of Nonpolar Liquids

The present section contains a detailed discussion of the influence on the Ps formation of the addition of mainly nonpolar solutes to nonpolar solvents, while we in the next section discuss the Ps inhibition in polar liquids. Section 7.6 will mainly deal with the measured Ps yields for mixtures of nonpolar and polar liquids. Opposite to the positron reaction field where only tens of papers have been published, over hundred papers discuss the Ps yields in mixtures of nonpolar liquids. Consequently, only a selection of Ps yield results can be discussed here. I have attempted to select those Ps yield results which give the best review of the whole field. A complete discussion of all Ps formation papers in the field could not be included, because of lack of space.

The discussion in this and the next two sections will mainly be in terms of results and concepts of the following disciplines:

a) The theories discussed the preceding sections.
b) Reaction kinetics, as briefly reviewed in Sect. 6.2.
c) Radiation and photo chemistry, in particular pulse radiolysis, flash photolysis, and steady state scavenging.
d) Studies of light particle states and motion in liquids, as discussed in Chap. 4.
e) Studies of electron affinities and attachment reactions in gases.

f) Electron spin resonance (ESR) and light absorption studies of irradiation created species in frozen matrices, mainly glasses.
g) General chemistry studies, such as investigations of electron states on molecules, bond strengths, solvation, and solubility.

In many cases knowledge is not available on the processes and properties which are used in the interpretation of the measured Ps yields, e.g., on the positron mobility and affinities and on electron properties at high concentration of scavengers. We can then normally use the Ps yields to predict new effects.

This is a very large field of science, and no scientist can understand in detail all that knowledge. Hence, nobody can critically review the whole field. I have attempted to the best of my ability to give a reasonably correct discussion here.

In general, the solutes reacting with Ps in nonpolar liquids inhibit also the Ps formation. The only exceptions seem to be certain small molecules where an electron trapped in an initially formed state cannot easily relax into the ground state of the formed radical anion. Two such cases, namely those of SO_2 and O_2 in nonpolar liquids, will be discussed below. Apparently, solutes reacting with Ps have normally so high electron affinities in the ground state of the formed anion that the positron cannot pick off the electron to form Ps, and thus, total Ps inhibition is measured, and the Ps yield in the pure scavenger liquid is zero.

At first we shall discuss the mixtures of nonpolar liquids and scavengers which do not react with Ps. Solutions of halogenated hydrocarbons are typical cases of such mixtures. In Table 7.1 are shown the Ps inhibition parameters according to (7.30) for a selection of solute/nonpolar-solvent systems. The parameters for other nonpolar systems can be found in the references given in the table.

The halogenated hydrocarbons can be divided into three categories with respect to the way the Ps formation is influenced. The iodonated and bromonated solutes inhibit the Ps yield to low or zero value, mainly because they react with the electron to form anions which dehalogenate rapidly, as for the CCl_4 case, (7.6,7). The fluoronated solutes form stable anion radicals, and the Ps yield depends on whether the solutes react also with positrons. Except for some cases such as CCl_4 and $CHCl_3$, the chloronated solutes constitute the "category of transition" between the two other categories, where the Ps yield depends on dechloronation time. Some "mixed compounds" containing, e.g., both fluoride and bromide groups, seem to belong to the transition category, too. The stability and the dehalogenation time of the anion radicals formed in nonpolar solvents depend mainly on the carbon-halogen bond energies, which are approximately 448, 326, 285, and 213 kJ/mol for the C–F, C–Cl, C–Br, and C–I bonds, respectively (see, e.g., [6.16,7.27]). Because of its higher bond energy compared to those of the other carbon-halogen bonds the C–F bond seems to be stable in the anions in nonpolar liquids. In polar solvents the C–Cl bond might in some cases break before, e.g., the C–Br bond because of the higher solvation energy of chloride compared to that of bromide in some protic media [7.28]. We shall now discuss the three categories in some detail.

Table 7.1. Positronium inhibition parameters (Eq. (7.30))

Solute/solvent	α [M^{-1}]	β	Ref.
CCl_4/tetradecane	20.5	0.88	[7.1]
CCl_4/hexane	24.2	1.0	[5.20]
CCl_4/isooctane	46.0	0.67	[7.1]
CCl_4/tetramethylsilane	59.0	0.58	[7.1]
CCl_4/cyclohexane	30.0	0.82	[7.19]
CCl_4/benzene	17.4	0.93	[3.7]
C_2H_5Br/hexane	5.3	0.57	[7.1]
C_2H_5Br/tetradecane	3.9	0.61	[7.1]
C_2H_5Br/isooctane	6.0	0.48	[7.1]
C_2H_5Br/neopentane	2.4	0.73	[7.1]
C_2H_5Br/tetramethylsilane	1.04	0.94	[7.1]
C_6H_5Cl/hexane	8.2	0.62	[7.20]
C_6H_5Cl/neohexane	9.1	0.64	[7.20]
C_6H_5Cl/isooctane	10.3	0.30	[7.20]
C_6H_5Cl/cyclohexane	4.0	0.41	[7.20]
C_6H_5Cl/benzene	4.2	0.50	[7.19]
$C_6H_5CH_2Cl$/cyclohexane	15.7	0.65	[7.20]
o-$C_6H_4Cl_2$/benzene	14.3	0.57	[7.19]
m-$C_6H_4Cl_2$/benzene	12.7	0.39	[7.19]
p-$C_6H_4Cl_2$/benzene	4.2	0.39	[7.19]
1,3,5-$C_6H_3Cl_3$/benzene	12.7	0.61	[7.19]
1,2,4,5-$C_6H_2Cl_4$/benzene	12.3	0.63	[7.19]
C_6HCl_5/benzene	3.8	0.48	[7.19]
C_6Cl_6/benzene	3.2	0.58	[7.19]
1,2,4,5-$C_6H_2Cl_4$/benzene	11.0	0.64	[3.7]
1,2,4,5-$C_6H_2Cl_4$/$C_6H_5CH_3$	7.0	0.68	[3.7]
1,2,4,5-$C_6H_2Cl_4$/o-$C_6H_4(CH_3)_2$	4.8	0.58	[3.7]
1,2,4,5-$C_6H_2Cl_4$/1,3,5-$C_6H_3(CH_3)_3$	1.4	0.43	[3.7]
1,2,4,5-$C_6H_2Cl_4$/1,2,3,4-$C_6H_2(CH_3)_4$	2.1	0.50	[3.7]
1,2,4,5-$C_6H_2Cl_4$/$C_6H_5NH_2$	3.2	0.93	[3.7]
C_6H_5Br/hexane	13.1	0.69	[7.21]
C_6H_5Br/neohexane	38.3	0.50	[7.21]
C_6H_5Br/isooctane	34.8	0.49	[7.21]
C_6H_5Br/cyclohexane	13.4	0.60	[7.21]
C_6H_5Br/benzene	13.9	0.72	[7.21]
o-$C_6H_4Br_2$/benzene	23.7	0.66	[7.21]
1,3,5-$C_6H_3Br_3$/benzene	19.7	0.73	[7.21]
1,2,4,5-$C_6H_2Br_4$/benzene	16.4	0.64	[7.21]
C_6Br_6/benzene	20.1	0.72	[7.21]
CH_3Cl/cyclohexane[a)]	1.13	0.48	[7.22]
CH_3Cl/benzene[a)]	0.49	0.75	[7.22]
CH_3Br/cyclohexane[a)]	12.8	0.56	[7.22]
C_2HCl_3/cyclohexane	8.2	0.45	[7.23]
C_2HCl_5/isooctane	100	0.43	[7.23]
C_2Cl_6/isooctane	122	0.53	[7.23]
SF_6/Hexane[b)]	20	0.9	[7.2]
SF_6/isooctane[b)]	100	0.7	[7.11]
SF_6/neopentane[b)]	216	0.53	[7.11]
CS_2/hexane[c)]	1.3	1.0	[7.24]
C_6F_{14}/hexane[c)]	1.6	0.67	[7.25]

Table 7.1. (*Continued*)

Solute/solvent	α [M^{-1}]	β	Ref.
CCl_4/hexane	23.6	0.77	[7.26] (45 C)
CCl_4/hexane	23.6	0.87	[7.26] (1 C)
CCl_4/hexane	23.2	0.96	[7.26] (−30 C)
1,2,3,5-$C_6H_2Cl_4$/$C_6H_5CH_3$	0.7	0.39	[7.26] (67 C)
1,2,3,5-$C_6H_2Cl_4$/$C_6H_5CH_3$	1.6	0.42	[7.26] (45 C)
1,2,3,5-$C_6H_2Cl_4$/$C_6H_5CH_3$	6.8	0.58	[7.26] (1 C)
1,2,3,5-$C_6H_2Cl_4$/$C_6H_5CH_3$	10.0	0.71	[7.26] (−30 C)
$C_6H_5NO_2$/benzene	18.0	0.85	[5.21]
$C_6H_5NO_2$/heptane	6.5	1.0	[5.21]

[a] Above $I_3 = 20\%$. [b] Above $I_3 = 30\%$. [c] Low concentrations

The Ps inhibition parameters of CCl_4 and C_2H_5Br in various nonpolar liquids are shown in Table 7.1. The solid curves through the measured points in Fig. 7.1 are the fits by use of (7.30) with the parameters in Table 7.1. Clearly, the measured points are well fitted by the curves. Similar qualities of the fits are normally obtained for other systems. In Fig. 7.4 the α-value and the rate constant of electron reaction, k, [4.13] for CCl_4 and C_2H_5Br versus V_0 [7.29–31] for several nonpolar solvents are plotted [7.1]. V_0 is the energy of the electron in its conduction band compared to vacuum (Table 4.1). These systems were studied in order to show the strong correlation between α and k. While both α and k decrease for increasing V_0 for CCl_4, they show a maximum for C_2H_5Br at the same V_0. In a rough way, but not in details, there is also a proportionality between α and k. For systems studied until now where strong Ps inhibition is found a similar degree of proportionality between the two constants has been found.

The rates of electron attachment to electron acceptors in nonpolar liquids have been interpreted by *Henglein* [7.31], who discusses also the CCl_4 and C_2H_5Br cases. The main idea of his interpretation is that a maximum in the rate of the kinetic reaction, i.e., apart from the diffusion controlled rate, is expected when there is a maximum of overlap of the energy distributions of the initial and final states. The initial state is the excess electron separated from the scavenger molecule in the solvent, while the final state is the state of the unrelaxed anion, i.e., the electron is added to the scavenger molecule but the molecule and the surrounding liquid have not yet had time to adjust their atomic coordinates to the presence of the electron. The electron tunnels into the acceptor from its ground state (about 0.2 eV below V_0 in hexane) in the liquid, or it reacts from the thermally excited "free state" of energy V_0 (−0.02 eV for hexane compared to vacuum). Hence, the rate constants, k, are strongly dependent on the properties of the excess electron in the liquids. The measured, strong correlation between α and k constitutes, therefore, an important argument in favour of the spur model of Ps formation.

It is important to note that the rate constants were determined for solute concentrations of normally 10^{-6} M, and in some cases up to about 0.2×10^{-3} M

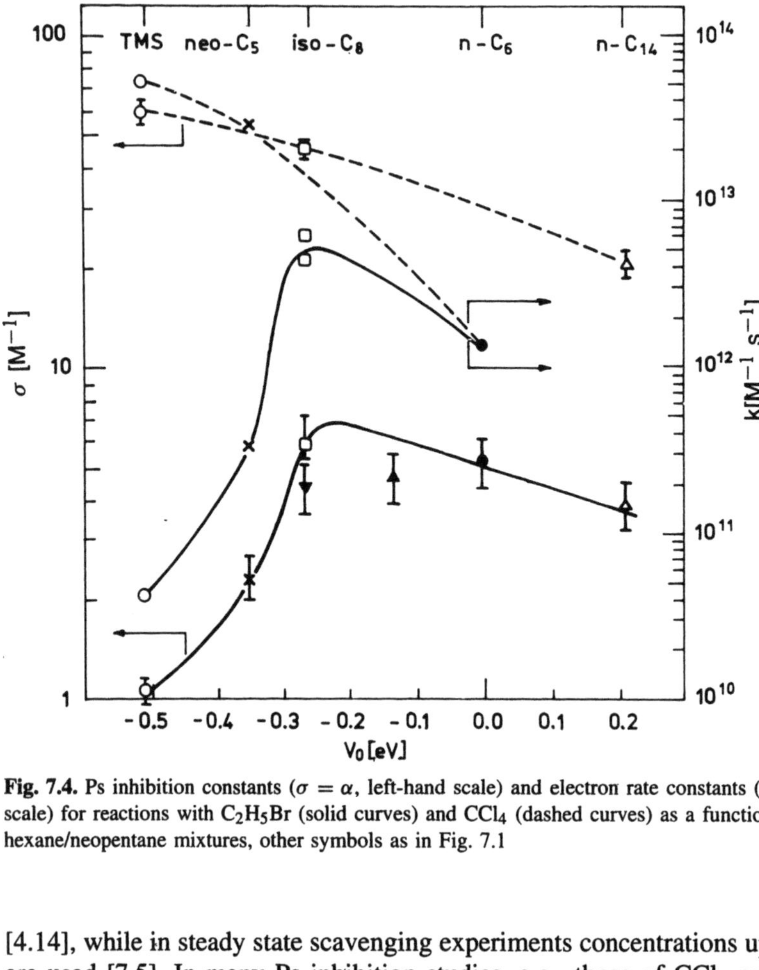

Fig. 7.4. Ps inhibition constants ($\sigma = \alpha$, left-hand scale) and electron rate constants (k, right-hand scale) for reactions with C_2H_5Br (solid curves) and CCl_4 (dashed curves) as a function of V_0: (▲) hexane/neopentane mixtures, other symbols as in Fig. 7.1

[4.14], while in steady state scavenging experiments concentrations up to 0.5 M are used [7.5]. In many Ps inhibition studies, e.g., those of CCl_4 and C_2H_5Br [7.1], scavenger concentrations up to the pure scavenger liquid are investigated. The initially formed encounters of electrons and scavenger molecules play an important role in steady state scavenging and Ps inhibition studies at high scavenger concentrations. Table 6.2 shows the percentage of encounter pairs for various concentrations of scavengers at a typical reaction radius of 1 nm, which is the value which can be reasonably assumed for the reactions of the partly delocalized excess electron in many nonpolar liquids. This must be taken into account if the α and k values are correlated. Furthermore, the rates are determined by the random diffusion of mainly the fast moving electron and the kinetic reaction process. The scavenging of spur electrons (and other spur reactants, such as the positive ions and the positrons) occurs in the spurs where it is influenced by the electric fields of the other charged particles, and where the effect of the drift of the electron in the field is stronger than that of random diffusion. Taking into account other effects, such as the field dependence of the dielectricity constant and electron mobility, and the dependence of the spur size on the scavenger concentration at moderate and high concentration, it is not surprising at all that the Ps inhibition constant is not proportional to the rate constant except in a semiquantitative way.

Some remarks on the positron reaction with CCl_4, (7.11), might be useful here. As discussed in Sect. 5.3, we have good reasons to believe that the "free" positron annihilation in benzene gives an angular correlation spectrum like 1 M nitrobenzene/benzene. We measured that this spectrum was 2% higher than that for positrons bound to chloride, but 10% lower than the spectrum for 2 M CCl_4/benzene, in the top, while that of 2 M CCl_4/benzene was fairly close to the pure CCl_4 spectrum. Hence, the addition of CCl_4 to benzene in a concentration which is high enough to inhibit all Ps formation results in a spectrum very different from the "free" positron spectrum, and the difference cannot be explained by positrons reacting with chloride in the spur (7.9). 1,2,3,5-$C_6H_2Cl_4$/benzene gives roughly the same result. Furthermore, CBr_4 in benzene reacts also with the positron in benzene if the same criterion is used [1.37]. These results speak strongly in favour of a positron reaction with the three molecules in benzene.

Let us now discuss the influence of adding C_6F_6 to certain pure nonpolar liquids and to mixtures of strong Ps inhibitors and nonpolar liquids. The addition of C_6F_6 at a high concentration of about 0.2 M traps all the electrons, i.e., the reactions which are fast because of the rapidly moving electron in the pure liquid, become slow. For example the electron ion recombination takes roughly 6 ps in pure hexane, but the recombination of $C_6F_6^-$ and the positive ion in hexane is expected to take several nanoseconds. The measured increase in the Ps yield on addition of C_6F_6 to hexane and a CCl_4/hexane mixture, as shown in Figs. 7.2,3, indicates that also the positron is rapidly moving in hexane. Similar results have been found for several other nonpolar solvents, including cyclohexane where the positive ion is fast moving, too. This shows that the positron must be much faster, i. e., have higher mobility and diffusion constant, than the positive ion in cyclohexane. The "antiinhibition" effect was first found by *Anisimov* and *Molin* [7.32], who found antiinhibition by C_6F_6 in iodobenzene/benzene, iodobenzene/hexane, nitrobenzene/benzene, and tintetrachloride/benzene solutions, by perfluoropyridine, 1,4-bis-trifluoromethylbenzene, pentafluorobenzene, difluorobenzene, carbondisulfide, naphthalene, α-naphthylamine, and carbondioxide in iodobenzene/benzene, and by carbondioxide in iodobenzene/hexane solutions. Many other antiinhibition cases have been reported, e.g., in [7.2,33–42], including various fluorine-substituted benzenes in CCl_4/benzene [7.35,42], C_6F_6 in various inhibitor/benzene solutions [7.33,35,39,42], and "antirecombination" by various fluorine-substituted benzenes in cyclohexane and hexane [7.38,42,43], biphenyl in hexane [7.2], and benzene in hexane, isooctane, and cyclohexane [7.2]. Benzene is not an electron acceptor in vacuum, but might trap electrons in some nonpolar liquids because of the polarization energy of the formed anion. In particular, carbondioxide and biphenyl are interesting electron acceptors as they bind the electron so weakly in some solvents that it has a significant detachment probability, see [7.44–46].

However, while many electron acceptors binding the electron weakly in anion states which are longlived compared to the lifetime of the positron, give the antiinhibition and antirecombination effects, other electron acceptors with very similar properties decrease the Ps yield on addition to nonpolar liquids. To understand these results, we must discuss the Ps yields for various other mixtures.

Figure 7.5 shows the Ps yields for CS_2 mixtures with several nonpolar liquids up to pure CS_2 [7.24]. Except for tetramethylsilane (TMS) a minimum in the Ps yield was found. CS_2/cyclohexane gave a minimum somewhat deeper than that found for hexane, while CS_2/benzene showed a nearly constant Ps yield for all compositions [7.2]. SF_6 mixtures up to roughly 2 M SF_6 give also minima, as shown in Fig. 7.6 [7.11]. The SF_6 mixtures were only studied below 2 M SF_6. In pure SF_6 at 20°C at its vapour pressure of 21 bar the o Ps yield is 44% [4.27]. Ps yields for the SF_6/hexane mixtures are very close to the SF_6/heptane results [7.2]. Furthermore, C_6F_{14}/hexane (see Fig. 7.2) and /isooctane mixtures give two minima, one at 0.3 M and the other at about 3.5 M

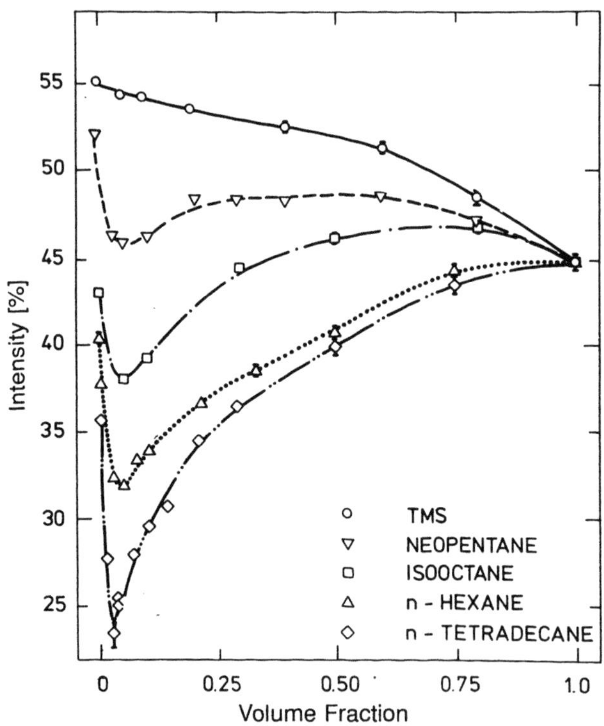

Fig. 7.5. The o Ps yield versus volume fraction of CS_2 in mixtures of CS_2 and various nonpolar liquids. TMS is tetramethylsilane. The curves are best fits to experimental points [7.24]

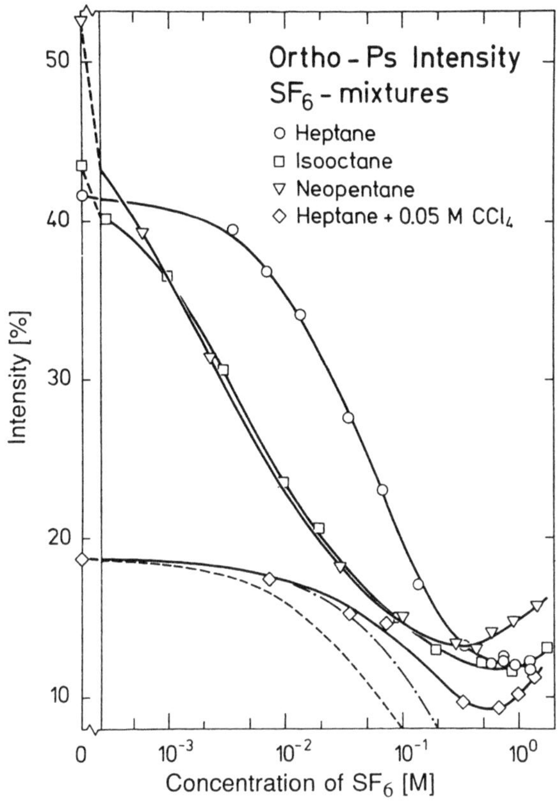

Fig. 7.6. The o Ps yield versus SF_6 concentrations in various nonpolar liquids. The solid curves for isooctane and neopentane for I_3 above 30% are (7.30) with the parameters of Table 7.1. Below 30% they are visual fits [7.11]. For the dashed and dash-dot curves see [7.11]

C_6F_{14} [7.3]. The first minimum corresponds to the ones found for CS_2 and SF_6 mixtures. Another result of interest in this context is the decrease of the Ps yield on the addition of CH_3Cl to benzene and cyclohexane up to pure CH_3Cl [7.22]. As discussed in Sect. 7.3, the steady state scavenging results showed that the formed anion, CH_3Cl^-, is longlived compared to the positron. It is at first sight difficult to understand that an increase in the Ps yield at low concentration is not found in these CH_3Cl mixtures. We shall now discuss these results in some detail.

In order to explain the results an improved spur model was proposed [7.22]. The new aspect was that the localization of the positron is also invoked in the interpretation. It was assumed that a decrease in the Ps yield on addition of a compound to a nonpolar liquid is only obtained if both the excess electron and the positron are localized on the compound, i.e., if the two particles get a mobility and diffusion constant in the order of magnitude of those of the massive ions. If any one of them is rapidly moving it might always be possible for it to pick off the other particle, localized on molecules, such as CS_2, SF_6, C_6F_6, CH_3Cl, etc., to form Ps. Localization of the positron was also included in the initially proposed spur model [1.35], but until the 1983 mainly the electron properties were used in the interpretations of the Ps yields, except for the clear cases of positron reactions.

The difference in the Ps yield for C_6F_6 and C_6F_{14} solutions can now be explained by the assumption that only C_6F_{14} scavenges the positron, while both molecules scavenge the electron, as is known from their positive electron affinities. At low C_6F_6 concentrations the antiinhibition effect explains the measured increase in the Ps yield, as discussed above. At low C_6F_{14} concentration the decrease in the Ps yield can be explained by two effects. First, part of the Ps formation in the pure liquid is caused by the higher mobility of the positron than that of the positive ion, which permits the positron to react with the electron before the ion. This difference disappears on positron scavenging. Second, Ps formation by the trapped particles, if this is possible, might take longer time than the lifetime of the positron. This effect explains the decrease in the Ps yield in the same way as the difference in the Ps yields between polar and nonpolar liquids is explained by positron annihilation in the polar, but not in nonpolar, liquids before Ps is formed.

Another effect must be taken into account in the interpretation of Ps yield data, namely that the particles can tunnel between, or delocalize on, the scavenger molecules at high scavenger concentration [7.24]. This effect explains the minimum found for some CS_2, and for the SF_6 and C_6F_{14}, mixtures. The tunnelling sets in when the concentration reaches 0.2–0.5 M, and at still higher concentration the particle becomes partly delocalized on the scavengers. As for pure nonpolar liquids the high mobility and diffusion constant of one or both of the particles in the mixture result in high Ps formation. It might be somewhat difficult to sort out whether it is the electron, positron, or both that become faster moving because of that effect. In the CS_2 mixtures, except for benzene, it seems to be most likely that it is the electron, because the depth of the minimum correlates well to the expected depth of the potential well trapping the electron. The electron is bound more strongly if the lowest electron energy in the pure liquid is high. Hence, we expect the strongest binding and the deepest minimum in the Ps yield in tetradecane, where V_0 is 0.21 eV (see Table 4.1), and the most shallow Ps yield minimum in tetramethylsilane having $V_0 = -0.51$ eV, as shown in Fig. 7.5. The lowest energy of the electron in the pure liquid is about 0.2 eV lower than V_0 for tetradecane and hexane, while it is roughly equal to V_0 for neopentane and TMS [7.46]. Of course, we cannot exclude that the positron trap depth on CS_2 fortuitously correlates to V_0 in the same way as that of the electron. This interpretation of the CS_2 results differs somewhat from that given in [7.24].

The inclusion of the scavenging of the positron in the interpretation of the Ps formation for nonpolar mixtures was done after careful considerations, which are discussed in [7.22]. It explains also the difference of the CS_2 and SF_6 mixtures, which now can be correlated to differences in the positron scavenging, and it need not be correlated to different electron affinities of the two molecules, an interpretation which was tried previously. The electron affinities in the literature are fairly uncertain. The decrease in the Ps yield on addition of CH_3Cl to cyclohexane and benzene can also be ascribed to the effect of positron scavenging by CH_3Cl [7.22]. Altogether, a more detailed study of the Ps yields measured until now shows that the improved spur model can account for the results. Without the positron scavenging this could not be done.

However, the improved model predicts new results, which in one case have been measured. It is predicted that the positron is trapped on C_6F_{14}, but not on C_6F_6, in hexane. The trapping of positrons on a fluorine containing molecule is easy to detect as the broad component of the angular correlation spectrum for such molecules is very broad compared with that of most other spectra, as illustrated for pure C_6F_6 in Table 5.2. The AC-spectra for C_6F_6 and C_6F_{14} mixtures with hexane were measured at Risoe [5.12]. In the C_6F_6 case the fwhm of the main gaussian of the broad component varied roughly linearly with the C_6F_6 concentration, which indicates that the positron does not preferentially annihilate in a bound state on C_6F_6. On the other hand, the fwhm of the main gaussian in the broad component of the AC-spectra for the C_6F_{14}/hexane mixtures increased strongly at very low C_6F_{14} concentration indicating an preferential annihilation of the positron in a bound state on C_6F_{14} molecules. The main gaussian in the broad component constitutes more than 79% of the broad components for the mixtures. Hence, the predicted effect was confirmed. At present, this seems to be the only detailed test of the improved spur model. Of course, it cannot be totally excluded that this nice agreement between the prediction and experiment is fortuitous.

Partly triggered by the Ps inhibition results the radiation chemists have studied C_6F_6 and CS_2 solutions in some detail. The group in Delft found that the excess electron mobility in pure C_6F_6 is 40 times larger than the value found for molecular ions in that liquid [7.47,48]. Furthermore, the electron did not react with SF_6, the standard electron scavenger, because it has a lower energy in C_6F_6 than on SF_6. However, it reacted with CCl_4, O_2, $CHBr_3$, CBr_4, and TCNE. C_6F_6 reacted with the electron in tetramethylsilane and isooctane with rate constants 1.2×10^{14} and 5.6×10^{13} $M^{-1}s^{-1}$, respectively. They also found that addition of, e.g., isooctane, cyclohexane, and benzene to pure C_6F_6 decreased the electron mobility very much. They looked for this decrease because of the decrease in the Ps yield found on adding hexane to pure C_6F_6, i.e., because of the second minimum found at 3 M for the C_6F_6 mixture curves, Figs. 7.2,3. There is normally a very strong correlation between the Ps yield and the electron mobility, as discussed for pure liquids in Chap. 5. Low Ps yield corresponds to low electron mobility. The cause of this decrease in the electron mobility and Ps yield seems to be unknown. As the electron has higher energy in, e.g., hexane than in C_6F_6 the results are difficult to understand. Also C_6F_{14} mixtures with hexane show that effect (Fig. 7.2). *Gee* and *Freeman* [7.49] measured the electron and positive ion mobilities for pure CS_2. They found a ratio between the two mobilities of 6–10 over a large temperature range. By means of optical detection of ESR spectra *Molin* and *Anisimov* [7.50] measured a strong broadening in the ESR spectra of $C_6F_6^-$ at higher C_6F_6 concentrations in squalane solutions with 10^{-3} M deuteroanthracene. This is very probably caused by the transfer of the electron between C_6F_6 molecules. Hence, radiation chemistry measurements fully agree with the interpretation of the Ps formation data for liquids containing the electron scavengers CS_2 and C_6F_6.

Concerning the stability of the anion radicals in nonpolar liquids some of the chloronated hydrocarbons constitute a "class of transition", where the dechloronation time of the anion is comparable to the "free" positron annihilation time. In particular, *Wikanders* Ps yield studies of halogenated hydrocarbons in nonpolar liquids show that clearly. To investigate this effect in detail, Ps yields in mixtures of tetrachlorobenzenes with various liquids were measured [3.7,7.26]. In Table 7.1 the α and β values for mixtures with benzene and methylsubstituted benzenes are given. Clearly, the strength of the Ps inhibition decreases with increasing number of methyl groups. 1,2,3,5- and 1,2,4,5-$C_6H_2Cl_4$ gave similar results. Furthermore, the strength of the Ps inhibition by 1,2,3,5-$C_6H_2Cl_4$ increased with increasing benzene concentration in benzene/hexane mixtures. The explanation of these effects is very probably that the dechloronation is faster in benzene and slower in the highly methyl-substituted benzenes and at high concentration of hexane, because the solvation energy of Cl^- is largest in benzene and lowest in hexane and in the highly methyl-substituted benzenes. As the dehalogenation time becomes faster compared to the time of the electron pick off from the anion by the positron when, e.g., the benzene concentration is increased, Ps formation by the pick off process, (7.8), becomes less important. Such fast dehalogenation times seem not to have been studied by other techniques.

Two examples of the effect of temperature variation on the Ps inhibition are given in Table 7.1 [7.26]. The effects of addition of CCl_4 to hexane and toluene and of C_2HCl_3 to hexane were fairly independent of the temperature. A similar result was obtained for 1,2,3,5-$C_6H_2Cl_4$ in hexane, for which, however, the Ps inhibition is very weak. On the other hand, the Ps inhibition by 1,2,3,5-$C_6H_2Cl_4$ in toluene, m-xylene, and mesitylene increased strongly with decreasing temperature, an effect which was opposite to that expected. We expected that the dechloronation process would be slower at lower temperatures, and hence, that the Ps inhibition would decrease with decreasing temperature. This problem is discussed in detail in [7.26]. Very few cases of the effect of temperature on Ps inhibition in nonpolar liquids have been published [7.26,51].

A discussion of various other Ps inhibition results might be appropriate here. The highest value of the inhibition constant, α, is that describing the initial decrease in the Ps yield on the addition of SF_6 to neopentane, Table 7.1. This is in agreement with the very large spur size in neopentane, Table 4.1, and with the fact that the rate constant for the electron reaction with SF_6 in neopentane (Table 4.2) seems to be the largest one ever measured in chemistry for a reaction where one of the reactants is uncharged. In general, the larger α-values are obtained for the solvents with the largest spurs, Table 4.1 and 7.1. *Ito* [7.41] has correlated the electron scavenging effects on Ps formation with the fluorescence intensity in cyclohexane. The studied fluorescence was mainly a product of the geminate recombination, and hence, it was appropriate to correlate it to the Ps yield. As expected, the inhibition of the Ps yield was somewhat stronger than that of the fluorescence, for scavengers such as CCl_4 and $CHBr_3$, reacting also with the positron, while the two inhibition processes were not correlated for CS_2 for which electron pick off from CS_2^- by the positron is very important for the Ps formation. *Wild* and *Ache* [7.52] measured the Ps yield versus scavenger concentration for various electron scavengers in benzene, including several "mixed" halogenated benzenes of which meta-chlorofluorobenzene gave a minimum at 0.1 M.

We have until now discussed only the Ps inhibition by compounds which do not react with Ps, i.e., which do not decrease significantly the o Ps lifetime when added to nonpolar liquids. The reactions of Ps will be discussed in detail in Chap. 8. In particular, it will be shown that the relative intensity of the long lifetime component, I_t, extracted by a computer analysis of the measured lifetime spectra in terms of exponentials, is not the ortho Ps intensity, $P_t(0)$. Equation 8.29 gives I_t in terms of $P_t(0)$. On taking into account the Ps inhibition, we see that I_t in a typical case might decrease at low concentration, then increase to infinity, followed by an increase from minus infinity. This is shown in Fig. 7.8 for various aqueous solutions and in Fig. 8.1 for nitrobenzene in benzene. This effect, and the fact that the difference between the o Ps and the "free" positron lifetimes is small at higher Ps quencher (inhibitor) concentrations, make a well defined determination of I_t, and hence $P_t(0)$, very difficult for many Ps inhibition systems. Actually, the importance of the difference between I_t and $P_t(0)$ was realized very late (in 1975, [7.10]), and many older $P_t(0)$ values are incorrect. Furthermore, partly because no good computer programmes which permit analyses taking into account the finite resolution of the lifetime setup, were used, $P_t(0)$ or I_t values were normally not given in the papers on Ps quenching. In summary acceptable o Ps yields are rare for nonpolar solute/solvent systems in which Ps quenching takes place. Because a detailed discussion of the Ps quenching and its influence on the possibility of determining the Ps yields will be given in Chap. 8, we shall here only give a qualitative treatment of the Ps inhibition by some selected Ps quenchers.

Actually, fairly few papers report results on the Ps yields if Ps quenching takes place in the solutions, and most of these are on polar systems treated in the next section. For nonpolar mixtures, in particular, *Jansen* et al. [7.53] describe the problems of obtaining the Ps yields for several fairly different solute/solvent systems. For SO_2 in heptane, where the Ps quenching rate constant was $3.7 \times 10^{10}\ M^{-1}s^{-1}$, the Ps yields determined from I_t were very dependent on the assumed lifetime of the positron in the product of the Ps quenching reaction, i.e., on λ_m or λ_f in (8.4,5). On the other hand, the Ps inhibition constant was well defined for the SO_2-TMA complex, for which the Ps quenching was small. Another small molecule, O_2, in heptane behaves similarly to SO_2 with respect to Ps inhibition and quenching. These two small molecules seem to be the only solutes for which the Ps inhibition is weak or missing although they react strongly with Ps. In [7.53] this was explained as caused mainly by an inefficient electron transfer from the initially formed excited state to the ground state in the anion formed by the reaction of the spur electrons with the molecules, resulting in a low rate constant of the reaction. This interpretation differs somewhat with that given in [7.31], and discussed above. Both the Ps inhibition and quenching depended strongly on the nature of the solvent and whether or not the molecule took part in complex formation. For SO_2 in dioxane the inhibition constant was difficult to determine.

The Ps inhibition constants for nitrobenzene in benzene and heptane are shown in Table 7.1, while the Ps quenching rate constants are given in Table 8.2. The inhibition constants were determined by use of the intensities of the narrow peaks in the angular correlation spectra. These intensities depend only little on the Ps quenching, and hence, the inhibition constants are much easier to determine than in the cases where only the lifetime results are used. Apparently, these are the only systems with Ps quenching where angular correlation data have been used for that purpose. Nitrobenzene in benzene has also been studied in [7.54].

We have until now not discussed the very recent results which have indicated that a second o Ps lifetime of 0.8–1.9 ns (2–8%) is measured for many liquids. Roughly speaking this fourth lifetime does not influence our treatment of the Ps inhibition very much as we have mainly used qualitative or semiqualitative arguments in the discussions. However, the fourth lifetime component is reflected in the inhibition results, mainly by a decrease in the second lifetime for increasing inhibitor concentration above roughly 0.5 M.

7.5 Positronium Formation in Solutions of Polar Liquids

We shall in this section discuss the changes in the Ps yield measured if solutes are added to polar liquids. The main topic will be ionic solutions exemplified by aqueous solutions. Apparently, this is the most difficult topic of the book. The main concepts and ideas discussed in the preceding sections apply as well for polar liquids. However, new processes taking place only in polar liquids complicate the Ps formation problem. The most important difference between nonpolar and polar solvents is that the electron and positron are strongly solvated in polar solvents in their equilibrium states. Because the solvated particles move slowly in an electric field, and diffuse slowly, the spur process of Ps formation takes much longer time in a polar liquid. For example, the spur recombination of the hydrated electron and positive ion in water takes about 5–10 ns [5.27], compared to a recombination time of roughly 10 ps in hexane [4.1], for the normal, mainly singlepair, spur, as discussed in Chap. 5. The lifetime of the free positron in liquids is about 400 ps. Hence, Ps cannot be formed by the fully solvated electron and positron separated by the normally found range of secondary electrons. Consequently, Ps is formed by electron-positron reactions before full solvation or in initial encounter pairs of solvated particles. Another complicating effect is that the spurs in polar solvents are smaller than in non-polar solvents, and higher concentrations of spur reactants are, therefore, found in the polar liquids.

As done in the preceding section we shall, at first, illustrate the basic principles by a simple discussion of typical cases, namely those of Ps inhibition by addition of solutes to water. The Ps yields for solutions of other polar solvents behave similarly. This will be followed by the main theoretical discussion. At last a short review of the available experimental results will be given.

Qualitatively, the inhibition of the Ps formation by ions in water and other polar solvents can be well rationalized in terms of the spur model. Strongly oxidizing ions, such as MnO_4^-, $Cr_2O_7^{--}$, and IO_4^-, reduce the Ps yield to zero and react with Ps, resulting in no LT and AC components due to Ps at about 0.5 M concentration. The result of the Ps reaction is oxidized Ps, i.e., a free positron; bound state formation; or a mixture of those two products (see Sect. 8.5). Moderately oxidizing ions, such as NO_3^-, inhibit the Ps yield to zero, but do not react with Ps, except if they are paramagnetic and react by spin conversion. Weakly oxidizing ions, such as Cd^{++}, Ag^+, and Tl^+, inhibit Ps formation only partly, and do not react with Ps except if paramagnetic (spin conversion). Very weakly oxidizing ions, such as Na^+, K^+, Mg^{++}, and Ca^{++} neither inhibit the Ps formation nor react with Ps. The correlation to the standard reduction potentials of the ions will be discussed below.

Let us illustrate the inhibition of Ps formation by strongly oxidizing ions by the LT results for aqueous solutions of MnO_4^-, $Cr_2O_7^{--}$, IO_4^-, and Cu^{++} shown in Figs. 7.7–9 [7.10]. All the lifetime spectra were analyzed by use of POSITRONFIT into three exponential components. The second lifetime τ_2 was fixed to 395 ps. Such constraint is necessary, if the o Ps lifetime gets shorter than about 1.2 ns. The Ps reactions and the analyses are discussed in detail in Chap. 8. We shall here only note that the lifetimes can be well fitted by (8.1) resulting in well defined rate constants. Exceptions are the Cu^{++} ions where complex formation with the counter ion Cl^- influences the Ps quenching.

The kinetic equations used normally in the interpretation of Ps reactions, are solved and discussed in details in Sect. 8.1, and a standard case, that of the Ps nitrobenzene reaction in benzene, is described in Sect. 8.2. It is shown that the solution of the equations is a sum of three (or four) exponentials if a Ps reaction occurs. In particular, the preexponential factor, I_t, of the longest lifetime component turns out not to be the intensity of o Ps decay, $I_3(= P_t(0)$ in Sect. 8.1). I_3 must be obtained from I_t by use of (8.29), if no spin conversion takes place. As shown in Fig. 7.8, I_t was fitted to the long lifetime intensities extracted from the LT spectra, by use of the inhibition constant, α, as fitting parameter, and for $\beta = 1$ in (7.30). The rate constants extracted from the lifetime, τ_3, versus concentration were used. The α-values and the rate constants are given in Table 7.2. I_3 is shown in Fig. 7.9.

These are standard cases of Ps inhibition in the presence of Ps reactions. The experimental and analytical difficulties will be discussed in detail below. We shall here only note that the determination of the inhibition constants, in

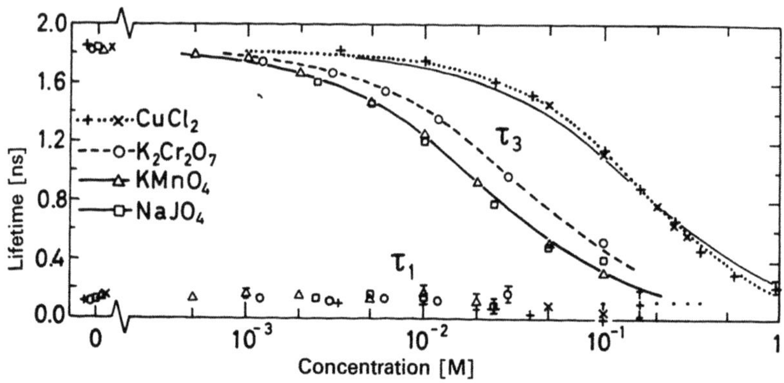

Fig. 7.7. Quenching of Ps. The short and the long lifetimes versus solute concentrations. The thin line for $CuCl_2$ and the curves for $NaIO_4$, $KMnO_4$, and $K_2Cr_2O_7$ are given by (8.1) with the rate constants given in Table 7.2. The dotted curve for $CuCl_2$ is calculated assuming that the three solutes: Cu^{++} and the complexes $CuCl^+$ and $CuCl_2$ react with Ps [7.10]

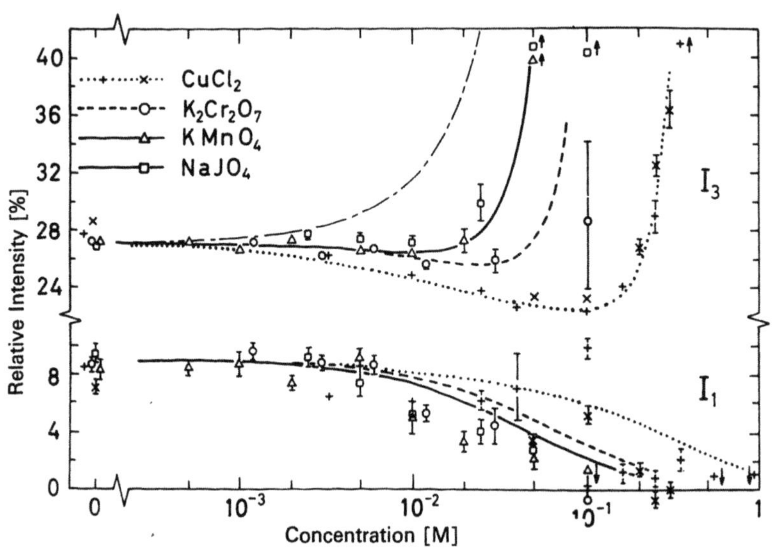

Fig. 7.8. The relative preexponential factors, I_t and I_s, versus solute concentrations. Points with arrows are for I_t but lying outside the frame of the drawing in positive and negative direction. The curves are drawn by use of the expressions (8.29) for I_t and (8.18) ($\gamma = 0$) for I_s, respectively, and the Ps yield expression (7.29,32) with the inhibition constants given in Table 7.2 [7.10]

particular the exponent β, is a difficult task (see Chap. 8), and the uncertainties in the extracted parameters should be carefully estimated before strong conclusions concerning the physics of the inhibition are stated.

In Fig. 7.10 is shown the Ps inhibition by a moderately oxidizing ion, NO_3^- in water [6.4,7.55]. NO_3^- does not react with Ps in any solvent. This

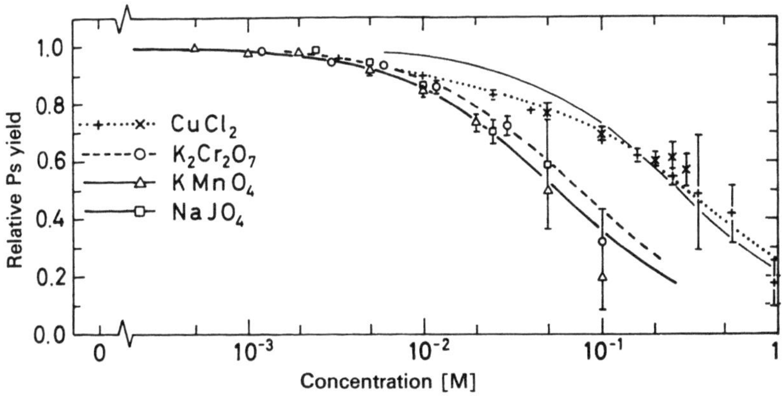

Fig. 7.9. Inhibition of Ps formation. The relative Ps yields, as determined from I_t and τ_3 by use of (8.29), versus concentrations. The curves for $NaIO_4$, $KMnO_4$, $K_2Cr_2O_7$, and the thin one for $CuCl_2$ are drawn according to (7.29,32). The dotted curve for $CuCl_2$ includes inhibition at low concentrations by Cl^- according to (7.33) for $K = 60 \pm 25\ M^{-1}$ and $f = 15\%$. Better K and f values are given in Chap. 6 [7.10]

typical case illustrates that the extraction of the inhibition parameters is easy. The parameters are given in Table 7.2.

If the relevant oxidizing (electron accepting) property of an ion is weak, Ps cannot be totally inhibited, as shown in Fig. 7.10 for Cd^{++} and Ag^+ [7.57]. These ions represent the transition from Ps inhibition to no inhibition. Approximately, the Ps yield can only be inhibited to a certain constant level, the saturating yield, as expressed in Table 7.2.

The very weakly oxidizing ions, e.g., Na^+, K^+, Mg^{++}, and Ca^{++}, which cannot trap electrons in water (see below), do not influence the Ps yield in polar liquids.

We shall now give a detailed and basic discussion of the Ps inhibition in polar liquids and its correlation to radiation chemistry results. At first, a treatment of the experimental situation and the spectrum analyses is appropriate. It is more difficult to determine the Ps inhibition parameters in polar than in nonpolar liquids, simply because the intensity of the long lifetime component due to o Ps annihilation in polar liquids is only about half of that found in nonpolar liquids. Furthermore, the long lifetime is normally shorter. In particular, this is the case for water. Many lifetime studies of aqueous solutions were performed before 1975. However, Ps inhibition was normally not detected, if Ps was also quenched by the added solutes. The main reason was that it was not realized that the correction of the preexponential factors, I_t, for the effect of the Ps reaction had to be performed to get the o Ps yields, as given by (8.29). The correction for the Ps reaction was not used in practice before 1975, although the formula, (8.29), was published ten years earlier by *McGervey* [7.58]. As shown clearly in Figs. 7.8 and 8.1, I_t, which was mistakenly thought to be the o Ps

Table 7.2. Positronium inhibition and quenching constants for solutes in water compared to their respective hydrated electron reaction rate constants, k_-, and C_{37} which characterizes the decrease in the initial (~ 30 ps) yield of hydrated electrons on adding the solute [7.55–57]. See text for further comments

solute	α [M^{-1}]	K [M^{-1}]	f	Expr.	k' [$M^{-1}ns^{-1}$]	$k_- \times 10^{-8}$ [$M^{-1}s^{-1}$]	C_{37} [M]
Ag^+		9.9	0.30	B		360	
$BrCH_2COO^-$	2.2			A		62	
BrO_3^-	1.44			A	0.036	34	1.3
BrO_4^-	14.3			A	5.85	70	
Cd^{++}		12.4	0.29	B		500	
$ClCH_2COO^-$	0.55			A		12	6.1
Cl_3CCOO^-	3.0			A	0.043	85	0.5
ClO_3^-	0.14			A	0.009	3.5	
ClO_4^-	0.084			A	0.0		
$Co(EDTA)^-$	10			A	11.3	290	
CrO_4^{--}	18.6			A	15.0	200	0.2
$Cr_2O_7^{--}$	13			A	16	600	0.12
Cu^{++}	2.5			A			0.9
$Fe(CN)_6^{3-}$	11			A			0.8
H_2O_2	1.28			A			1.44
Hg^{++}	21			A			0.062
$HgCl_2$	15			A			0.08
H^+		6.0	0.26	B		230	
ICH_2COO^-	5.1			A	5.79	120	
IO_3^-	3.45			A	0.075	85	0.39
IO_4^-	18			A	26	110	0.14
MnO_4^-	18			A	26	220	
NO_2^-	1.4			A		45	1.6
NO_3^-	3.9			A			0.42
Ni^{++}	0.3			A			2.6
Pb^{++}	2.4			A	0.011	390	0.27
$S_2O_8^{--}$	2.7			A	6.1	106	0.6
SeO_4^{--}	5.6			A	0.041	11	0.42
$Te(OH)_6$	10.5			A		320	0.075
Tl^+		8.6	0.26	B	0.01	300	
Acetaldehyde		2.13	0.31	B		35	
Acetic acid	0.06			A		1.8	
Acetone	0.045	2.3	0.31	D		79	
2.3-butanedione	1.7			A	1.43	100	1.12
Dimethyloxalate	2.3			A			0.76
Ethanol	0.0					0.0	
Formate	0.0					0.0	
Formic acid	0.1			A		1.5	
Fumerate (dian.)	4.7			A			0.62
Glycine	0.07			A		0.0	
Maleate (dian.)	0.6			A		17	1.1
Maleic acid	4.6			A	0.612	120	0.41
Malonic acid	0.43			A		13	2.51
Nitromethane	6.0			A	5.62	220	0.30
Oxalic acid pH1	2.2			A			0.7
Oxalic acid pH2	1.1			A			2.0
Oxalic acid pH3.5	0.75			A			3.5
Succinic acid	0.18			A		3.7	
Tart. acid(D.L.M.)	0.29			A		5.9	

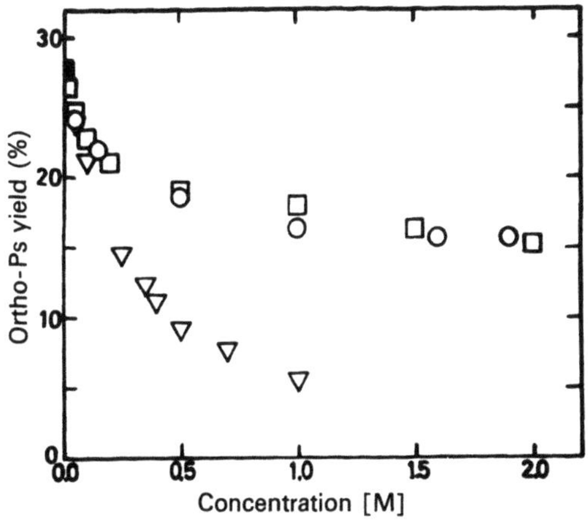

Fig. 7.10. Inhibition of Ps formation. The preexponential factor, I_t, versus concentrations. Except for small corrections at high concentrations in the $AgClO_4$ case, I_t is equal to the o Ps yield, $I_3(c)$, in a good approximation. (□) $Cd(ClO_4)_2$ [7.57], (○) $AgClO_4$ [7.57], and (▽) NO_3^- [7.55]. ClO_4^- inhibits the Ps yield weakly ($\alpha \simeq 0.08\ M^{-1}$). The inhibition parameters are given in Table 7.2

yield, is roughly constant up to high concentration of the solute, and hence, it was incorrectly assumed that Ps was not inhibited by the ions reacting with Ps.

The results of earlier lifetime studies of Ps inhibition were also partly incorrect because a good computer program for the analysis of the LT spectra was not available. The first good computer analyses seem to be those reported in [7.10]. In particular, the long lifetime intensities determined by use of a ruler on semilogarithmic plots, or by use of programs which could not take into account the finite resolution of the lifetime set-up, seem to be somewhat uncertain, see Sect. 3.1.

Another problem is the validity of fixing the second lifetime in the analyses of the LT spectra. The purpose is to make possible the extraction of a reasonable intensity of the long lifetime for long lifetimes below about 1.2 ns, where the analysis cannot separate well this lifetime from that of the free positron annihilation of roughly 400 ps. Actually, LT spectra are not just a sum of three exponentially decaying components, in general. We discussed this problem in Chap. 5, and we shall encounter it in Chap. 8. Part of the discussion of the physics of the Ps formation, given below, concerns this problem, too.

The basic problem in our discussion is, of course: Which processes in the positron spur determine the Ps formation? As mentioned previously, an electron-positron pair which are completely solvated, cannot form Ps before the positron decays, except if they are solvated very close to each other. Essentially, this means that Ps is formed before both particles are solvated, or if the particles are solvated in close encounter positions, i.e., where the distance between the particles is less than, or roughly equal to, the reaction radius of the fully solvated particles. We discussed this in Chap. 5 for the pure polar liquid cases, and in Chap. 6 on the solvated positron reactions. How is the Ps formation influenced if an electron or positron acceptor is present in the positron spur? At first, we discuss the electron trapping. We shall see that positron trapping is very similar. A trapping of the electron on the acceptor after the positron has annihilated, e.g., after ~ 400 ps, will not, of course, influence the Ps yield. Hence, on account of the low mobility of the nonsolvated particles, we expect that only the trapping of the electron before it is fully solvated, or as a solvated electron in an initial close encounter with the acceptor, can change appreciably the Ps yield. A solvated electron which has to move a distance larger than roughly the reaction radius of solvated electron acceptor reaction, has a very low probability to be trapped before the positron decays. However, presolvated electrons might get trapped on the acceptor before solvation, as they move much faster and, therefore, have much larger mobility and diffusion constant, than the solvated electron. Probably, there is in most liquids a distribution of presolvated states which can be occupied by the thermalizing electrons, i.e., the electron can be in many states ranging from subelectronic; over "free", also called "dry" states; and trapped but not solvated; to fully solvated states.

Information on the behavior of the electrons in these states has mainly been obtained by radiation chemists. Of course, we cannot in this book review this information in any detail. However, it is very important to understand certain aspects of the radiation chemistry methods, before the information is applied in the interpretation of the Ps formation. Almost all rate constants of reaction of the solvated electron have been determined by a measurement of the light absorption of the solvated electron in an appropriate wave length interval [4.2,8,9]. A reaction is indicated by a decrease in the absorption on addition of the electron acceptor, or as function of the time in studies of intraspur reactions. In certain polar liquids, e.g., alcohols, some of the nonsolvated electron states can be followed if they have lifetimes against solvation which are longer than the finite time resolution of the used apparatus, e.g., roughly 30 ps for the best pulse radiolysis [4.9], and about 100 fs for the best laser flash photolysis, set-ups [4.10]. Information on the nonsolvated electrons that have been studied directly, have been obtained by a measurement of their infrared absorption [4.9,10]. However, the chemical reactions of the short-lived nonsolvated electrons having lifetimes smaller than about 30 ps, have only been investigated by pulse radiolysis measurements of the initial yield of solvated electrons, at about 30 ps and longer, except in very few cases [7.59]. In most studies of the reactions of the nonsolvated electrons the initial yield decreases exponentially with the electron acceptor concentration, and the strength of the reaction is reported as the concentration where the yield is reduced by a factor of e, C_{37},i.e., where 37% of the yield is not scavenged [5.5,7.59]. The initial yield is usually corrected for the reactions during the finite resolution time of the set-up by use of certain assumptions about the solvated electron reactions.

Clearly, the best correlation between the Ps inhibition and radiation chemistry is expected to be between the strength of the inhibition and C_{37} [7.59]. The inhibition constant, α, may be correlated to C_{37} in the cases of full Ps inhibition by an electron acceptor. However, which inhibition constant should be chosen if the Ps yield is only partly inhibited, and hence, (7.30) fits the Ps yields very badly?

It is appropriate here to discuss the Ps inhibition studies of the Strasbourg group and their interpretation of the measured results, as they have made the by

far largest studies of Ps formation in polar liquids. They were able to express the measured Ps yields, $I_3(c)(= P_t(0))$, versus concentration, c, of the solute by the following expressions [7.56]:
A) Complete inhibition was defined as a case where (7.30) for $\beta = 1$ could be used, i.e.,

$$I_3(c) = I_3(0)/(1 + \alpha c). \tag{7.32}$$

B) Ps inhibition down to a plateau value, I_p, as given by

$$I_3(c) = I_3(0)\{f(1 + Kc)^{-1} + 1 - f\}, \tag{7.33}$$

where $f = (I_3(0) - I_p)/I_3(0)$, see also [7.10].
C) Ps enhancement up to a plateau value, I_{max}, described by

$$I_3(c) = I_3(0) + (I_{max} - I_3(0))\{1 - (1 + \beta' c)^{-1}\}. \tag{7.34}$$

D) A combination of cases A) and B), namely

$$I_3(c) = I_3(0)(1 + \alpha c)^{-1}\{f(1 + Kc)^{-1} + 1 - f\}. \tag{7.35}$$

E) A combination of B) and C), given by

$$I_3(c) = I_3(0)(1 + \gamma' c)(1 + \delta' c)^{-1}\{f(1 + Kc)^{-1} + 1 - f\}. \tag{7.36}$$

It is important to realize that these expressions are mainly empirical. The extracted fitting parameters should only be used after a careful analysis of the actual physical processes of the Ps formation has been performed.

In Table 7.2 the fitting constants and the used expressions are given for 49 ions and molecules in water [7.55–57]. Table 7.2 shows also the rate constant of a reaction with Ps, if a reaction takes place, and the rate constant of the reaction with the hydrated electron, if such reaction occurs. Furthermore, C_{37}, the solute concentration at which the initial ($\sim$ 30 ps) yield of hydrated electrons has decreased to 37% of that in pure water, is shown in the cases where it is known. Figure 7.11 gives the Ps inhibition constants (expressions A) or B)) versus the hydrated electron rate constant in a double logarithmic plot. Additionally, a double logarithmic plot of α versus $(C_{37})^{-1}$ is shown in Fig. 7.12 [7.59]. It is also shown in [7.59] that a plot of α versus the rate constant for hydrated electrons measured at high concentration gives a much poorer correlation than that shown in Fig. 7.12, as expected. Hence, the experimental data on the inhibition of Ps in water are in good agreement with the spur model and the expectation that the Ps formation occurs mainly by nonsolvated particles, apart from the initial close encounter effect, as discussed above.

The Strasbourg group has proposed a more detailed model, which, however, seems to be questionable. They have used the model in many papers. They suggest that Ps can be formed via two pathways involving either quasifree ("dry") or localized (presolvated) particles. Furthermore, on the basis of a simple scheme of the reactions they suggest the following. A solute reacting only with localized particles will lead to limited inhibition whereas a solute reacting effectively with the quasifree particles will result in complete inhibition at high

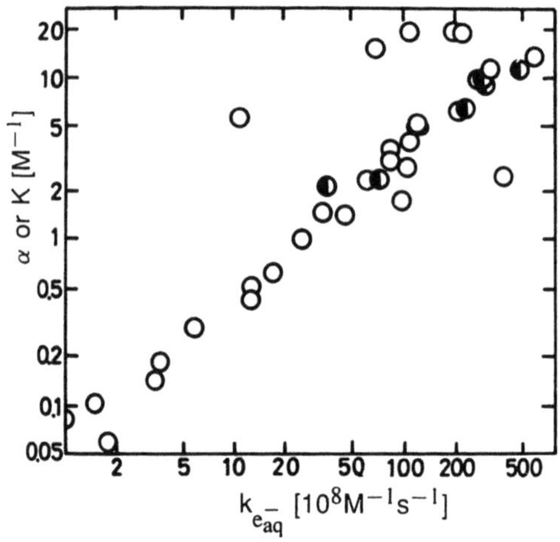

Fig. 7.11. The inhibition constants, (○) α (7.29,32) or (◐) K (7.33), versus hydrated electron rate constants for 36 of the solutes in water in Table 7.2 [7.56]

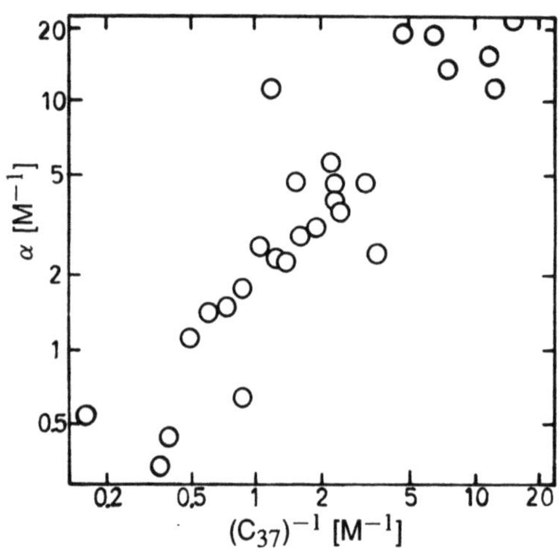

Fig. 7.12. The inhibition constants, α (7.29,32), versus $(C_{37})^{-1}$ for 27 solutes in water [7.59]

enough concentration. Thus in their model the inhibition constants α and K (7.32–35) would be correlated with the rate constants of the reactions of the solutes with either quasifree or localized electrons (positrons), respectively. The enhancement parameters β' and γ' are correlated with the scavenging ability for H_2O^+ (OH) of a solute.

Several objections to the Strasbourg model can be raised. The most important are: First, it is difficult to believe that the two nonsolvated states assumed to exist can be reasonably well distinguished from all the other states in the continuous distribution of states of nonsolvated particles. With few exception [4.9], no particular states in the full distribution are thrown into relief by the available experimental data including the Ps inhibition results. Second, there are no reasons to prefer $\beta = 1$ in (7.30), as the inhibition results for nonpolar liquids show very clearly. Hence, a deviation from the case of $\beta = 1$ needs not necessarily indicate that a second nonsolvated particle state is involved in the scavenging. Third, initial close encounters of the fully solvated particles with the scavenger might form before the reaction occurs. This will depend on which process is the fastest one, solvation or reaction. The model does not include this possibility, which follows directly from reaction kinetics as discussed in Chap. 6. Fourth, the model does not explain why, in particular, the weakly oxidizing ions, such as Ag^+ and Cd^{++}, are supposed to scavenge only the localized electrons (see below).

In principle, the ions in a polar liquid can be arranged in a series according to their electron accepting behavior with the strongly accepting ions first and the very weakly accepting ions at last in the series. Within that series there must be a transition from Ps inhibition to no Ps inhibition. Concerning Ps inhibition, electron accepting means the ability of the ions to scavenge nonsolvated electrons, apart from the above mentioned close encounter effect determined by the ability to scavenge solvated electrons. In the cases of nonpolar liquids we used mainly the electron affinity of the molecules measured in vacuum, and we took into account the polarization energy of the anion radicals formed by the electron scavenging, in our discussion. In the case of polar liquids one normally uses the electrode potentials to indicate the electron acceptor strength of ions, and the electron affinity in vacuum is very seldom used. This is, of course, also possible in Ps inhibition studies. However, one must realize that the normally used standard reduction potentials of the ions express the ability of the ions to cause an electric potential in a standard galvanic cell, and provided certain conditions are fulfilled, see, e.g., [6.16,7.60]. For example, the standard reduction potentials are, in general, caused by multi-electron transfer processes at the electrodes, and they may be influenced by the binding energy of metal atoms in solid metal. The electrode potentials used in the discussion of the scavenging of electrons causing Ps inhibition, must be calculated in agreement with the actual process taking place, as in the discussion of hydrated electron reactions [4.2,7.61].

H.J. Ache and coworkers used the potentials calculated in [7.61] for divalent ions in water in the discussion of the Ps inhibition results [7.62–64]. They used the "Ore plus hot Ps reaction" model of Ps inhibition to explain their results (see Sect. 7.6). In 1972 [7.62] they discovered the partial Ps inhibition to roughly a plateau, I_p, by certain ions: Pb^{++}, Sn^{++}, Cd^{++}, Zn^{++}, Tl^+, and Ag^+. Their f values (7.33), shown in Table 7.3, are somewhat incorrect because they lacked a good computer program, and corrections for effects of counter ions: ClO_4^- giving partial inhibition [7.65] due to electron scavenging and Cl^- giving partial Ps inhibition by positron scavenging, were not performed. The Strasbourg group have more recently remeasured the partial Ps inhibition by some of the ions, as shown in Table 7.2. In Table 7.3 are also shown the electrode potentials for single electron attachment for divalent ions estimated in [7.61] and the standard reduction potentials [5.29],. Furthermore, the electrode potentials given in [7.63] for monovalent and divalent ions for the single electron reactions: $A^{n+}+e^- \rightarrow A^{(n-1)+}$, in water are shown. In the approximate

Table 7.3. The transition from Ps inhibition to no Ps inhibition for ions in water. E' [7.63] and E'' [7.61] are specially calculated electrode potentials for single electron reactions, while E_{SRP} is the standard reduction potentials. f is the fraction of the Ps yield which is inhibited at high concentrations, as given in [7.63], except for Ag^+ and Sn^{++} for which f is estimated here by use of the data in [7.62–64]

Ion	E'[V]	$-E''$[V]	$-E_{SRP}$ [V]	f
Cu^{++}	0.2		−0.16	1.0
Pb^{++}	1.4	1.1–1.5	0.13	0.8
Sn^{++}	1.8		0.14	0.35
H^+	2.1		0.00	0.28
Cd^{++}	2.1	1.9–2.5	0.40	0.24
Ag^+	2.2		−0.80	0.28
Tl^+	2.3		0.34	0.0
Zn^+	2.6	2.5–3.2	0.76	0.04
Na^+	3.5	3.6	2.71	0.0
K^+	3.6	3.6	2.92	0.0
Mg^{++}	4.4	4.0–4.8	2.4	0.0
Mn^{++}		3.1	1.03	
Fe^{++}		3.1	0.41	
Co^{++}		3.1	0.28	0.0
Ni^{++}		2.7	0.23	
Ca^{++}		4.8–5.5	2.8	
Ba^{++}		4.6–5.1	2.9	
Cs^+		3.5	2.9	

calculations in [7.61] the electrode processes are divided into several steps and the Gibbs free energies of the steps are calculated to be used as input in the equation for the potential of the standard galvanic cell (e.g., (11.27) in [6.16]). In the case of a monovalent ion, such as Ag^+, a contribution of the Gibbs free energy of sublimation to the electrode potential is subtracted from the measured standard reduction potential. In this way the fact that the metal atoms do not get bound to the solid metal in the positron spur reaction is approximately taken into account. As seen in Table 7.3, these calculated potentials are well correlated to the plateau in o Ps yield, I_p. As the potential decreases towards more negative values I_p increases to the o Ps yield for pure water. This transition from Ps inhibition to no inhibition by ions in water is a strong argument in favour of the spur model.

A more straightforward way would be to correlate the Ps yield to the actual change in Gibbs free energy of the electron trapping reaction itself. This has not been reported in the literature, and probably the calculation of this energy is fairly difficult. The above mentioned, easier calculation of the special electrode potentials uses experimental numbers in the estimation of certain energies. Of course, the correlation with the special electrode potential expresses only a necessary condition for the electron scavenging reaction in the given approximation. It is well known that electron transfer reactions do not always take place even if the energy conditions permit the reactions. This is clearly illustrated by the discussion of the hydrated electron, e^-_{aq}, reactions with ions in [4.2]. e^-_{aq} has an electrode potential of about −2.7 V, as discussed also in [7.61], and energy conservation allows a reaction if the other reactant has an electrode potential, calculated as in [7.61], above that value. However, the rate of the e^-_{aq} reactions is not determined solely by the gain in free energy between the initial and the final states. More important is the availability of a vacant orbital in the reactant. Henglein's discussion of excess electron reactions in nonpolar liquids [7.31] illustrates this point clearly. Only few rates of reactions of the nonsolvated electron in polar liquids have been reported compared to the e^-_{aq} case, and hence, the interpretation is more uncertain at present. Of course, it must be expected that nonsolvated electron reactions follow the general trend found for electron transfer reactions in general.

At high concentrations special effects occur. Ps inhibition studies are performed up to 1–3 M, nonsolvated electron reactions are studied up to 0.5–1 M, while the typical e^-_{aq} rate constant studies are performed at concentrations below 1 mM, typically 0.05 mM. At 0.05 mM the concept of a rate constant is well defined, and the approximations used in the theoretical derivations of the rates are valid. At 1 M no theoretical calculations of the rates seem to be reasonably correct, as the normally used approximations are invalid (see Sect. 6.2). For example, a reaction radius of 0.6 nm implies that the main number of the reactions occur in close encounter pairs, and only a small percentage of the electrons need to diffuse before the reaction takes place. The discussions of the Ps inhibition in polar liquids must take the high concentration effects into account. This concerns, in particular, the partial inhibition cases.

Let us discuss the Cd^{++} case in some detail. Cd^{++} inhibits the o Ps yield to a plateau of roughly 29% of the yield in pure water [7.56,57] at 1–2 M. According to the "Risoe interpretation" of the partial Ps inhibition, the electron trapped on Cd^{++} (i.e., in the Cd^{+} ion), which is weakly bound, can be picked off again by the positron to form Ps. This cannot happen, if the electron is trapped on more oxidizing ions, such as NO_3^-. As for Ps formation in pure liquids, this pick off process is limited by the lifetime of the positron. Hence, it must happen mainly before the positron is fully solvated or in close encounters of the solvated positron and Cd^{+}, depending on the distances and rate of the processes involved. Some of the electrons, which would have formed Ps in the pure liquid, are trapped on Cd^{++} and are not picked off by the positron. Thus part of the Ps formation is inhibited. Actually, the exact cause of the inhibition only to a plateau is not known. Probably, the size of the wave function of the trapped electron on Cd^{++} is smaller than that of the nonsolvated, or solvated in close encounters, electron, and its mobility is lower. This reduces its chance of forming Ps with a positron. One of the causes of the plateau in I_3 might be that the electron may tunnel from one Cd^{++} ion to another in the field of the positron at higher Cd^{++} concentrations. The effect is similar to the tunnelling of electrons in weakly bound anion states at higher concentration of, e.g., CS_2 in nonpolar liquids, as discussed in Sect. 7.4. This will increase the probability of Ps formation by that electron before the positron decays, if the Cd^{++} concentration is increased to high concentrations. Other causes are possible, of course. It is important to realize that the properties of a nonpolar solution of high concentration, e.g., its excess electron and positron properties, are different from those of a pure liquid, and hence, it is not surprising at all if the Ps inhibition by certain ions is found to be different from the pure liquid case. Actually, it is more surprising that the simple expressions for the Ps inhibition (e.g., (7.30) for $\beta = 1$) can be applied reasonably well up to concentrations of several moles/liter, in most Ps inhibition cases.

The interpretation given here is well illustrated by two particular inhibition results. First, Co^{++} does not inhibit the Ps formation in water [7.65], although e^-_{aq} and Ps react with Co^{++} with rate constants $k = 1.2 \times 10^{10}$ $M^{-1}s^{-1}$ [4.2] and 2.2×10^{10} $M^{-1}s^{-1}$ [7.65], respectively. Apparently, Mn^{++} [5.21] and Ni^{++} [7.65] do not (or only very weakly) inhibit Ps formation either, but they react also with Ps. Clear information on the inhibition properties of Fe^{++} seems to be missing, but Fe^{++} reacts with Ps. This seems to be the only cases in a polar liquid where Ps quenchers do not inhibit Ps formation. As discussed in Sect. 8.4, this does not speak against our interpretation. For example, Co^{++} (like Mn^{++}, Ni^{++}, and Fe^{++}) reacts with Ps by spin conversion and not (or maybe very weakly) by oxidizing or compound formation. A spin conversion reaction can take place even if the paramagnetic ion is weakly oxidizing. The single electron reduction potentials of Co^{++}, Mn^{++}, Ni^{++}, and Fe^{++} are given in Table 7.3. They are below that of Cd^{++}, and hence, the lack of Ps inhibition agrees well with the correlation shown in Table 7.3. Their reactions with e^-_{aq} are discussed in [7.61].

A second interesting result is that Cd^{++} does not give antiinhibition if added to aqueous solutions of $NO_3{}^-$ [7.66]. Again this is easy to understand as the reason for the antiinhibition in nonpolar liquids is the much higher mobility and diffusion constants of the positron compared with those of the other reactants. In polar liquids the slowly moving, solvated positron has no greater chance of picking off the trapped electron than the positive ion. Similarly, antiinhibition of muonium formation by, e.g., C_6F_6 in nonpolar liquids is not possible because the muon is strongly solvated and hence, slowly moving [7.67].

The inhibition of Ps formation in polar liquids by positron scavenging seems to be very similar to the cases of electron scavenging, as discussed in detail in Chap. 6. However, all the reported cases belong to the class of weakly bound positrons on the ions, where the positron can be picked off the by the excess electrons, similarly to the Cd^{++} case for the electron. Hence, inhibition to a plateau is measured. Furthermore, few experimental cases have been reported. Our detailed discussion of the electron scavenging by Cd^{++} in water covers the positron scavenging as well, if electron is exchanged by positron and vice versa.

Inhibition of the Ps formation by ions in other polar liquids than water seems to follow the same trends as in water. In particular, electrons (positrons) are scavenged by the same electron (positron) scavenger ions as in water.

Furthermore, the ions giving inhibition to a plateau in water seem also to give inhibition to a plateau in other polar liquids. However, fairly few cases of Ps inhibition in other polar liquids have been reported (see below). It is important to note that the degree of dissociation of ions might be much lower in other polar liquids than water.

The Ps inhibition by electron and positron acceptors different from ions in polar liquids seems to be qualitatively very similar to the inhibition by the same acceptors in nonpolar liquids, which was discussed in details in Sect. 7.4. However, polar liquids have not been studied as much as nonpolar liquids. In general, polar solvents differ from nonpolar solvents with respect to electron reactions by a) the stronger solvation of the formed anions, b) possible influence of hydrogen bonding, and c) other reactions of the formed anions. For example, in some mixed halogenated hydrocarbon anions containing both bromide and fluoride groups, bromide seems to split off in nonpolar and fluoride in hydrogen bonded polar solvents. As the positron reacts with the bromide resulting in a narrowing of the AC and DB spectra this difference might be detectable.

Until now we have discussed Ps inhibition in polar liquids in general using particular cases as illustrating examples. Let us finish this section by discussing shortly some other interesting experimental results. As usual in this book we do not attempt to discuss all the available data on our present topic, which is published in more than hundred papers, because of lack of space.

The *Strasbourg group* have published more data on the Ps inhibition in polar liquids than any other group. The main part of their studies of aqueous solutions was treated above (e.g., Table 7.2, Figs. 7.10–12, and (7.32–36)).In [7.68] they report on the partial Ps inhibition by CN^-, $Fe(CN)_6{}^{4-}$,and $Fe(CN)_6{}^{3-}$ in water. Interestingly, the initial decrease in I_3 at low concentrations is more than six times stronger for two last ions than for CN^-, indicating that the positron scavenging, which is the cause of the inhibition, is roughly independent of whether or not CN^- is in a complex. Very few results of Ps inhibition by ions in compounds have been reported. Future studies, e.g., on positron reactions with halides and pseudohalides in complexes, would be useful. They found also a plateau, I_p, of about 35% of I_3 for pure water. They state that this result indicates that f in (7.33) is not constant in a given solvent, contrary to their previous assumption [7.56]. This conclusion agrees with Risoe results and those of [7.62]. A comment on Risoe's pseudohalide results [7.69] and the answer from the Risoe group [7.70] illustrate the difference between the interpretations of the two groups.

In a series of papers results of Ps inhibition in polar liquids other than water were reported by the *Strasbourg group*. In particular, the temperature effects on the Ps yield and inhibition were studied (see [7.71] and ref's therein). For example, they have studied iodide in water, methanol, glycol, a 40–60% glycerol-water mixture, and 1,3-propanediol. They found partial inhibition by iodide due to positron scavenging. In some of the liquids enhancement of the Ps yield caused by positive ion scavenging by iodide at higher concentrations was found, too. Parameters characterizing the Ps yield behavior

(7.32–36) were given. In general, the influence of temperature on the Ps yield and inhibition is fairly moderate, and the change of polar solvents results in rather small changes in the character of the Ps inhibition mechanism. They have studied mainly positron scavengers, weakly to moderately oxidizing electron scavengers, and fairly weak molecular electron acceptors, using solvents, such as H_2O, D_2O, glycerol/water mixtures, ethyleneglycol, methanol, ethanol, N-ethylacetamide, dimethylsulphoxide, hexamethylphosphorotriamide, n-propylamine, triethylamine, allylamine, aniline, N-N-diethylamine, dimethylaniline, hydroquinone, and 1,3-propanediol.

Lazzarini and coworkers measured the Ps inhibition by the ions NO_3^- and Cl^- in H_2O, D_2O, methanol, ethanolamine, ethylenediamine, H_2O-ethylenediamine mixtures, liquid NH_3, and other solvents [7.72,73]. They studied also the temperature effects on the Ps inhibition by NO_3^-, Cl^-, and Cd^{++} in water [7.74]. Apparently, Lazzarini and his group used the "Ore plus hot Ps reaction" model longer than other chemists (see [7.73]). His opinion in 1986 is explained in [7.75].

Smith and his group have studied the influence of solutes on the Ps yield in aqueous solutions ([7.76–78] and ref's therein). In particular, the difficult task of fitting the inhibition parameters of (7.32–36) if several solutes are simultaneously present in water, is discussed in [7.76].

Byakov and coworkers interpret the Ps inhibition somewhat differently form the interpretations discussed in this book [7.79]. In particular, they correlate Ps formation to radiolytic hydrogen formation. Their interpretation in terms of only "dry" (quasifree) electron scavenging in the positron spur is probably too simple.

Anisimov and *Molin* [7.80] reported on Ps inhibition by several ions in tert-butanol/water mixtures, interpreted by use of the spur model in 1975.

Information on Ps inhibition by strongly oxidizing ions can be found in the literature published before 1975. However, because of the above mentioned difficulties in the analysis of the lifetime spectra, the given Ps yields are fairly uncertain. In particular, Ps inhibition by Co^{3+} complexes is reported in [7.81], by Fe^{3+} complexes in [7.82] and by Hg^{++} complexes and Sn^{4+} in [7.83].

7.6 Miscellaneous Ps Yield Effects

Until now we have in this chapter discussed the changes in the Ps yields on spur-electron scavenging. Positron scavenging was treated in Chap. 6. In this section we shall discuss various other Ps yield effects. The discussion will include Ps formation in mixtures of nonpolar and polar liquids. In particular, the effects of clustering of the polar molecules and of scavenging the positive ion will be treated. Ps formation in solids will be discussed in Chap. 9.

The Ps yields in mixtures of two nonpolar liquids, which are hydrocarbons, seem to show fairly small deviations from a linear dependence on the concentration of one of the components [7.2,24]. Exceptions are the cases where the

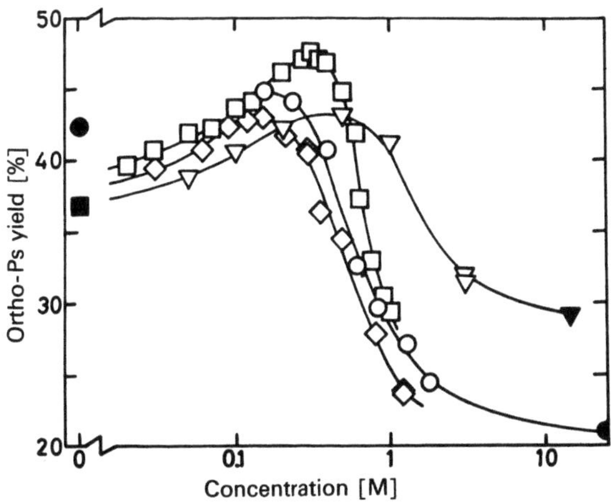

Fig. 7.13. The o Ps yield during transition from nonpolar to polar liquids. The Ps yield versus concentration of the polar component for mixtures of: (▽) neopentanol/cyclohexane [5.21], (◇) pyridine/pentadecane [7.84], (□) NH_3/cyclohexane [5.26], and (○) methanol/hexane [7.87]. Filled marks are for pure liquids. The lines are guide to the eyes

electron is trapped on the molecules of one of the liquids, e.g., in benzene mixtures [7.2]. However, only few cases have been reported.

On the other hand, if certain nonpolar liquids are mixed with particular polar liquids strong, characteristic dependencies on the concentration of the polar molecules are measured. Figure 7.13 shows the Ps yield versus the concentration of the polar component for four selected mixtures. The shown Ps yields are influenced by two effects: A) At low concentration of the polar component Ps formation increases for increasing polar solute concentration due to the effect of scavenging of the positive ion by the polar solute. B) At 0.5–1.5 M clustering of the polar component sets in, the positron and electron are trapped by the clusters, and the Ps formation processes change to those typical of the polar liquids. We shall now discuss these two effects in some detail.

Let us start by discussing the positive ion scavenging effects [7.85]. If a solute, S, which reacts with the positive ions in the positron spur, is added to a pure solvent, the following extra equations describe the nonhomogeneous spur kinetics apart from (7.1–5):

$[S + M^+] \rightarrow [M + S^+]$	M^+ scavenging	(7.37)
$[S^+] \rightarrow$ [products]	S^+ reactions	(7.38)
$[e^- + S^+] \rightarrow [S]$	recombintion	(7.39)
$[e^- +$ products$] \rightarrow [?]$	recombination	(7.40)

The products of the two reactions (7.38,40) depend on the solute. If the recombination reactions (7.39,40) are slower than the recombination reaction in the pure liquid (7.2), the excess electrons in the positron spur will have a longer lifetime. This means that more electrons will be able to form Ps through reaction (7.3) than in the pure solvent, and hence, the Ps yield increases on the addition of the positive ion scavenger, S.

Experimentally, an increase in the Ps yield on addition of positive ion scavengers has been measured in many reported cases. However, certain important exceptions were found. For example, dioxane and triethylamine in heptane [7.85]; dimethylamine, triphenylamine, and tetramethylparaphenylenediamine in cyclohexane [7.86]; and wellknown OH scavengers in water [7.56] give a Ps yield increase. They give no decrease in the Ps yield at higher concentrations because the clustering effects are absent for those solutes. On the other hand, the positive ion scavenger, cyclopropane, gives no extra Ps formation when added to cyclohexane and trans-decalin, although the hole is fast moving in these two solvents.

Before we give a more comprehensive discussion of the positive ion scavenging effect, it is appropriate here to describe the Ps yield decrease due to clustering of certain solutes in particular nonpolar liquids. The Ps yield versus concentration of methanol in hexane, shown in Fig. 7.13, is very similar to that found for ethanol/n-hexane, n-propanol/n-hexane, and n-propanol/cyclohexane mixtures [7.87]. The decrease in the Ps yield from the value of 44% at 0.3 M to 24% at 1.8 M is very probably caused by the trapping of the electrons and the positron by methanol clusters in the positron spur. The character of the Ps formation processes changes thereby from that typical for nonpolar liquids (enhanced somewhat due to positive ion trapping on methanol) to that typical for polar liquids. The polar liquid type of processes are dominating if only 10% of the molecules are methanol.

This explanation is well supported by other experimental Ps yield results. The positive ion trapping effect is less pronounced in n-propanol/isooctane mixtures [7.87], and it is absent in neopentane mixtures with methanol, ethanol, and cyclohexanol [7.24] and in n-propanol/tetramethylsilane (TMS) mixtures [7.88]. The reason is probably that positive ion trapping effect is counterbalanced by the decrease of the Ps yield due to a reduction of the spur size and electron mobility on the addition of the alcohols. The strong correlations between the Ps formation and the spur size and electron mobility were discussed in Chap. 5, where it was concluded that an obvious explanation of these correlations has not been found, as yet. However, as the spurs and the electron mobility are very large for isooctane, neopentane, and TMS it is expected that the Ps yield will decrease on addition of polar solutes for those liquids, in particular. In [7.24] it was reported that the Ps started to decrease at the concentration where the alcohol solvated electron absorption started to occur in methanol/neopentane mixtures at increasing methanol concentration. However, in the ethanol and cyclohexanol mixtures with neopentane the Ps yield decrease started at a lower concentration than the solvated electron absorption. The only available temperature effect study of the clustering effect on the Ps yield [7.87] shows that at 248 K the decrease in the Ps yield occurs at a concentration of ethanol which is only 40% of that found at 298 K. This is in agreement with the expectation that the clustering occurs at a much lower concentration at lower temperatures. Figure 7.13 shows that the decrease in the Ps yield sets in at a higher alcohol concentration in the neopentanol/cyclohexane mixture. This is expected, as neopentanol is an almost spherical molecule, for which the clustering is more difficult than for linear alcohols. Maybe a cluster of only four molecules is the largest possible cluster for neopentanol.

Of course, it is expected that the clustering effect on the Ps yield is found if the polar molecules form hydrogen bonds, in particular. The mixtures of NH_3 in cyclohexane in Fig. 7.13 show such a case. The solubility of NH_3 in cyclohexane is limited. The first mixtures found to show the clustering effects were the H_2O/dioxane mixtures [7.89] (explained in [1.35]). This study was followed by the report on mixtures of pyrrolidine in heptane [7.85].

A more comprehensive discussion of the clustering effect of the Ps yield necessitates that an overview of the available knowledge on the clustering of polar molecules in nonpolar liquids is given. Because of lack of time and space this cannot be done here. Apparently, only fairly scattered investigations of these phenomena have been published. For example, NH_3 mixtures with nonpolar liquids seem not to have been studied at concentrations where clustering is important, in radiation chemistry. Some references are found in [1.35,7.24,85].

The semi-quantitative theory given in the preceding section on Ps formation in polar liquids applies probably also in the mixtures where the clustering effects on the Ps formation occur. In particular, similar explanations of the low Ps yields seem to work in the mixtures. This is a very interesting field of research for the future.

Let us return to the discussion of the positive ion scavenging effects on the Ps yield. Also in this case it is obvious that a deeper understanding of the Ps yield data necessitates a more comprehensive treatment of the positive ion reactions, studied mainly in radiation chemistry. A complete research of the literature could not be done because of lack of time, and hence, only some general comments shall be given here.

At first it must be noted that the reactions of the positive ions have been studied less than those of the anions in radiation chemistry. However, interests in the positive ions have risen recently. We again encounter the problem that the cations have been studied mainly at times which are large compared to the lifetime of the positron of roughly 400 ps and the time of the electron reactions in the spur of the nonpolar liquids of about 10 ps in linear hydrocarbons. Cation reactions taking place after the positron has decayed do not, of course, influence the Ps yield. However, they may have been very important for the results on cation chemistry reported by radiation chemists. Hence, an important problem is: Can we reliably extract information on the chemistry of the positive ions in the positron spur from the reported results on the cation reactions at much longer times? Very few picosecond reactions of cations have ever been published compared to what is known about picosecond electron reactions. On the other hand the situation seems to be improving at present. This cannot be said about the available information on the clustering of polar molecules.

We can illustrate the uncertainties in the interpretation of the positive ion scavenging effects by a discussion of particular Ps yield results. Pyridine mixtures with nonpolar liquids were studied in [7.90,84]. In [7.90] it was reported that the Ps yields increased on addition of pyridine at low concentrations in those liquids: cyclohexane, cyclohexane-D12, trans-decalin, and cis-decalin, where the positive ions have a high mobility, i.e., where they are somewhat delocalized. Further, no increase was found in the liquids: CS_2, isooctane, hexane, and benzene, having slowly moving positive ions. This was explained by the effect of the slowing of the fast positive ion on trapping, a very reasonable explanation it seemed. However, it was soon realized that a "classical" spur diffusion theory would not predict these results, as the holes in, e.g., cyclohexane are much too slow compared with the electrons. Hence, the slowing of the holes was expected not to change significantly the Ps yields. This was recently substantiated by the results for pyridine/pentadecane mixtures, where an increase in the Ps yield was found, see Fig. 7.13 [7.84], although no fast hole is expected in pentadecane. It was then decided to study the "classical" hole scavenger, cyclopropane, [4.1] in cyclohexane and trans-decalin. Cyclopropane gave no Ps yield increase at all! Recently obtained new results for other hole scavengers do not clarify the situation [7.84]. A good explanation of the positive ion scavenging effect on the Ps yield must await further investigations in Ps yield as well as radiation chemistry research.

Once more it is important to realize what is actually measured in radiation chemistry. In the case of cyclopropane in cyclohexane a reduction of the mobility of the fast hole to that obtained for normal massive ions was measured at nanosecond times [4.1]. We expected that a hole trapping before the main part of the spur reactions has finished, i.e., before roughly 10 ps, will cause an increase in the measured Ps yield, provided the trapping slows down the electron positive ion recombination reaction. This Ps yield increase was not obtained, although other positive ion scavengers cause the expected effects. This seems to indicate that the hole trapping on cyclopropane either does not influence appreciably the recombination, or that a shortlived, nonrelaxed state of the trapped hole on cyclopropane is formed which does not influence the Ps formation, but which decays to the relaxed hole on cyclopropane before the mobility of the nonrelaxed hole can been measured. This hypothesis can then be tested by, e.g., very fast radiation chemistry, which maybe is too slow, or fast flash photolysis experiments. Further search for such cases would be useful.

7.7 The Older Ps Formation Models

The spur model of Ps formation has been used in all the discussions of the measured Ps yield in condensed matter in this book. Apparently, it is the mostly accepted model today. However, some scientists still use the older models of Ps formation, and hence, it is necessary that these models are shortly described and critically discussed in this section. Of course, we cannot discuss all the arguments against the older models, and for the spur model, here. The good agreement between the spur model predictions and explanations and the experimental facts, described in the other sections and chapters, is, by far, the most important argument for the spur model. We shall mainly discuss selected Ps yield results which cannot be explained by the older models except if very unlikely assumptions are used.

In gases of very low density Ps is formed in single collisions of the positron with one atom or molecule at the time, and Ps will decay intrinsically before it undergoes a collision with other gas atoms or molecules. The Ps formation can, in principle, be calculated by the methods usually used in scattering theory of quantum electrodynamics (QED). However, only the very simple problems, such as positron collisions with small single atoms, can apparently be handled.

At higher densities, i.e., around 0.0001–10 amagat (1 amagat $= 2.704 \times 10^{19}$ cm^{-3}) in polyatomic gases, Ps is still mainly formed in single collisions but it can split up again in collisions with the gas atoms or molecules. These are the densities at which the *Ore model* seems to apply. Energy conservation demands that Ps cannot be formed if the positron energy is below $V - E_b$, where V is the ionization energy of the atoms or molecules and E_b the binding energy of Ps, i.e., 6.8 eV. Furthermore, it is assumed in the Ore model that Ps breaks up during collisions if its kinetic energy after the formation is larger than the ionization energy of the gas. To get a lower estimate of the Ps yield, it is sometimes assumed that Ps breaks up if its kinetic energy is above the lowest electronically excited states of the gas, E^*. This leads to an estimate of the Ps yield, P, that it should be within the interval:

$$E_b/V > P > (E^* - (V - E_b))/E^* \tag{7.41}$$

The "Ore gap" is the positron energy interval where Ps is mainly assumed to be formed. Hence the Ore gap has a lower boundary of $V - E_b$ and an upper boundary of V or E^* depending on the assumptions used. In the Ore gap Ps formation is assumed to compete with other positron processes, such as positron slowing down by collisions and possible positron bound state formation.

At densities above 5–10 amagat in polyatomic gases spur effects start to be significant for Ps formation. This is in agreement with radiation chemistry results. Consequently, the distances between the thermalized positron and one or more of the thermalized electrons, created during the slowing down of the positron, are not large compared to the Onsager radius at the given temperature, as they are at lower densities. Hence, the positrons not forming Ps by the Ore model process have a significant probability of forming Ps by the spur process. In a very simple model [7.91] the Ps yield has been expressed as:

$$P = P_0 + (1 - P_0)\exp(-\tau_{Ps}/\tau_+)(1 - \exp(-r_c/R)). \tag{7.42}$$

P_0 is the fraction of Ps formed by the Ore mechanism, τ_{Ps} the Ps formation time, τ_+ the positron lifetime, r_c the Onsager radius, and R the electron-positron distance at thermalization. The last term in (7.42) gives the probability that Ps is formed by the reaction of the positron and an electron, while $\exp(-\tau_{Ps}/\tau_+)$ is the probability that Ps is formed before the positron decays, and $\exp(-r_c/R)$ is the probability that the positron and electron escape the reaction. τ_{ps} and r_c

are given by (5.8) and (6.19), respectively. Of course, (7.42) is a very rough approximation. However, it illustrates that two main processes determine the Ps formation at those densities in the simplest possible case. At increasing density the Ps yield increases at first because the average distance between the positron and the electrons at thermalization decreases. However, at higher density the mobilities of the particles decrease, and hence, the positron has a greater probability to decay before Ps forms, i.e., the Ps yield saturates, or even decreases somewhat, at the highest densities. The few available experimental results of Ps formation in moderately dense gases of SF_6 (1–22 amagat), CO_2 (1–25 amagat), C_2H_6 (1–60 amagat), and N_2 (1–150 amagat) could be semi-quantitatively fitted by this qualitative model of a mixture of the Ore and spur models [7.91], by use of reasonable values of the parameters. The main problem in this model is that the physics of the processes involved is fairly unknown, and hence, only very uncertain results can be derived.

Jacobsen ([7.91] and unpublished) has estimated the densities at which the spur process of Ps formation becomes significant. As p Ps has a much shorter lifetime than o Ps, the breaking up of p Ps seems not to be fully developed at the density where the spur process of Ps formation becomes significant. Hence, the Ore model might not be the dominant process of p Ps formation at any gas density. Furthermore, at certain densities o Ps, but not p Ps, might break up, and the positron might form Ps again by a spur process. This would change the o Ps to p Ps ratio away form the normally assumed 3 to 1. Apparently, this effect has never been looked for. We shall not in this book discuss the Ps formation in gases in more detail (see, e.g, [1.50]).

Typical liquid densities are: neon, 1375 amagat; H_2O, 1226 amagat; benzene, 252 amagat; and naphthalene, 170 amagat at the melting points. In normal polyatomic liquids the spur model can account for all the Ps formation. In atomic liquids the spur model seems to apply too, although detailed Ps yield results are rare, and very few studies of the effects of addition of solutes have been performed. There are several experimental facts which indicate that all Ps formation is caused by the spur processes in normal polyatomic liquids. For example, the Ps yield can be inhibited totally by addition of many molecules or ions not reacting with Ps to the liquids, e.g., CCl_4 to nonpolar liquids and NO_3^- to water. If some of the Ps formation was caused by the Ore process one would expect that part of the Ps yield could not be scavenged. We shall discuss the possibility of "hot Ps reactions" below.

At the start of Ps chemistry up to about 1965 the Ore model was used to explain the Ps formation in condensed matter. However, it was soon realized that this model could not account for the measured Ps yields [1.31]. Hence, McGervey proposed the *"Ore plus hot Ps reaction"* (OhPr) model, where it is assumed that Ps is formed by the Ore process, but that it can undergo reactions with the molecules of matter before it is thermalized. As the Ps thermalization times were reasonably assumed to be short compared to the o Ps lifetime of several nanoseconds, such reactions would show up as a reduction of the intensity of the long lifetime component of the lifetime spectrum if Ps was destroyed by the hot Ps reaction. The OhPr model was practically the only alternative to the spur model when the latter was proposed in 1973, although other models were used, too (see below).

Let us first explain why the Ore model alone cannot explain the available facts. We shall only refer to selected cases. The total Ps yields for pure liquids (see Tables 5.1–3) vary from zero to almost 80% for liquids with very similar Ore gaps. It is almost impossible to "explain" those yields by use of the Ore model alone. To do this one has to assume very unreasonable changes in the positron slowing down properties from one liquid to another. Furthermore, the changes in the measured Ps yields on the transition between solid and liquid phases are almost impossible to explain, too (see Chap. 9). For example, for polar substances, such as H_2O and alcohols, typical Ps yields, P, are: in crystalline solids 60–72%, in amorphous solids about 45%, and in the liquid phases 26–36%. As there is less "free volume" in the crystalline solid (except for ice) compared to the other phases, one would expect the Ore gap, corrected for the negative Ps affinity of the phases, to be narrowest, and, therefore, that the Ps yield would be lowest, in this phase, in clear disagreement with experiments. Many other Ps yields speaking against the Ore model alone could be mentioned. Consequently, the Ore model by itself is invalid. Chemists interested in Ps formation have accepted this fact many years ago. However, physicists, in particular "slow positron researchers", occasionally use incorrectly the Ore model alone to explain the Ps yield in the bulk of solids (see Chap. 9).

It is less obvious that the OhPr model is also invalid. Actually, if a fairly limited field of Ps formation research is considered it might even be somewhat difficult to show that this model is less applicable than the spur model. The reason is mainly that the hot Ps reactions are totally unknown experimentally, and hence, many processes can be assumed to take place in the theoretical "explanations". However, if all the available Ps formation results are considered the OhPr model is clearly invalid. This is generally accepted today, mainly because the spur model is in very good agreement with experiments, and not so much because the OhPr model can be easily shown to be incorrect.

Let us mention some of the measured Ps yields which cannot be reasonable well explained in terms of the OhPr model. The changes in the Ps yield on transitions between phases constitute such cases. They were shown above to be inexplicable in terms of only the Ore processes. However, it is also very difficult indeed to argue for some hot Ps reactions which could explain the results. Hot Ps reactions with polar molecules, such as H_2O and alcohols are unlikely, but it is still more unlikely that the rate of such reactions should vary so that the changes in the measured Ps yield at the phase transitions could be explained. Experimental results concerning hot Ps reactions are totally unavailable, but even within the "freedom of speculation" which this fact allows, very few have tried, as yet, to apply the OhPr model to explain the Ps yield changes at phase transitions. Other Ps yield results which are almost inexplicable in terms of the OhPr model are the antirecombination and antiinhibition effects discussed in details in Sect. 7.4. It is simply very difficult to imagine an explanation of the strong increase in the Ps yield on addition of a very low concentration of the antiinhibitor in terms of the OhPr model. In [7.92] this was not attempted, but another explanation was proposed instead. However, this other explanation has not been used further. The specific effects of clustering and positive ion scavenging, discussed in Sect. 7.6, cannot be reasonably explained by the OhPr model either. Important attempts to avoid the use of the spur model can be found in [7.75,92,93]. The spur model explains well the effects which cannot be explained in terms of the OhPr model. Some of these effects serve also as strong tests of the validity of the spur model. If these effects were not found the spur model would be in trouble. In Chap. 9 we shall discuss other unsuccessful attacks on the spur model based on the interpretation of slow positron results for the surface of ice and the measured electric field effects on the Ps yields in polymers.

We can conclude that the main number of Ps atoms are formed by the spur process. However, can we exclude that a small percentage, e.g., 1–3%, are formed by the Ore process? The answer is that this cannot be totally excluded, but that the hot Ps reactions must be included and assumed to have very exceptional properties in order that the measured facts could be explained. As discussed above, the main point is that the Ps yield can be inhibited to zero by many solutes, and that this inhibition occurs smoothly as a function of solute concentration. There are no properties of the measured Ps inhibition curves, such as humps, etc., which indicate a possible change from mainly a spur model inhibition process to a hot Ps reaction inhibition process. It is very unlikely that the Ps inhibition curves would not show humps for all the reported inhibition cases, if the inhibition was caused by two different processes. Of course, this is only correct in so far as the hot Ps reaction and the scavenging process in the spur can be separated. In few cases of spur reactions the electron positron state is expected to be partly a Ps state and partly "free" particles in the positron spur, and such a separation might be questionable. But apart from such rare cases, i.e., apart from the uncertainties in the definition of a Ps state, the experimental facts seem to indicate that even a low percentage of Ps formation by the Ore process is fairly unlikely.

It is important to realize that the physics of the positron, electron and Ps processes at the energies typical for the spur reactions is understood only in a very qualitative way. Hence, it is almost impossible to argue in favour of the spur model, or any of the other models, in terms of the physical processes alone. Of course, experimental facts, such as spur properties measured in radiation chemistry, can be used in the arguments.

8 Positronium Reactions in Solutions

The reactions of Ps in solutions, also called Ps quenching, have been a very important topic of the chemical physics of positron annihilation. In particular, Ps reactions were much studied up to 1975 and after 1986, while Ps formation was an important topic in the interval 1975–1986. The rate constants of the Ps reactions, k, are approximately determined from the measured o Ps lifetimes, τ_3, using the equation

$$\lambda_3 = \tau_3^{-1} = \lambda_p + kc, \tag{8.1}$$

where $\lambda_p = \tau_p^{-1}$ the pick off decay rate and c the concentration of the reactant reacting with Ps. Many Ps reactions are diffusion controlled, and hence, they belong to the class of fast reactions which can normally be studied directly only in radiation and photo-chemistry. For example, positronium seems to react with MnO_4^- and Mn^{++} in water by oxidation and spin conversion, respectively, and with nitrobenzene in benzene.

At first we shall discuss the kinetic equations describing the Ps reactions in general. Solutions of the equations will be given in typical cases, and the approximations used in setting up the equations will be evaluated. This is followed by a discussion of the Ps reactions in selected cases. A treatment of all published papers on Ps reactions is certainly too large a topic for this book.

8.1 Kinetic Equations of Positronium Reactions

In this section we shall formulate the kinetic equations describing the reactions of Ps which take place after its formation in the thermalized or very nearly thermalized state and before its annihilation. We refer to the liquid state although the equations are valid in gases and solids, too. A correct and complete solution of these equations for reasonably general cases is not available in the open literature (see e.g. [8.1,7.58]). We shall first give the equations and the solutions and then discuss in some details the validity of the equations.

We denote the probabilities of observing the positrons as p Ps, o Ps, free positrons, or Ps in a bound state at a given time, t, by P_s, P_t, P_f, and P_m. The kinetic equations are then:

$$dP_s/dt = -(3\gamma + \lambda_s + \lambda_p + \lambda_0 + \lambda_c)P_s + \gamma P_t, \tag{8.2}$$

$$dP_t/dt = -(\gamma + \lambda_t + \lambda_p + \lambda_0 + \lambda_c)P_t + 3\gamma P_s, \tag{8.3}$$

$$dP_f/dt = -\lambda_f P_f + \lambda_0(P_s + P_t), \tag{8.4}$$

$$dP_m/dt = -\lambda_m P_m + \lambda_c(P_s + P_t), \tag{8.5}$$

Here $\lambda_s = \tau_s^{-1} = 7.98 \times 10^9\ \mathrm{s}^{-1}$ is the p Ps 2γ-annihilation rate, and $\lambda_t = \tau_t^{-1} = 7.04 \times 10^6\ \mathrm{s}^{-1}$ is the o Ps 3γ-annihilation rate. The spin conversion rate can normally be written as $\gamma = k_{sc}c_{sc}$, where k_{sc} is the rate constant and c_{sc} is the concentration of the converter. Similarly, λ_0 and λ_c can normally be written as $\lambda_0 = k_0c_0$ and $\lambda_c = k_cc_c$, respectively, where $k_0(c_0)$ and $k_c(c_c)$ are the corresponding rate constants (concentrations), respectively. $\lambda_p = \tau_p^{-1}$ is the pick off rate of the solution. λ_f and λ_m are the 2γ-annihilation rates of the free positron and the Ps bound states, respectively.

We wish to find the solution of these equations assuming that the initial probabilities: $P_s(0)$, $P_t(0)$, $P_f(0)$, and $P_m(0)$ are known. This task can easily be performed by use of standard calculation techniques of mathematics. To avoid too complicated expressions we have not presented the solutions in terms of only the primary constants of equations (8.2–8.5), i.e., we use secondary constants: A to D, A' to D', E, F, G, H, N, λ_1, λ_2, λ_s', and λ_t'. These probabilities versus the time, t, are given by:

$$P_s = A\exp(-\lambda_1 t) + B\exp(-\lambda_2 t), \tag{8.6}$$

$$P_t = C\exp(-\lambda_1 t) + D\exp(-\lambda_2 t), \tag{8.7}$$

$$\begin{aligned} P_f = {} & (\lambda_0(A + C)/(\lambda_f - \lambda_1))(\exp(-\lambda_1 t) - \exp(-\lambda_f t)) \\ & + (\lambda_0(B + D)/(\lambda_f - \lambda_2))(\exp(-\lambda_2 t) - \exp(-\lambda_f t)) \\ & + P_f(0)\exp(-\lambda_f t), \end{aligned} \tag{8.8}$$

$$\begin{aligned} P_m = {} & (\lambda_c(A + C)/(\lambda_m - \lambda_1))(\exp(-\lambda_1 t) - \exp(-\lambda_m t)) \\ & + (\lambda_c(B + D)/(\lambda_m - \lambda_2))(\exp(-\lambda_2 t) - \exp(-\lambda_m t)) \\ & + P_m(0)\exp(-\lambda_m t), \end{aligned} \tag{8.9}$$

$$A = C'(-P_s(0)B' + P_t(0)D')/N, \tag{8.10}$$

$$B = D'(P_s(0)A' - P_t(0)C')/N, \tag{8.11}$$

$$C = A'(-P_s(0)B' + P_t(0)D')/N, \tag{8.12}$$

$$D = B'(P_s(0)A' - P_t(0)C')/N, \tag{8.13}$$

$$A' = \lambda_1 - \lambda_s', \quad B' = \lambda_2 - \lambda_s', \quad C' = \lambda_t' - \lambda_1,$$

$$D' = \lambda_t' - \lambda_2, \text{ and } N = (\lambda_1 - \lambda_2)/(\lambda_t - \lambda_s).$$

Furthermore,

$$\lambda_s' = \lambda_s + \lambda_p + \lambda_0 + \lambda_c,$$

$$\lambda_t' = \lambda_t + \lambda_p + \lambda_0 + \lambda_c,$$

$$\begin{aligned} \lambda_{1,2} = {} & 2\gamma + \lambda_t/2 + \lambda_s/2 + \lambda_0 + \lambda_p + \lambda_c \\ & \pm (4\gamma^2 + \lambda_s\gamma + (\lambda_s - \lambda_t)^2/4 - \lambda_t\gamma)^{1/2} \end{aligned} \tag{8.14}$$

A lifetime experiment determines in a good approximation the 2γ-annihilation rate:

$$L(t) = -\frac{d}{dt}(P_s + P_t + P_f + P_m), \tag{8.15}$$

and ignoring λ_t, which will not contribute to the 2γ-decay:

$$L(t) = \lambda_s P_s + \lambda_f P_f + \lambda_m P_m + \lambda_p(P_s + P_t). \tag{8.16}$$

This gives:

$$L(t) = I_s\lambda_1 \exp(-\lambda_1 t) + I_t\lambda_2 \exp(-\lambda_2 t) + I_f\lambda_f \exp(-\lambda_f t) + I_m\lambda_m \exp(-\lambda_m t), \tag{8.17}$$

$$I_s = A(\lambda_s + \lambda_p)/\lambda_1 + C\lambda_p/\lambda_1 + E\lambda_f/\lambda_1 + F\lambda_m/\lambda_1 \tag{8.18}$$

$$I_t = B(\lambda_s + \lambda_p)/\lambda_2 + D\lambda_p/\lambda_2 + G\lambda_f/\lambda_2 + H\lambda_m/\lambda_2 \tag{8.19}$$

$$I_f = P_f(0) - E - G, \tag{8.20}$$

$$I_m = P_m(0) - F - H, \tag{8.21}$$

$$E = (A + C)\lambda_0/(\lambda_f - \lambda_1); \quad F = (A + C)\lambda_c/(\lambda_m - \lambda_1);$$

$$G = (B + D)\lambda_0/(\lambda_f - \lambda_2); \quad H = (B + D)\lambda_c/(\lambda_m - \lambda_2).$$

The o Ps 3γ-annihilation is given by $M = \lambda_t P_t$. Hence,

$$M = (C\lambda_t/\lambda_1)\lambda_1 \exp(-\lambda_1 t) + (D\lambda_t/\lambda_2)\lambda_2 \exp(-\lambda_2 t) \tag{8.22}$$

The total intensity of o Ps 3γ-annihilation is

$$M_{tot} = C\lambda_t/\lambda_1 + D\lambda_t/\lambda_2. \tag{8.23}$$

Similarly we get that the intensity of the narrow component in the angular correlation spectrum due to p Ps intrinsic 2γ-annihilation is:

$$I_n = A\lambda_s/\lambda_1 + B\lambda_s/\lambda_2, \tag{8.24}$$

and the total pick off component from o Ps and p Ps decay:

$$I_{po} = (A + C)\lambda_p/\lambda_1 + (B + D)\lambda_p/\lambda_2. \tag{8.25}$$

Furthermore, the AC component due to free positron annihilation has an intensity:

$$I_{fr} = P_f(0) + E\lambda_f/\lambda_1 + G\lambda_f/\lambda_2 - E - G, \tag{8.26}$$

while that of the Ps bound state is:

$$I_{bs} = P_m(0) + F\lambda_m/\lambda_1 + H\lambda_m/\lambda_2 - F - H. \tag{8.27}$$

The solution of the equations is normalized to give:

$$I_s + I_t + I_f + I_m + M_{tot} = 1.0, \tag{8.28}$$

$$I_n + I_{po} + I_{fr} + I_{bs} + M_{tot} = 1.0.$$

It must be realized that this normalization is different from those of, e.g., the POSITRONFIT and PAACFIT computer programs, for which the sum of the intensities of the exponentials (gaussians) in the lifetime (angular correlation) spectra is normalized to 1.0. For example, for benzene, $I_s + I_t + I_f + I_m$ is equal to 0.99 in the given solution, while $I'_s + I'_t + I'_f + I'_m = 1.0$ in a POSITRONFIT output.

It is strongly recommended that a computer program is made which calculates the measurable quantities given by the expressions above. The input parameters in a calculation are the rates, rate constants, and initial intensities. The Ps intensities can be given by the α and β parameters of (7.30). In Sect. 8.2 a standard case of fitting the measured numbers by use of some of the input parameters as fitting constants is discussed.

If there is no spin conversion, i.e., for $\gamma = 0$, (8.14) gives the o Ps decay rate: $\lambda_2 = \lambda_t + \lambda_p + \lambda_0 + \lambda_c$, which for λ_0 or λ_c equal to 0 and by disregard of λ_t reduces to (8.1). For $\lambda_t = \lambda_0 = \lambda_c = 0$, $\lambda_2 = \lambda_p + \gamma$ in first nonvanishing order in γ/λ_s only, and a more complex expression applies for larger γ.

The peculiar behavior of the intensity of the long lifetime is best illustrated in the case where $\gamma = \lambda_0 = 0$ and by disregard of λ_t and the contribution to I_m from p Ps reactions. Hence, we get $\lambda_2 = \lambda_p + \lambda_c$ and:

$$I_t = P_t(0)(\lambda_m - \lambda_p)/(\lambda_m - \lambda_p - \lambda_c), \tag{8.29}$$

$$I_m = P_m(0) + P_t(0)(-\lambda_c)/(\lambda_m - \lambda_p - \lambda_c). \tag{8.30}$$

Typically, $\lambda_m = 2.5\times10^9\ s^{-1}$, $\lambda_p = (2-5)\times10^8\ s^{-1}$, and $k_c = 3\times10^{10}\ M^{-1}s^{-1}$ for strong reactions, and hence λ_c can easily become larger than $\lambda_m - \lambda_p$. Clearly, $I_t \to \infty$ for $\lambda_c \to \lambda_m - \lambda_p$. For $\lambda_c > \lambda_m - \lambda_p$, $I_t < 0$. The very large positive and negative values of I_t are counterbalanced by an opposite variation in I_m. We shall discuss typical Ps quenching cases in the other sections of this chapter.

Equations (8.29,30) describe also the kinetics of Ps trapping in vacancy-type defects in insulating solids (Chap. 9), where, however, $\lambda_m < \lambda_p$, $I_t > 0$, and the strong variation in I_t is absent.

Let us now discuss the validity of (8.2–5) and the given solutions. The 3γ-annihilation of the free positron, Ps during pick-off, and the Ps bound state on molecules have been disregarded. This is a fairly good approximation if only the 2γ-annihilations are measured, as the 2γ-rate is expected to be 371 times larger than the 3γ-rate for these states. However, if the 3γ-annihilations are measured the three contributions cannot be neglected! The positron reactions were discussed in details in Chap. 6, and they can easily be included in the kinetic equations if necessary. The equations (8.2–5) can also be modified to account for, e.g., bubble collapse effects [8.1], possible Ps transfer reactions, and/or Ps attachment/detachment processes.

A more fundamental question is: Why are the expressions (8.2–5) valid in general? Parts of the treatment in Sect. 6.2 can be used to answer this question. At first, we note that the Ps states must be an incoherent superposition of the o Ps and p Ps states, i.e., the Ps state must be either a p Ps or a linear superposition of the o Ps states. Hence, (8.2,5) cannot be used if a magnetic field is present. It is normally assumed that the Ps states are formed with equal probability and random phase in the spin space, and experiments seem to show that this assumption is valid. In other words, "rapid phase randomization" occurs among the spin states of the electron positron pair during or shortly after Ps is formed. As discussed in Chap. 2, the reason for the validity of this seems to be unknown. If a magnetic field is present, equal probability and random phase for the energy eigen states in the spin space are used, and are in agreement with experiments.

If the above discussed assumptions are correct or good approximations, (8.2–5) are valid. They are the type of kinetic equations used normally to describe, e.g., electron trapping in defects, etc. in solids, and in reaction dynamics in general. Our discussion in Sect. 6.2 is then applicable. The rate constants: λ_s, λ_t, λ_f, and λ_m account for first order reactions, namely the annihilations of the four states. They are justified by QED calculations, as described in Chap. 2. The other terms describe bimolecular reactions in the pseudo first order approximation. As the concentration of the solutes reacting with Ps is very nearly constant during a typical experiment, this is a very good approximation.

The use of a rate constant in the bimolecular terms must be justified, too. We discussed this in some details in Sect. 6.2, where it was realized that the "classical reaction kinetics" theories were the best theories to apply in this justification, even if the light reactants were somewhat delocalized, which could be the case for the positron in some systems. The Ps atom is neutral, and hence, a charge on a reactant will only influence the rates by the rather small polarization effects. However, it is necessary to discuss the effects of nonhomogeneity and high reactant concentrations on the Ps reactions.

It is normal practice to look away from nonhomogeneity in the interpretation of Ps reactions. Because of the presence of the positron spur around or close to the formed Ps this is only approximately correct. It is appropriate to discuss this problem in some detail here.

As discussed in Chap. 5, we have good reasons to believe that o Ps participates in secondary reactions with the other species in the positron spur. It was experimental results which forced us to accept this, namely an extra o Ps lifetime component of (0.8–1.8) ns ((2–8)%) in high quality LT spectra, and a too large narrow peak in the AC spectra compared to the o Ps lifetime intensity. The interpretation is that Ps reacts with the positive ion and the other spur reactants, such as excess electrons. This results in many comparatively shortlived o Ps lifetime components, which are approximated by one, shorter than normal, o Ps component in the POSITRONFIT analysis. Furthermore, the narrow peak in the AC spectrum is enhanced because p Ps is shortlived, and hence, it has little chance to react with the spur reactants, and also because the product of an o Ps reaction with electrons contributes to the narrow peak in the AC spectra. In principle, the extra o Ps components could be taken into account in the kinetic equations by introduction of an extra equation for a new o Ps component and by giving it the measured average lifetime and intensity by choice of rate constants and initial intensity. We have refrained from that here as we know much too little about these o Ps processes. No experiments sensitive enough to see directly the reactions and formation probabilities of those shortlived Ps states have been made. We have only measured weak, but significant, changes in the middle lifetimes of roughly (390–500) ps in three components fits of standard quality LT spectra, which seem to show directly the inhibition and quenching of the "intermediate" o Ps states, as discussed in Chap. 5. New experiments are urgently needed! At present, the best procedure is to base the discussion of the Ps reactions on the three term fits of the LT spectra, and assume that the shortlived o Ps components, because of their relatively low total intensity, have caused only small errors in the analyses. Consequently, we look away from these states in the kinetic equations.

Strictly speaking, the spur species will compete with Ps in reactions with the added solutes in some cases. For example, many solutes reacting with Ps react also with electrons. However, this normally leads to Ps inhibition, which is taking into account in the specification of the initial probabilities of the Ps and positron states in the kinetic equations.

As even the o Ps lifetime is fairly short, high concentrations of the Ps quencher must be used. The reduction of the o Ps lifetime starts to be significant at quencher concentrations of 0.001 M, and all o Ps have been quenched at about 0.5 M, for an o Ps lifetime in the pure solvent of 2–4 ns and diffusion controlled rate constants of normally $(2\text{–}3) \times 10^{10}\ \mathrm{M}^{-1}\mathrm{s}^{-1}$. Hence, the high concentration problem in Ps reactions must be discussed. However, somewhat unexpectedly (8.1) seems to describe the measured o Ps lifetimes fairly well in the Ps quenching studies reported until now. Exceptions are the Ps reactions where the quencher changes structure as the concentration varies, such as the changes due to complex formation for the Cu^{--} ions in Fig. 7.7. Nevertheless, a discussion of the Ps reactions in general and, in particular, the high quencher concentration limit is an important contribution to the understanding of the physics of the reactions.

In Chap. 6 the high concentration corrections to the "classical reaction kinetics" theories in the normally used approximations were discussed. The formation of initial encounter pairs could be described by (6.27), while reactions close to but outside the reaction spheres were taken into account by the time dependent term in the expressions for the rate constants, such as (6.13,15,21). However, Ps reactions are complicated by the presence of the Ps bubble, which was discussed in Chap. 4. As an example, let us discuss the Ps reactions in a solution of a strong Ps quencher and a nonpolar liquid, such as a nitrobenzene/benzene solution.

As mentioned in Chap. 7, Ps quenchers nearly always inhibit Ps formation by reacting with the spur electrons. At 1 M concentration the average distance between the inhibitor molecules is roughly 1.2 nm. Hence, during the Ps formation, and the probably simultaneous Ps bubble creation, the electron, positron, and Ps atom will be very close (~ 0.6 nm) to the nearest Ps inhibitors. We can expect the bubble to form within about 1–10 ps, while a typical lifetime against Ps reaction, $t = (kc)^{-1}$, is roughly 50 ps at 1 M and for $k = 2 \times 10^{10}\ \mathrm{M}^{-1}\mathrm{s}^{-1}$. This is short compared to the annihilation lifetime of the product of Ps reactions (~ 400 ps). A bubble radius of 0.4–0.6 nm seems to be typical for standard nonpolar liquids. Consequently, at 1 M concentration the quencher molecules will normally be situated in the first layer of molecules on the bubble surface, provided the quenchers are not attracted to or repelled from the bubble surface. For example, C_6F_{14} might be repelled by the Ps bubble surface in hexane (Sect. 7.4). As discussed in Chap. 5, the bubble surface is probably illdefined, and its size is strongly varying due to the thermal motion of the molecules. Furthermore, Ps, being a light particle, may not be completely confined in the bubble but penetrates somewhat the first few layers. In some solids and liquids a larger part of the Ps wave function might be outside the cage/bubble. In discussions of reactions between two massive reactants an "encounter" of the two reactants in a solvent "cage" is often referred to [6.12]. It is important to realize the corresponding "encounter" of Ps and a reactant is the state described above, i.e., a Ps bubble with a reactant on the illdefined bubble surface.

At low quencher concentrations Ps and the quencher may form such an "encounter" during their random diffusion in the solvent. This is the diffusion controlled step of the reaction. At high concentrations Ps will normally be in an "encounter" with one or more reactants during and after its formation. In the "encounter" Ps and the reactant will undergo repeated "collisions". Actually, as Ps is a light particle the "collisions" cannot be reasonably well described except by quantum mechanical calculations. It might be more useful to refer to concepts, such as wave function overlap, tunnelling reaction into states on the quencher, "wiggles" in the Ps wave function in the potential of the quencher, transitions to lower states, etc. For example, these concepts are used in treatments of the trapping of electrons in defects in solids, see, e.g., [6.22]. The theory of Ps trapping is even less developed than that of the electron. During "collisions" a Ps reaction might take place, the so called kinetic step of the reaction. If the kinetic step is slow the rate constant is given by (6.15–17). The quencher and the Ps might also separate, and a spin conversion reaction might have occurred if the quencher is a free radical. In practice, the measured Ps rate constants are nearly always interpreted in terms of "classical reaction kinetics", which was discussed in Sect. 6.2. For the Ps nitrobenzene reaction in benzene $k = 2.3 \times 10^{10}\ \mathrm{M^{-1}s^{-1}}$ is measured. As this value is amongst the largest ones obtained for benzene as a solvent we can assume that the reaction is approximately diffusion controlled. On assuming that the sum of the diffusion constants is twice the self-diffusion constant for benzene ($D = 2.2\ \mathrm{cm^{-2}s^{-1}}$) we obtain a reaction radius of about 0.7 nm from (6.14), which is a reasonable value as the Ps bubble radius is 0.4–0.6 nm. The time dependent term in the expression for the rate constant, $R/(\pi D't)^{1/2}$ (6.13), is 0.19 and 0.11 at 1 ns and 3 ns, respectively. Hence, the correction due to this term is much less important than in solvated positron reactions, as the o Ps lifetime is about 8 times longer than that of the positron, while the diffusion constants are similar. We can thus disregard the time dependence of the rate constants in (8.2–5). Further, it seems in agreement with experiments to assume that Ps reacts very fast with nitrobenzene in benzene when nitrobenzene is included in the first layer of the Ps bubble surface. In general this seems to be true for most rapid Ps reactions, as the cases discussed below will show.

In papers reporting measured rate constants of Ps reactions many different problems in the interpretation of the results have been discussed. The relatively simple ones include, for example, how to separate the diffusion and kinetic steps of the reaction, and what is the expected correlation between the rate constant and solvent viscosity, self diffusion constants, dielectricity constant, etc.? Of course, the temperature dependence of the rate constants have been interpreted in terms of the Arrhenius equation:

$$k = A \exp(-E_a/kT), \tag{8.31}$$

where A is the preexponential factor and E_a the activation energy. In certain cases this is a very bad approximation, indeed. Reactions with aromatic molecules have been explained by reference to the properties of the molecules, such as the electron with-drawing effect of its substituents and the content of carbonium ions in its electron states. Empirical correlations of the Hammett equation type have been invoked too. We shall see below that the rates of Ps reactions with ,e.g., diamagnetic molecules and ions correlate mainly with their electron accepting properties and the solvent surface tension. Correlation to the "special" redox potentials might be invoked in aqueous solutions. Furthermore, complex formation of the reactant with solvent and solute molecules and/or ions influences normally the Ps reactions very much.

In summary, a large part of the theories and explanations used normally in chemical reaction dynamics – the heart of chemistry – is applied in Ps reaction dynamics, too. This constitutes a very large field of chemistry, and it cannot be thoroughly and completely reviewed by only one scientist. A similar statement was written concerning the large field of science used in the interpretation of Ps inhibition, and it can be repeated in Chap. 9, where the Ps and positron properties (e.g., trapping in defects) in solids, such as molecular crystals, shall be discussed. In the next three sections selected cases of Ps reactions will be discussed to the best of my ability.

8.2 An Example of Ps Quenching and Inhibition

The Ps quenching and inhibition by nitrobenzene in benzene can be used as a good example to illustrate the practical problems encountered in the Ps chemistry studies. Actually, it was the only really suitable case I could find, where both the Ps quenching and inhibition are strong and for which angular correlation spectra are available. Although nearly one thousand AC spectra for liquids have been measured at Risoe since 1970, very few strong Ps quenching cases have been studied. Furthermore, very few AC spectra for liquids have been reported by other groups.

The lifetime spectra were measured in 1974. Leading edge discriminators were used, and the resolution was about fwhm = 400 ps. The quality of the measurements was poorer than what can be obtained today. The angular correlation spectra were of normal quality. The resolution was roughly fwhm = 1.2 mrad.

The spectra were analyzed in the following way. The LT spectra were fitted by three lifetime components and by use of a fixed free positron lifetime, $\tau_2 =$ 400 ps. We found $\tau_2 = 430-490$ ps in free three term fits at low nitrobenzene concentrations, while 1 M nitrobenzene/benzene gave $\tau_2 = 390\pm10$ ps and 1 M CCl_4/benzene gave $\tau_2 = 400\pm10$ ps. The most important component in the pure CCl_4 LT spectrum is a 458 ± 5 ps term. The higher τ_2 value at low nitrobenzene concentrations is probably caused by the distribution of shorter o Ps lifetime components of fairly low intensities in the spectra, as discussed in Chap. 5. It is impossible to get reliable intensities of the long lifetime component for $\tau_3 <$ 1.2 ns, if the middle lifetime, τ_2, is not fixed. The choice of the fixed τ_2 value constitutes a main difficulty in the analyses of the LT spectra in Ps quenching cases. As discussed in Chap. 5, the free positron lifetimes are close to 400 ps in most "normal" liquids. CCl_4 seems to be a "special" liquid, probably because the spur electrons are rapidly converted into Cl^- which trap the positrons resulting in a lifetime different from that of the positron bound in CCl_4. The determined lifetimes and preexponential factors are shown in Fig. 8.1.

The angular correlation spectra were fitted by a sum of the three benzene components due to p Ps intrinsic, o Ps and p Ps pick off, and free positron 2γ-annihilation, the fitting parameters being the intensities of the components. The shapes of these components were given in Chap. 5, where the assumptions used in their determinations are discussed too. The most questionable

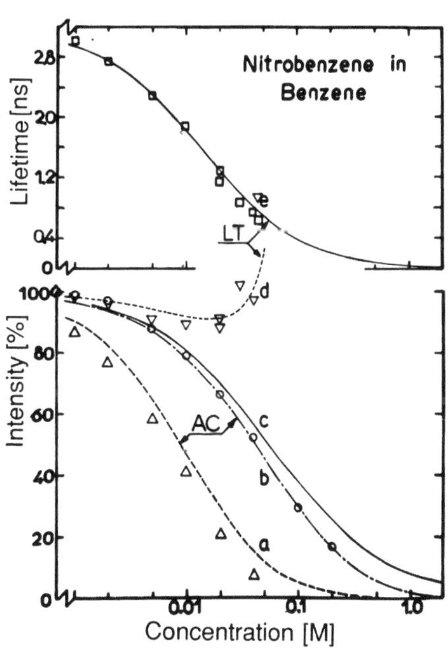

Fig. 8.1. Positronium quenching and inhibition by nitrobenzene in benzene. The marks show: (□) o Ps lifetime; (▽) preexponential factor, I_3'; (O) p Ps intrinsic AC component (narrow); and (△) pick off AC component. The intensities are normalized to 100% in pure benzene. Curves *a*, *b*, *d*, and *e* are the corresponding calculated curves, while curve *c* shows the Ps inhibition alone

approximation was probably that the free positron annihilation in benzene was assumed to give the same shape as that of the 1 M nitrobenzene/benzene solution. It cannot be excluded that the positron annihilates in some bound state with nitrobenzene, and/or the nitrobenzene anions formed by spur electron reactions, in that solution. However, the used shapes are the best ones available at present.

The fitting of the results was done as follows: First, the rate constant of the Ps nitrobenzene reaction was determined to be $k = 2.3 \times 10^{10}$ $M^{-1}s^{-1}$ by fitting of (8.1) to the measured lifetimes by use of $\lambda_p = 0.32 \times 10^9$ s^{-1}. In particular, a good fit at the lower concentrations was preferred, as shown in Fig. 8.1, curve e. Second, the Ps inhibition parameters (7.31) were determined to be $\alpha = 18.0$ M^{-1} and $\beta = 0.85$ by fitting (8.24), with $P_s(0)$ and $P_t(0)$ given by (7.30) and $P_s(0)/P_t(0) = 1/3$, to the measured p Ps intrinsic annihilation intensities of the AC spectra for $k = 2.3 \times 10^{10}$ $M^{-1}s^{-1}$. We assumed that the product of the Ps nitrobenzene reaction had $\lambda_m = 2.5 \times 10^9$ s^{-1}. Furthermore, the relative intensities of the p Ps, o Ps, and free positron states in pure benzene were taken to be 14.5%, 43.6%, and 40.9%, respectively, as the pure benzene gives $\tau_3 = 3.15$ ns and $I_3 = 43\%$, and 1.0% of the positrons decay by 3γ-annihilation of o Ps. A "trial and error" fitting method by use of a computer program was applied. The fitting curve (*b*) is shown in Fig. 8.1, where curve *c* shows the decrease in intensity caused by only the Ps inhibition. To test the results of the fitting and the used assumptions the preexponential factor, I_3', and the pick off AC components were calculated by use of the determined k, α, and β values. Small systematic differences between the measured and calculated numbers show that one or more of the used assumptions are somewhat incorrect. Intuitively, I feel that the shapes of the pick off and free positron AC components are somewhat wrong, for the reasons discussed above. This is, of course, a very preliminary interpretation. The example discussed here, and many other reported cases, show that the kinetic equations (8.2–5) and the inhibition expression (7.30) seem to describe the Ps quenching and inhibition fairly well. It is important to note that we cannot decide whether the Ps reaction with nitrobenzene results in a bound state or an oxidation of Ps by use of the analyses described here.

The used method of determining the parameters seems to be the best one, if both LT and AC spectra have been measured for solutions where both Ps quenching and inhibition are strong. However, in most works only the LT spectra have been used for the determination. The Strasbourg group have used extensively results of Doppler broadening measurements to test the parameters obtained by fitting the LT spectra for polar solutions. In the future, a very good investigation of the Ps quenching and inhibition should include LT measurements of high quality (e.g., fwhm $\simeq$ 200 ps, low background, and good statistics) and reasonably good AC measurements. Four term fits of the LT spectra should be used to get information on the quenching and inhibition of the shorter o Ps lifetime components, too. Uncertainties in the shapes of the AC components of the pure solvent could be reduced, if AC spectra for many

Ps quenchers and inhibitors in that solvent were included in data used in the determination of the shapes. In a similar way the decay rates of the free and bound state components could be determined with a smaller uncertainty than in the cases studied until now. Such high quality studies have not been reported yet.

8.3 Positronium Quenching by Mainly Organic Molecules in Solutions

We shall now discuss selected results of studies of o Ps reactions with mainly organic molecules. The criterion for the selection is that some of the most important features of the Ps chemical reactions should be reasonably well illustrated by the discussion. The characteristic Ps reactions with diamagnetic molecules in nonpolar liquids will be included. This has been the mostly investigated topic in Ps chemistry in organic liquids.

The reactions of Ps are mainly caused by three of its properties. First, Ps is a free radical, and hence, it can react by spin conversion, and, in principle, it can form covalent "bonds" with other free radicals. Secondly, the ionization energy of Ps is only 6.8 eV, which is low compared to that of most molecules. Consequently, it is an electron donor reacting strongly with electron acceptors. Thirdly, Ps is repelled by molecules which are not electron acceptors. This property is the cause of: a) the Ps reaction with, i.e., trapping at, hole type defects, e.g., vacancies, in certain molecular crystals, b) the Ps bubble formation in liquids, and c) the self-trapping of Ps in soft molecular solids.

Let us begin with a discussion of the reactions of Ps with electron acceptors, which we, at first, assume are diamagnetic molecules. The electron affinity, EA, is the amount of energy (in eV) which is evolved when a molecule or atom adds one electron in vacuum. An electron acceptor can be defined as a molecule or atom with a positive electron affinity. In Table 8.1 the electron affinities of 33 selected molecules are shown. Until quite recently there could be significant disagreement between electron affinities determined by use of different methods [8.4,5]. Apparently, the values in Table 8.1 are the best available, at present Electron affinities of molecules are of interest in many areas of chemistry [8.5].

Next, let us discuss shortly charge transfer (CT) complexes. In its own strange way Ps seems to form "CT-complexes" with electron acceptors, as we shall discuss below. A CT-complex is a complex between an electron acceptor (A) and an electron donor (D). An electron donor is a molecule or atom with a low ionization potential, such as Ps. The binding energy of a CT-complex between ordinary, heavy molecules arises from a partial transfer of charge from the highest occupied molecular orbital (HOMO) of the donor to the lowest empty molecular orbital (LEMO) of the acceptor. The binding molecular orbitals (MO) of the CT-complex are often approximated by linear combinations of the unperturbed MO of the acceptor and the donor extended to include the MO of the negatively charged acceptor and the positively charged donor.

Table 8.1. Electron affinities of selected molecules. References: benzene and the chlorobenzenes [8.2], nitrogenoxide [8.3], aromatic molecules [8.4], and all other molecules [8.5]

Molecule	EA [eV]	Molecule	EA [eV]
benzene	−1.1	fluorobenzene	−0.89
chlorobenzene	−0.75	1,4-dichlorobenzene	−0.28
1,3,5-trichlorobenzene	0.15	nitrogenoxide	0.1
naphthalene	0.15	phenanthrene	0.31
oxygen (O_2)	0.5	nitromethane	0.48
carbondisulphide	0.51	hexafluorobenzene	0.52
anthracene	0.56	2.6-dimethylnitrobenzene	0.81
p-nitroanisole	0.91	o-nitrotoluene	0.92
p-nitrotoluene	0.95	m-nitroaniline	0.95
m-nitrotoluene	0.99	nitrobenzene	1.01
sulphurhexafluoride	1.05	o-nitrobiphenyl	1.07
sulphurdioxide	1.1	p-nitrobiphenyl	1.2
p-chloronitrobenzene	1.26	9,10-anthraquinone	1.59
o-dinitrobenzene	1.65	m-dinitrobenzene	1.65
p-nitrobenzonitrile	1.72	p-benzoquinone	1.91
p-dinitrobenzene	2.0	chloranil	2.78
tetracyanoethylene	3.17		

The donors and acceptors are normally named according to the type of MO donating and accepting the electrons, e.g., σ^- and π-donors or -acceptors.

When an electron donor and an acceptor collide, i.e., when they take part in an "encounter" in a solvent "cage" [6.12], they may form a CT-complex. This CT-complex may split up and form A and D. In equilibrium $[AD]/([A][D] = K$, where K is the complexity constant, and [AD],[A], and [D] are the CT-complex, acceptor, and donor concentrations, respectively. The light absorption spectrum of the CT-complex is not the same as the sum of spectra of the two individual components, a new band, the CT-band, is produced. A recording of this band is the most important method of studying CT-complexes [8.17,18]. In particular, most of the acceptors with the electron affinities above 1 eV in Table 8.1 are known to form CT-complexes with various donors. We shall see below that the Ps inhibition and quenching by an acceptor can change very much if the acceptor forms a CT-complex with a donor. Positronium reacts normally with electron acceptors having EA > 0.5−1.0 eV.

The most interesting property of the Ps reactions with electron acceptors is the strong temperature dependence of the rate constant of the reaction, k, found for most acceptors. Normally, k increases with the temperature at low temperatures, reaches a maximum, k_m, at t_m, and then decreases with temperature above t_m. In Fig. 8.2 the variations of $\log(k)$ with the reciprocal temperature for six acceptors in toluene [8.7] are shown. The electron affinities are given in Table 8.1. Similar dependence of k on temperature has been found for many acceptors in various solvents. Table 8.2 gives the results for all the acceptor/solvent systems found in the literature. The rate constants, k, at the lowest

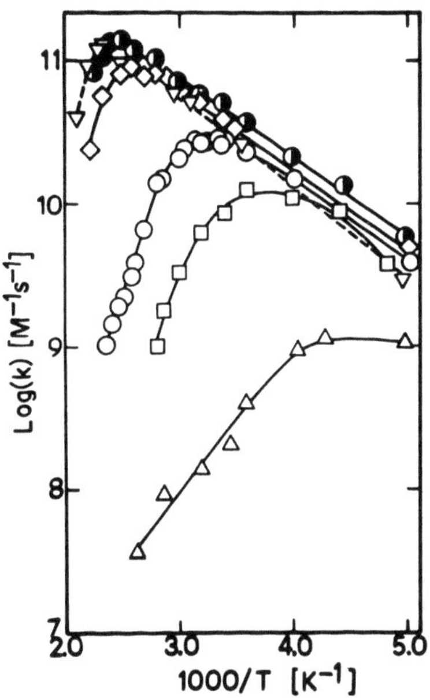

Fig. 8.2. Rate constants (log(k)) for thermal o Ps reaction with various substrates in dilute toluene solutions as a function of the reciprocal temperature (K^{-1}) [8.7]. △ 2,6-dimethylnitrobenzene, □ p-nitroanisole, O nitrobenzene, ◇ p-nitrobenzonitrile, ▽ p-benzoquinone, ◑ p-dinitrobenzene

and highest temperature investigated are given. Furthermore, the rate constant, k_m, and temperature, t_m, at the maximum of the rate constant versus temperature are shown, too. The first cases of such temperature dependence were reported by *Goldanskii* et al. [8.12] in 1972.

Some short remarks on the quality of the measurements are appropriate here. Most available Ps quenching data were measured before 1975, and hence, the quality of the results is poorer than what can be obtained today. All the results of Table 8.2 are extracted from lifetime spectra, measured with a resolution of 0.4–1.2 ns. In most cases good computer programmes for the analyses of the spectra were not used. The o Ps intensities are not mentioned, with the exception of a few papers where it is stated that the intensities were roughly constant for the acceptor concentrations studied (e.g., [8.7]). Furthermore, in some cases only few, dilute acceptor solutions were studied. As shown in the typical case of nitrobenzene in benzene in Fig. 8.1, the preexponential factors, I_3', which before 1975 were assumed to be the o Ps intensities, are, actually, roughly independent of the concentration, if the second lifetime, τ_2, is fixed to be 400 ps in three-term fits. However, in many measurements before 1975 only two-term fits were used. The rate constants were calculated from (8.1) and the measured longest lifetimes. Angular correlation results were not used. However, as the longest lifetimes are fairly easy to obtain, the reported rate constants are reasonably well determined. In this context, it should be realized that the interesting variations in the rate constants are very large, and they are normally much larger than the uncertainties of the rate constants. On the other hand, high quality LT and AC studies of Ps reactions with electron acceptors might give new and unexpected results in the future.

Table 8.2. Temperature dependence of the Ps rate constants of reactions in various solvents. The rate constants at the lowest and highest temperature studied, and at the maximum of k versus t (k_m, t_m), are given. E_c is the Ps bubble energy at t_m calculated by use of the infinite square well model

Reactant /solvent	Temp. int. [°C]	k [$M^{-1}ns^{-1}$]	k_m [$M^{-1}ns^{-1}$]	t_m [°C] E_c [eV]	Ref.
oxygen	−74	15	58	7	[8.6]
/n-hexane	74	9		1.08	
oxygen	8	24	33	57	[8.16]
/n-octane	104	19		1.03	
oxygen	−74	14	56	35	[8.6]
/isooctane	74	18		1.01	
2,6-dimethylnitrobenzene	−73	1.0	1.0	−51	[8.7]
/toluene	112	0.035		1.48	
p-nitroanisole	−70	3.5	11	−10	[8.7]
/toluene	84	1.0		1.38	
o-nitrotoluene	5	0.5	0.5	5	[8.8]
/n-heptane	100	0.13		1.13	
o-nitrotoluene	−50	10	15	−17	[8.8]
/toluene	85	3.2		1.39	
m-nitroaniline[a)]	25	0.09	0.17	60	[8.9]
/n-octanol	108	0.1		1.20	
nitrobenzene	−100	6	15	−50	[8.6]
/n-hexane	67	1		1.23	
nitrobenzene	−73	5	17	−17	[8.10]
/n-heptane	84	0.2		1.10	
nitrobenzene	−154	0.02	2.4	−23	[8.15]
/isopentane	30	0.2		1.08	
nitrobenzene	−113	0.5	16	−38	[8.6]
/isooctane	67	0.5		1.19	
nitrobenzene	20	21	21	20	[8.11]
/benzene	85	2		1.31	[8.12]
nitrobenzene	−73	4.5	26	30	[8.7]
/toluene	160	1		1.27	
nitrobenzene	−73	0.4	17	100	[8.10]
/n-pentanol	127	9		1.05	
nitrobenzene	10	1	8	72	[8.11]
/n-octanol	160	2		1.17	[8.12]
nitrobenzene	13	10	13	25	[8.10]
/water	97	1.5		2.06	
o-nitrobiphenyl	−20	3.1	5.1	35	[8.13]
/xylene	74	1.8		1.28	
p-nitrobiphenyl	−15	1.4	2.5	35	[8.13]
/xylene	78	1.6		1.28	
p-chloronitrobenzene	20	2	7	97	[8.9]
/n-octanol	148	3		1.12	
9,10-anthraquinone	20	53	60	60	[8.9]
/benzene	112	35		1.18	
p-nitrobenzonitrile	−73	4.5	90	112	[8.7]
/toluene	180	30		1.02	
p-benzoquinone	20	52	100	127	[8.9]
/benzene	127	100		0.95	

Table 8.2. *(Continued)*

Reactant /solvent	Temp. int. [°C]	k [$M^{-1}ns^{-1}$]	k_m [$M^{-1}ns^{-1}$]	t_m [°C] E_c [eV]	Ref.
p-benzoquinone	−73	2.5	120	150	[8.7]
/toluene	198	45		0.88	
p-dinitrobenzene	−70	10	105	84	[8.10]
/n-heptane	162	6		0.91	
p-dinitrobenzene[a)]	20	50	100	84	[8.9]
/benzene	84	100		1.11	
p-dinitrobenzene	−73	5	125	127	[8.7]
/toluene	182	80		0.97	
p-dinitrobenzene	−73	1	90	127	[8.14]
/n-pentanol	127	90		0.99	
chloranil	18	20	150	152	[8.9]
/n-octanol	152	150		1.00	
2,2′-dinitrobiphenyl	−18	15	37	50	[8.13]
/xylene	90	31		1.24	
2,4′-dinitrobiphenyl	−18	17	46	50	[8.13]
/xylene	88	40		1.24	
4,4′-dinitrobiphenyl	5	33	50	60	[8.13]
/xylene	90	39		1.21	

a) Uncertainties in [8.9].

Before we attempt to interpret the results in terms of the physics of the interaction of Ps with the acceptor and solvent let us state some facts which are obtained by a closer study of the reported data.

A) A comparison of the data of Tables 8.1,2 shows clearly that the rate constant, k, is not a simple function of the electron affinity. For example, the free radical oxygen (EA $= 0.5$ eV) has much higher k-values than acceptors with EA $\simeq$ 1 eV, e.g., nitrobenzene. On the other hand there is a clear tendency that high EA corresponds to a higher k_m-value found at a higher temperature, t_m, other things being equal.
B) Other acceptor properties, such as possible hydrogen bonding [8.19] and steric hindrances, seem to play a role, too.
C) In particular, the nine nitrobenzene solutions show that the rate constant also depends on the solvent. High surface tension seems to correspond to higher k_m-value. Furthermore, higher viscosity of a solvent corresponds, generally, to lower k-values at temperatures below t_m. For example, the low k-value in n-octanol can be correlated with the high viscosity of octanol. However, k is probably low in isopentane because of its low surface tension.
D) A plot of $\log(k)$ versus the reciprocal temperature (K), T^{-1}, below t_m is roughly linear, corresponding to an activation energy which always seems to be lower than that expected from the temperature dependence of the viscosity, typically a factor of two lower [8.7,10]. At temperatures above t_m, $\log(k)$ versus T^{-1} varies roughly according to a negative activation energy [8.7].

In the literature these experimental results have normally been interpreted as due to a complex formation of Ps and the electron acceptors, (Ps,A), according to the equation:

$$\mathrm{Ps} + \mathrm{A} \rightleftarrows (\mathrm{Ps,A}) \tag{8.32}$$

At low temperature the complex is assumed to be stable, and the variation with temperature is assumed to be caused mainly by the temperature dependence of the diffusion and kinetic processes of the forward reaction in (8.32), as is found for many chemical reactions. At temperatures above t_m the complex is assumed to be increasingly unstable for increasing temperatures, because it splits up by thermal activation. Actually, this model is more restrictive than necessary. The available experimental facts seem only to demand that the positron and electron leave the state(s) together to form Ps on thermal excitation at higher temperatures, but this is assured by the strong Coulomb attraction in nonpolar liquids with low dielectric constants. Hence, states where the positron and electron are totally separated, e.g., where the positron (electron) is mainly on the acceptor and the electron (positron) is mainly in the solvent, are not against experimental fact. Furthermore, many Ps-acceptor bound states might, and are very probably, involved. Hence, this interpretation is somewhat incorrect, as we shall see below.

Let us first formulate the probably best explanation, and later discuss it in some details. Further, we use the Ps reaction with nitrobenzene in hexane as a typical example. At higher temperatures k is low. Ps forms "collision cages" with nitrobenzene, where nitrobenzene takes part in the illdefined surface of the Ps bubble, at a diffusion controlled rate, but the shortening of the o Ps lifetime corresponds to a much lower rate constant if (8.1) is assumed to be valid. It cannot be excluded that the lifetime decrease is caused by a faster pick off rate in the "collision cage" and not by a "direct reaction", as assumed in (8.2–5). On lowering the temperature the surface tension increases, and hence, the size of the Ps bubble decreases, i.e., Ps is increasingly "squeezed" in the bubble. Consequently, the energy of the free Ps bubble increases, while that of the nitrobenzene-Ps "bound" states in smaller "bubbles" increases less, i.e., energy conditions begins to be favourable for bound state formation, and the k-value increases. At temperatures somewhat below t_m Ps might even be mainly "squeezed" out of the associated "bubble", and a bound state of the positron, electron, and nitrobenzene with little or no Ps attributes is formed. Depending on the nature of solvent the positron (electron) might even be more or less outside, and the electron (positron) on, the nitrobenzene molecule in that state. Physically, the temperature dependence of the rate constant reflects the "squeezing" of the Ps out of the bubble in a way very similar to that encountered in pure CS_2, as discussed in Sect. 5.5, and in some aromatic crystals treated in Sect. 9.3. As discussed in Chap. 5, it is the internal pressure, as given by the combination of the dispersion and polar, but not the hydrogen bond, components of the solubility parameter, which determines the inwards "squeezing" on Ps in the bubble. Hence, the surface tension correlates to the size of the Ps bubble only for the solvents where hydrogen bonds are absent or of little importance.

After thus having formulated the explanation of the temperature dependence of k, let us discuss the experimental and theoretical facts speaking in favour of the model. At first, it must be realized that the Ps-molecule complexes mentioned above should be called a Ps state only in certain cases. This is partly a nomenclature problem. Except for the Ps molecule complexes, states are only designated "Ps states" in the literature if certain Ps-attributes are measured. Normally, annihilation of Ps states results in a long lifetime due to o Ps pick off annihilation and a narrow peak in the AC-spectrum due to p Ps intrinsic annihilation. In some solids only one of these attributes is detected, as discussed in Chap. 9. None of these attributes are found if the Ps-molecule complexes annihilate in many cases. The situation is very similar to the annihilation from positron bound states with halides, for which the designation "Ps-state" is not used. Hence the designation "Ps complex" is misleading and should be avoided, if the two attributes are not found. Clearly, in states where the positron and electron are separated the two attributes are absent.

Very recently, *Kobayashi* [1.45,46] measured the rate constant, k, versus external pressure, p, at $t_r = 20°C$ for Ps reactions with several diamagnetic acceptors in nonpolar liquids. His results agree very well with our simple explanation of the temperature dependence of k. For example, for nitrobenzene solutions of benzene and cyclohexane k is diffusion controlled at all pressures. Hence, k is about $3 \times 10^{10}\ M^{-1}s^{-1}$, and it decreases with increasing pressure, i.e., k is roughly inversely proportional with the viscosity, as expected. On the other hand, in nitrobenzene/hexane solutions k increases strongly with p up to pressures of about 500 bar, where it reaches the diffusion controlled rate. At $p > 500$ bar k in the nitrobenzene/hexane solutions behaves roughly similarly to k in the benzene and cyclohexane solutions. The explanation is, of course, that the Ps bubble state has a higher energy than the positron-electron-nitrobenzene bound state in benzene and cyclohexane at all pressures. In hexane the Ps bubble has lower energy than the reaction product at pressures below roughly 500 bar, while its energy is above that of the reaction product for $p > 500$ bar. All the other acceptor/solvent combinations studied in [1.45,46] give results in full agreement with this simple explanation. In particular, the results for nitrobenzene solutions of pentane, hexane, heptane, octane, benzene, and cyclohexane are very strong arguments in favour of the model.

Concerning the explanation given above, the following AC results [8.20,21], measured at t_r, are of interest. Apart from fortuitous resemblance in shapes, AC spectra are "fingerprints" of the state at annihilation, as discussed in Chap. 3. High concentration (~ 1 M) benzene solutions of nitrobenzene, SO_2, dibenzoyl, p-benzoquinone, and DPPH give almost identical AC spectra within uncertainty, if the spectra are properly corrected for very small components due to annihilation of remaining Ps. These spectra are very different from those of high concentration solutions of nitrobenzene in hexane, heptane, and methanol and for pure nitrobenzene. Apparently, this indicates that the AC spectrum for 1 M nitrobenzene in benzene is a property of benzene only, and it has been ascribed to the annihilation of "free" positrons, as mentioned in Sect. 8.2. Clearly, this state should not be called a Ps state. However, nitrobenzene/heptane gives an AC spectrum very different from that of SO_2/heptane, if Ps is totally quenched, inhibited, or properly corrected for. Nitrobenzene/heptane and -/hexane give very similar AC spectra, apart from a somewhat higher k-value in heptane at t_r [8.6]. This indicates that benzene and heptane, and hence hexane, behave very differently with respect to annihilation of Ps and the positron. Furthermore, a 2 M nitrobenzene/methanol solution gives an AC spectrum which is very close to that of pure nitrobenzene [8.20], indicating that the annihilation is only a property of nitrobenzene at high concentrations in methanol. Few other AC spectra [8.20] for organic solutions of electron acceptors are available. In summary, the few available AC-spectra seem to show that the positron annihilates from states in the solvent, on the acceptor, or in some other states, depending on the properties of the solvent and acceptor.

The Ps bubble state in liquids was discussed in Sect. 4.4. In particular, several models of the Ps bubble were described. It was shown that a simple square well potential gave a Ps bubble of radius 0.4–0.6 nm in many liquids, e.g., hexane, but an energy of kT changed the radius by about 25%, Hence, the radius is illdefined, if this simplified model is used. It might be even more uncertain in reality. In (4.4) the total energy of the Ps bubble versus temperature, as calculated by use of the infinitely deep square well model, is given. By decreasing the temperature from 80°C to −100°C we can increase this energy by roughly 0.5 eV in hexane. If Ps encounters an acceptor which provides a lower energy for Ps, it will normally react with the acceptor. Apparently, Ps reacts normally at a roughly diffusion controlled rate, if the energy condition favours the reaction. For example, for nitrobenzene in hexane the energy of "bound" Ps (see below) seems to be comparable that of the Ps bubble at roughly $t_m = -50°C$, as the rate constant begins to decrease at t_m, if the temperature is increased. Table 8.2 shows the Ps bubble energy in the solvent, calculated by use of the infinitely deep square well model at the temperature t_m for selected acceptor/solvent systems. Excluding water, the energy at t_m for the nitrobenzene as acceptor is fairly constant, as it varies only between 1.31 and 1.05 eV for eight solvents. Probably this is speaking in favour of the assumed model of the reaction, taking into account the rough models used in the calculation. For example, we have neglected that n-pentanol, n-octanol, and water make hydrogen bonds with nitrobenzene, thereby probably changing somewhat the "Ps affinity" of the acceptor, and that benzene, toluene, and water have hydrogen bond components of the solubility parameter and surface tension, resulting in higher calculated Ps bubble energy than that expected from a more realistic calculation, in particular in the water case. Furthermore Table 8.2 shows clearly that the "Ps affinity", i.e., the effective binding energy of Ps on the acceptor, is not a simple function of the EA-value, provided, of course, that the model is reasonably correct. On taking into account the rather poor quality of the available data, a further interpretation of the data in terms of the model seems not to be useful. Future experiments to elucidate these points will be very important. Spin conversion of Ps is expected to occur, if Ps and a free radical which is also an electron acceptor, encounter, stay together for a short time thereby mixing their spin states, and then split up again. Hence, it is expected to occur at temperatures above t_m, in agreement with experiment, as discussed below.

It can be reasonably assumed that many states of the positron exist between that of free Ps in the bubble at very high temperatures and the fully bound positron-electron-acceptor state, giving no Ps attributes, at lower temperatures. The reason is, of course, that the states involved normally consist of many molecules which behave "classically". Well-separated quantum mechanical states, which are important for the description of a single light particle in a potential well, cannot be expected to play a role for the molecules in the surface of the Ps bubble. Hence, many bubble states, populated according to statistical physics at the given temperature, exist in the pure solvent, and also if the Ps bubble is associated with the acceptor, i.e., if Ps is only partly bound to the acceptor.

The Moscow group first proposed the "two-square-well" model to describe approximately the Ps acceptor interaction [8.11,22]. Apart from the normal, shallow, and broad potential well of the Ps bubble in the solvent, they introduced a deeper and narrower potential well to account for the Ps acceptor interaction. Their calculation showed that the Ps reaction depended on the bubble shrinkage rate. However, *Kobayashi and Ujihira* [8.23] revealed that the model was incorrectly used in [8.11,22]. Furthermore, they showed by calculation for some solvents that the energy which is necessary for the activation of Ps out of the bound state on the acceptor, is largest in the solvents with the largest surface tension. Plots of $\log(k)$ versus bubble shrinkage rate show no correlation, in general. Clearly, the "two-square-well" model assumes that Ps is present, and hence, it is a bad approximation if the positron and electron are separated in the bound state with the acceptor. However, it is expected to be a fairly good approximation for states where Ps is weakly bound. If the Ps bubble state has a lower energy than that of Ps associated with, or totally bound to, the acceptor, the model is inapplicable.

Recently, *Rochanakij and Schrader* [8.24] discovered an anomalous magnetic quenching of the o Ps ($m = 0$) in nitrobenzene/hexane, but not in -/benzene and -/cyclohexane, CCl_4/hexane, and biphenyl/hexane solutions, and in the pure solvents. Except for water and its solutions very few other liquids have been studied earlier. The weighted average, τ_3, of the o Ps ($m = 0$) and o Ps ($m = \pm 1$) lifetimes for a 0.1 M nitrobenzene/hexane solution decreased from 2.6 ns to 2.33 ns for a magnetic field increase from 0 kG to 2 kG, which is about ten times more than normally found. The effect has been reproduced by *Billard* et al. [8.25,26] by LT- and DB-measurements. Three attempts to explain the effect have been published in [7.4,8.26,27].

In addition, it has been proposed that the effect is caused by magnetic quenching of certain Ps-nitrobenzene states existing at the temperatures where they are just stable energetically, i.e., where they are not far from being transferred into non-Ps states [1.44]. Apparently, very strong magnetic quenching of Ps has only been reported in one other case, namely for solid naphthalene at t_r. As discussed in Chap. 9, Ps makes small cages by selftrapping in naphthalene at t_r. In the temperature interval $-20 - -80°C$ Ps is squeezed out of the cage state, and below $-80°C$ Ps is not found. Hence, Ps is just stable at t_r, as in the nitrobenzene/hexane case. The main idea of the explanation in [1.44] is not that the effect is found if Ps is squeezed, as it is in most normal liquids and solids. It is found if Ps is not far from being squeezed out of a Ps-like state and into a state which is not Ps-like.

In naphthalene the magnetic quenching of Ps can be explained by use of a model in which two Ps states exist having the relative density of electrons at the positron equal to roughly $\alpha' = 0.1$ and 0.25, respectively (see Sect. 9.3), and of about equal intensity. The nitrobenzene/hexane results can be roughly explained if we assume that about 15% of Ps is "swollen" Ps with $\alpha' \simeq 0.1$, and 85% is normal Ps with $\alpha' = 0.9$. This indicates that only about 15% of Ps is in a strongly "swollen" state in nitrobenzene/hexane at 0.1 M, a reasonable result as we shall discuss below. The dependence of the stability of Ps on the temperature explains why hexane ($t_m = -50°C$), but not benzene and cyclohexane, give the effect at t_r. Benzene has t_m close to t_r, (see Table 8.2). This seems to be the case also for cyclohexane, as the energy of the Ps bubble in cyclohexane at t_r is 1.23 eV, calculated as in Table 8.2. Hence only in hexane the just stable Ps is found at t_r. Also the pressure effects [1.45,46] speak in favour of the explanation. CCl_4 and biphenyl are unreactive towards Ps, and hence, the effect is absent for these acceptors. Probably, Ps and nitrobenzene in the "encounter cage" in nitrobenzene/hexane having "bubbles" a little smaller than that of free Ps, are the "swollen" Ps states, which give long o Ps lifetimes and anomalous magnetic quenching. These states can, therefore, be called "Ps acceptor complexes", and they might be the Ps complexes which resemble the normal, heavy CT-complexes. The proposed model explains the very few available results on the anomalous effect, the AC spectra, in particular the difference between hexane and benzene as solvents for nitrobenzene, and Ps in solid naphthalene.

Inhibition of Ps formation by an electron acceptor can give the same positron-electron-acceptor states as Ps quenching in certain solutions. The excess electrons (positron) will react with the acceptor and the formed product might react with the positron (an excess electron) giving these states. Of course, not all mechanisms of Ps inhibition give such states, as the positron might annihilate before it can react further, in some spurs because all electrons have recombined, for example. Clearly, at temperatures where the formed products are unstable and break up because of thermal activation, the states formed as results of Ps inhibition processes break up, too. The particular states assumed to give anomalous magnetic quenching will, of course, also be present in the cases where they are results of Ps inhibition processes. Hence, the processes in the positron spur which in normal systems give Ps inhibition, might give Ps formation, i.e., a long lifetime, in stead. Consequently, Ps formation cannot be theoretically separated from Ps inhibition in certain solutions!

We have until now discussed mainly the acceptor nitrobenzene. However, experiments and theory indicate that Ps reacts with the other electron acceptors in a way which is roughly similar to that of nitrobenzene. Apart from the strong influence of temperature, we expect that the reactions depend on the structure of the acceptor in the details. For example, if the acceptor is polar such as nitrobenzene, "swollen" Ps might be more probable. As found for nitrobenzene, the positron-electron-acceptor states at temperatures below t_m will probably depend very much on the solvent and acceptor properties. Hydrogen bond formation between the acceptor and the solvent might play a role. In polar solvents solvation of the particles might influence the structure of the states. For instance, the positron seems to be on nitrobenzene, and the electron solvated in methanol, in nitrobenzene/methanol.

The new, simple explanations of the temperature and pressures effects in the Ps reactions with acceptors and the anomalous magnetic quenching have just been published [1.44]. The models must be tested against experiments before elaborate theoretical predictions are worthwhile. We shall here only mention some obvious predictions of the models. Concerning the anomalous magnetic quenching of Ps, we can predict many experimental results. For example, a plot of the "strength" of the effect versus temperature will probably be shifted about 30°C to higher temperatures in heptane than in hexane, as t_m is -17°C in heptane compared with -50°C in hexane. In general, the effect will be correlated to t_m and the pressure effect of the solute/solvent system. The effect of changing the acceptor should be correlated to the structure of the acceptors, as done for the electron affinities [8.5]. As concerns the anomalous magnetic quenching effect in general, it can probably be expected, whenever Ps is close to be unstable, as shown by the naphthalene case. Furthermore, its variation, if parameters, such as temperature, pressure, electric field, and acceptor structure are changed, is expected to be explainable mainly in terms of the dependence of the stability of Ps on these parameters. In particular, studies of the pure electron acceptor liquids, such as CS_2, C_6F_6, and SF_6, and electron acceptor solids, such as naphthalene and the polyphenyls, will be interesting in that context.

Some comments on the older interpretations might be useful. Twenty years ago when the studies of Ps reactions with diamagnetic electron acceptors started, the rate constants were correlated mainly with the properties of the acceptor, such as the probabilities of structures containing carbonium ions in the molecule, Hammett constants, half-wave potentials, and lowest empty molecular orbital. These correlations were first studied by the Moscow group, see [8.11] and refs. therein. The Ps acceptor properties correlated with the electron affinities, which were much more uncertain twenty years ago [8.4,5]. As the strong temperature dependence of the k-values was discovered, it was realized that the solvent properties were very important too, and the mentioned correlations became less useful ([8.7], ref. 32). Because of lack of space, I have chosen to use mainly the electron affinity in the discussion. However, other acceptor properties might be of importance in the future to explain details in the formed bound states.

At last in this section let us recall two important facts, encountered many times in the preceding discussion. First, the physical processes of the Ps acceptor reactions are very unique in chemistry because of the particular structure of free Ps, namely the Ps bubble, in a liquid. Secondly, concerning the problems of experiment and analysis, we frequently were faced with a basic problem, namely: "Where is the positron, if Ps has been inhibited and/or quenched?" This is perhaps the most important unsolved problem in positron and Ps chemistry at present.

8.4 Paramagnetic Quenching of Positronium in Solutions

In the previous section we have mainly discussed Ps reactions with diamagnetic molecules. An exception was the Ps O_2 reaction. For paramagnetic molecules and ions spin conversion reactions of Ps, i.e., reactions where o Ps is changed into p Ps and vice versa, are possible, too. Actually, Ps reactions with the free radical, DPPH, seem to have been the first reactions studied in a liquid [1.27,7.58]. We shall in this section discuss the spin conversion reaction of Ps. At first a simple, basic discussion will be given. The details and problems of interpretation will then be discussed with reference to a special case, namely that of TMOPO in methanol, which is the mostly studied solution. The main part of the treatment concerns paramagnetic molecules as well as ions.

As stated in Sect. 8.3, a spin conversion might occur if Ps and a paramagnetic molecule or ion form an "encounter cage" in a liquids, where they stay together for some time, and then split up again resulting in a free Ps bubble state. Hence, if any stable, bound state is formed, the positron will not be in a state with Ps attributes after the reaction, and, consequently, spin conversion is not detected. In principle, a measured spin conversion of Ps might also be caused by annihilation of Ps during its stay in the "encounter cage". As discussed in Sect. 8.3, an "encounter cage" of Ps and a reactant might be fairly complicated involving many different states of Ps, of which some might be bubble states with smaller bubbles than that of the free Ps state. We expect spin conversion to occur at temperatures above t_m, the temperature where the rate constant is highest (see Sect. 8.3). At room temperature, t_r, we, therefore, expect that a spin converter is a fairly weak electron acceptor, as strong acceptors will very probably form stable, bound states on Ps reactions at t_r. Furthermore, we expect that spin conversion happens in solvents of low effective surface tension. The effective surface tension which is caused by only the dispersion and polar attraction between the molecules, but not the hydrogen bond part of the attraction, determines mainly the energy of the Ps bubble, and hence the possibility of reaction (see Chap. 5).

At present, only few results on spin conversion of Ps have been reported, and detailed studies are very rare. However, the available experimental results seem to be in full agreement with the above mentioned general expectations of the model discussed in Sect. 8.3. The measured roughly 8–10 times lower conversion than diffusion controlled rate constants can, apparently, by explained by the efficiency of the conversion process during the collisions, see below. The important conclusion is, therefore, that we seem to understand the spin conversion reactions in general. At the same time we shall see below that many detailed problems of these reactions are certainly not solved yet.

Three experimental methods are mainly used in the studies of spin conversion reactions. The lifetime (LT) measurements determine the long lifetime and its intensity. However, it is, in general, very difficult to separate between spin conversion and the other Ps reactions by use of only lifetime measurements. A "fingerprint" of a spin conversion is a deviation from linearity of the decay rate versus reactant concentration, but this deviation is normally too small to be give an unambiguous determination of the rate constants of the reaction. Figure 8.3 shows a case discussed in detail below. The Ps inhibition parameters (7.30) can be determined from the corrected long lifetime intensities. The angular correlation (AC) and Doppler broadening (DB) measurements can separate clearly between the spin conversion and other reactions. The intensity of the narrow component due to intrinsic p Ps annihilation increases at low concentration due to conversion of o Ps to p Ps, as shown in Fig. 8.4. At higher concentration the intensity decreases again because of Ps inhibition and/or because oxidation and/or other chemical reactions of p Ps begin to be important. A standard case of spin conversion is that of TMOPO in methanol, illustrated in Figs. 8.3 and 8.4. In principle, spin conversion can also be detected by use of an external magnetic field to mix the p Ps with the o Ps ($m = 0$) states. Collisions with paramagnetic reactants cause spin conversions of all the Ps states, and hence, the other two o Ps states, which are not influenced by the magnetic field

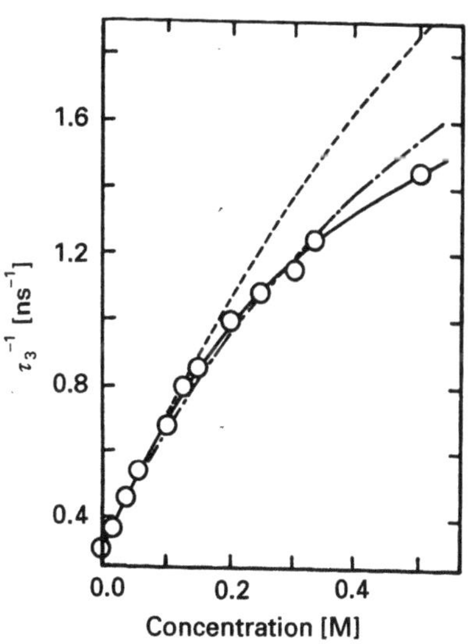

Fig. 8.3. Reciprocal long lifetime versus concentration of TMOPO in methanol. Measured points are from [5.21]. The chemical and conversion rate constants for the theoretical curves are: 0.0 and 4.8×10^9 $M^{-1}s^{-1}$ for the full, b) 1.10×10^9 and 3.70×10^9 $M^{-1}s^{-1}$ for the dashed, and c) 0.43×10^9 and 3.87×10^9 $M^{-1}s^{-1}$ for the dot-dashed line, respectively

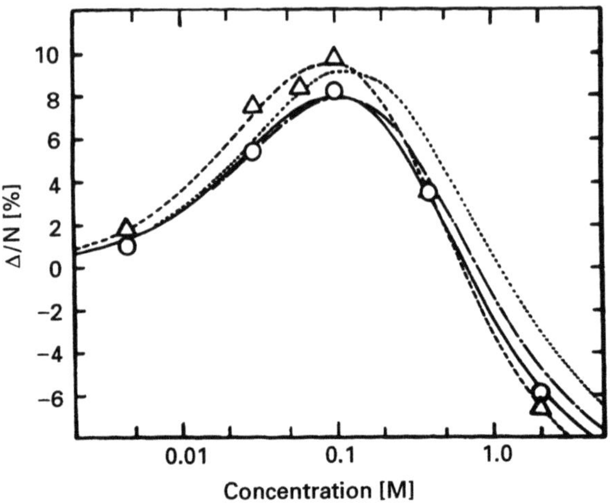

Fig. 8.4. Change in narrow peak intensity, $I_N(c) - I_N(0)$, versus TMOPO concentration, c, in methanol at 25°C (O) and 50°C (Δ). The four theoretical curves are determined by the following parameters (see Sect. 8.1). For all curves λ_s, λ_t, λ_p, λ_f, λ_m, and β are 8.0, 0.00714, 0.272, 2.5, 2.5 $(ns)^{-1}$, and 1.0. The rate constants, k_0, k_c, k_{sc}; the yields at $c = 0$, $P_s(0)$, $P_t(0)$, $P_f(0)$, $P_m(0)$; and the inhibition constant, α, are: 0.0, 1.1×10^9, 3.7×10^9 $M^{-1}s^{-1}$; 0.091, 0.273, 0.631, 0.0; and 1.9 M^{-1} for the full, 0.0, 1.4×10^9, 4.69×10^9 $M^{-1}s^{-1}$; 0.1, 0.3, 0.5945, 0.0; and 1.9 M^{-1} for the dashed, 0.0, 0.43×10^9, 3.87×10^9 $M^{-1}s^{-1}$; 0.091, 0.24, 0.664, 0.0; and 2.0 M^{-1}, for the dot-dashed, and 0.0, 0.0, 4.3×10^9 $M^{-1}s^{-1}$; 0.091, 0.24, 0.664, 0.0; and 2.0 M^{-1} for the dotted line, respectively. Data from [8.33], reanalysed

if no paramagnetic collisions occur, can also decay rapidly after having been converted into the o Ps ($m = 0$) state (see below).

Let us now discuss in some detail the Ps spin conversion by TMOPO, 2,2,6,6-tetramethyl-4-hydroxypiperidine-1-oxyl (HTMPO, [8.28]) in methanol. The discussion will be very complicated involving nearly all the aspects of Ps chemistry and inhibition, and, in particular, the problems of the measurement, analyses, and interpretation. It was decided to include this discussion because it is a very good example of the present stage of development, and furthermore, it may serve as a final test case on which a reader can examine his/hers skill in many of the disciplines treated in this book. A non-specialist might prefer to skip the next pages.

The stable organic free radical TMOPO seems to be one of the standard free radicals in chemistry, see e.g. [8.29,30]. Ps reactions with TMOPO were first studied by the Moscow group [8.31,32], followed by Risoe [8.33] and Strasbourg [8.28,34] works. Our discussion will be based on [8.28,33], which constitute the most detailed studies of a spin conversion case, TMOPO in methanol. *Abbé* et al. [8.28] measured LT and DB spectra for TMOPO/methanol at ten different temperatures between 203 and 333 K. Results of TMOPO in water and hexamethylphosphorotriamide and two other radicals in methanol were also given. The decay rate, $\lambda_3 = \tau_3^{-1}$, versus TMOPO concentration, c, at 294 K is shown in Fig. 8.3. The other temperatures gave similar LT and DB results. The Ps inhibition constants, $\alpha = 1.9 \pm 0.2$ M^{-1}, $\beta = 1.0$, were independent of temperature. In [8.34] 2.0 ± 0.5 M^{-1} and 1.0 were measured at 294 K.

The first problem we encounter is that nothing is said in [8.28] about whether the second lifetime, τ_2, has been fixed in the used three term analyses. It is wellknown that the Ps yield, in particular, but also the long lifetime, cannot be determined well if τ_2 is not fixed for long lifetimes below about 1.0 ns. As discussed in Chap. 5, the reason is partly that a distribution of shorter o Ps lifetimes is present in the lifetime spectra, and partly that the second and third lifetime components cannot be separated well in the analyses. As the inhibition parameters agree well with those of [8.33] at t_r, they are probably acceptable. The decay rates at high concentrations might be somewhat wrong though. Furthermore, for pure methanol $\tau_3 = 3.35$ ns in [8.28], which seems to be too short compared to the normal value of $\tau_3 = 3.58$ ns, Table 5.2. One encounters usually such problems if one wants to used LT data reported in the literature. Hence, better information on the analyses of LT spectra ought to be given, in general.

The second problem is that the DB results in [8.28] seem to give peculiar results after the computer analyses. The Strasbourg group use usually a DB detector of resolution, fwhm = 1.4 keV, corresponding to 5.48 mrad in AC measurements. Furthermore, they normally use the LT results to derive the fwhm of the DB spectra using the fwhm's of the components of the DB spectra obtained from previous, independent measurements and by solving the kinetic equations (8.2–5). The DB components are approximated by only one gaussian each. The fwhm of the DB spectra could not be fitted for TMOPO/methanol. Tentatively, they interpreted this fact by introducing an extra DB component due to positron bound state formation with TMOPO, having fwhm = 2.47 mrad, and got acceptable fits. Later [8.34] they found similar results for TMOPO in benzene, N,N-dimethylacetamide, and N-methylformamide, which were now, together with the TMOPO/methanol results, interpreted in terms of a new idea. They showed that the results could be well fitted by assuming that the p Ps formed through the spin conversion reaction has a much narrower momentum distribution, roughly by a factor of two, than the p Ps formed in the positron spur, indicating a more complete thermalization of the converted p Ps. However, this idea is in very clear disagreement with AC spectra measured at Risoe [8.21]. For example, the following intensity (fwhm) of the narrowest gaussian in a three gaussian fit of the AC spectra at 0.1 M TMOPO: benzene 27.93% (2.84 mrad), diethylether 31.56% (2.57 mrad), and methanol 16.98% (2.66 mrad) were obtained at Risoe. The fwhm's are identical to those of the pure liquids within the uncertainty (see Table 5.2), and the shapes of the remaining part of the spectra were also almost identical to that for the pure liquids. Hence, the proposed idea in [8.34] is wrong. Probably, the problems of interpretation of the DB spectra stem from the poor fit of the DB components in terms of only one gaussian. For the fitting of the more precise AC components more gaussians are normally needed, as discussed in the preceding chapters.

We are now left with the old AC spectra for TMOPO/methanol (Risoe, 1974, [8.33]) for a further discussion of the spin conversion reactions. DB data for any spin conversion case is only available from Strasbourg, and the above discussed DB data, which cannot be used, is probably the best available. The old AC results from Risoe seem to be the only AC data available for TMOPO/methanol, and furthermore, it is probably the best AC data measured for any spin conversion case. Hence, these, admittedly rather few, AC spectra were the only reasonable data to use in our discussion. In general, high quality AC results on spin conversion are, therefore, urgently needed!

The AC spectra for TMOPO/methanol were fitted by the sum of three gaussians by use of PAACFIT. Only five concentrations were studied at 298 K and six at 323 K. Of course, more concentrations ought to have been investigated in a thorough study. In [8.33] only the normalized counts in the top of the AC spectra were fitted by solutions of the kinetic equations (8.2–5), and the given changes in I_N were too small. The changes in I_N, shown in Fig. 8.4, are the changes in the intensity of the narrowest gaussian, $I_N(c) - I_N(0)$, where $I_N(0) = 8.80\%$ at 298 K and 9.67% at 323 K. In [8.28] the decay rate, $\tau_3{}^{-1}$, versus c at 294 K was well fitted assuming only spin conversion with a rate constant, $k_{sc} = 4.8 \times 10^9$ $M^{-1}s^{-1}$, as shown by the full line in Fig. 8.3. However, this gives an unacceptable fit of the AC data. This forces us to consider several explanations.

One possibility is that the assignment of the narrowest gaussian to the p Ps intrinsic decay is not a good approximation. The fwhm of the three gaussians are very similar to those of pure methanol, Table 5.2., and the intensities are roughly as expected. One question is: Can a part of the narrowest gaussian be caused by pick off, free positron, and/or the distribution of shorter o Ps lifetimes components? The measured roughly 3.0% intensity (fwhm = 2.62 mrad at 298 K and 2.55 mrad at 323 K, not deconvoluted for resolution as in Chap. 5) at 2 M excludes that it is caused by pick off, as any reasonable rates for quenching and inhibition of Ps give only about 0.6% pick off AC component at 2 M TMOPO. As the distribution of shorter o Ps lifetimes contains only about 4–6% even in a pure liquid, this component can reasonably be excluded, too. Furthermore, CCl_4, CBr_4, nitrobenzene, and quinone at high concentrations in methanol give no narrow component at all, i.e., if a 3.0 mrad gaussian is assumed a negative intensity is normally obtained [8.21]. However, we do not know whether any of these solutions give the free positron AC spectrum. It cannot be excluded that they all are due to bound states of the positron with the acceptor, although that seems to be rather unlikely. Another possibility is that we do not understand important aspects of the Ps physics and/or reactions, which cannot be totally excluded. It seems to be reasonable though to assume that the narrow gaussian at 2 M, and at all the other concentrations of TMOPO are due to p Ps annihilation.

We then are faced with the usually problem of a too large p Ps yield compared to 1/3 of the o Ps yield. For example, for pure methanol a correction for pick off of the $I_N = 8.8\%$ gives 9.09% p Ps, which results in 27.3% o Ps at 298 K, assuming a factor of 1/3 between p Ps and o Ps. We measure only 21% in the long lifetime component if a free three term fit is used. A four term fit would give a slightly different intensity. Can we reasonably assume that the missing 6.3% is roughly the intensity of o Ps reacting with positive ions, excess electrons, etc. in the positron spur to give the distribution of shorter o Ps lifetimes? One argument in favour of this large yield is that we expect these processes to be more important if the spurs are small and the o Ps lifetime is long, which is the case for methanol, in particular. High quality LT spectra for methanol and other similar liquids are not available yet. A chemical reaction of o Ps in the spur must compete with the spur reactions. This is not included in our kinetic equations (8.2–5). Hence these equations are only approximate.

We cannot get full agreement between the solutions of (8.2–5) and the combined LT data of [8.28] and the AC data of [8.33] for any rate constants, etc., as shown in Figs. 8.3–4 by the fitting curves. The full line at 298 K and the dashed line at 323 K are the resulting curves for the total rate constants given in [8.28] of 4.8×10^9 and 6.1×10^9 $M^{-1}s^{-1}$, respectively, but on assuming that 77% of these constants describe spin conversion compared with 100% in [8.28], and further, that all o Ps take part in the "normal", but not in the spur, reactions. The fits of the AC data are perfect, but they disagree markedly with the LT data, as illustrated by the dashed curve in Fig. 8.3 for $T = 298$ K. The best overall agreement at 298 K are the dot-dashed curves in Fig. 8.3,4, for which $k = 4.3 \times 10^9$ $M^{-1}s^{-1}$ with 90% conversion and 24% of o Ps undergo "normal" reactions. The dotted curve in Fig. 8.4 shows that 100% conversion, other parameters being equal, gives an unacceptable fit. 100% conversion can only be counterbalanced by an inhibition constant of over 4 M^{-1}, which would have been detected. The remaining disagreements between the fitting curve and the points in Fig. 8.3,4 can easily be explained as being caused by some of the effects mentioned above. We emphasize that other effects might also be important. For example, the concept of a rate constant is not expected to be a good approximation at higher concentrations, as discussed in Chap. 6. At 2 M the distance between the TMOPO molecules is only 0.95 nm, and initial encounter pairs play a strong role in the reactions! Another problem is: Can a saturation in the Ps inhibition be excluded?

In summary, this detailed discussion of the best available set of data on Ps spin conversion shows, in particular: 1) The experiments and data analyses resulted in few and rather uncertain numbers to be compared with the theory. 2) Many aspects of the theory are uncertain. In particular it is difficult to include the effects of the distribution of shorter o Ps lifetime in the theoretical model. 3) Little is known about the state of the positron after Ps inhibition and/or quenching. 4) The field is partly open, i.e., very much work needs to be done to sort out all the problems discussed above. 5) However, the main principles of spin conversion reactions are known today. Semi-quantitatively we understand the processes.

The determined values of the rate constants, inhibition constants, and yields give information on the physical processes of Ps, the positron, and spur electrons in the solution. Ps inhibition and the yields of the components of the LT, AC, and DB spectra have been discussed in greater parts of this book. However, the spin conversion process needs to be discussed in more detail. Theories of the spin conversion of Ps have been reported in [1.32,8.35–38]. The formation of the "reaction cage" of Ps and a reactant was shortly discussed above and in detail in Sect. 8.3. We expect that Ps spin conversion reaction at t_r on a paramagnetic reactant, which is a rather weak electron acceptor if the conversion dominates at t_r, changes to bound state formation at low temperatures. The reason is, of course, that the energy of the free Ps bubble increases on lowering the temperature due to increasing effective surface tension, as discussed in detail in Sect. 8.3, and hence, the stable, bound state becomes more favourable. At increasing temperatures above t_r the spin conversion rate constant is expected to decrease strongly for the same reactant, because the energy gained by the "reaction cage" formation decreases and maybe becomes negative, as the energy of the free Ps bubble decreases due to the lowering of the effective surface tension. Consequently, the determined rate constants, k_{sc} and k_c (or k_O), give information on the effective surface tension and the effective "Ps affinity" of the free radical, as for diamagnetic reactants.

Let us discuss shortly the Ps spin conversion process itself. According to [8.35,36] the Ps spin conversion processes are very similar to the spin exchange processes of heavy reactants. Apparently, spin conversion occurs because the spin of the Ps electron is not a good quantum number, when its wave function overlaps that of the spin of the unpaired electron on the radical. The exchange energy, which follows from the Pauli principle, is the driving energy of the spin conversion. Apparently, the theory explains why the conversion rate constants are always lower than the diffusion controlled ones by a factor of 8–10 [8.35,36]. I tried to understand the calculations on Ps spin conversion in [8.29,35–38], but had to realize that this would take too long time. I encountered, however, one problem worth mentioning here, namely: Are the spin conversion rate constants, γ_{ik}, independent of the initial and final states of Ps, in general, and if a magnetic field is present, in particular. γ_{ik} are the rate constants for conversion on the i'th Ps into to k'th Ps state [8.38]. For example, is the rate of conversion of o Ps ($m = 1$) into the p Ps state identical to that of o Ps ($m = 0$), in general, and if the two last mentioned states are perturbed in a magnetic field. In [8.38] γ_{ik} is given for, what seems to be, collisions in vacuum, and the values depend on i and k. Apparently, the exchange repulsion effect giving lower p Ps pick off rate than that of o Ps (see Chap. 2) will be dependent on the relative spin orientation of the Ps and unpaired electrons, and thereby give different annihilation rate and electron overlaps for different spin orientation during the collisions. Theoretical papers on that topic are most welcome. A sufficient, but not necessary, condition for the spin conversion described by (8.2–5) is that γ_{ik} is independent of i and k. Again we see that the usual description of the Ps chemistry in terms of the kinetic equations (8.2–5) needs not be correct, in general. A deeper understanding of the physical processes of the Ps reactions should be an important goal of future research.

Finally, let us mention some results, which illustrate the complexity of the spin conversion reactions of Ps. Ps reacts mainly by spin conversion with TMOPO in methanol, hexamethylphosphorotriamide, benzene, N,N-dimethylacetamide, and N-methylformamide, but by chemical reaction, i.e., oxidation and/or bound state formation, in water and m-cresol [8.28,32–34]. [8.39] gives the recent opinion of the Moscow group. Ps reacts with oxygen by spin conversion in gases and liquid nitrogen [8.40] and mainly by chemical reaction in hexane [8.41,42]. LT data of Ps reactions with O_2 and NO was reported in [1.33]. Oxygen in silicagel will by discussed in Chap. 10. The free radical diphenylpicrylhydrazil (DPPH) reacts with Ps by chemical reaction in benzene ([7.58] and refs. therein). Spin conversion of Ps on paramagnetic ions will be discussed in Sect. 8.5.

8.5 Positronium Reactions with Ions in Aqueous Solutions

One of the most studied topics in Ps chemistry has been the reactions of Ps with ions in polar liquids, in particular water. Opposite to the Ps formation case, where a large, qualitative difference between polar and nonpolar solvents is found, the chemical reactions of Ps in polar solvents are very similar to those in nonpolar solvents. Hence, the main part of the discussions in the preceding sections is relevant for polar liquids, too. In this section we shall, therefore, not discuss fundamental problems in details, but mainly describe typical cases of Ps reactions in aqueous solutions to give a short review of the field. At first a discussion of the experimental and analytical difficulties will be given. The influence of the structure of the ions on their reactions with Ps is then treated, and selected results are reviewed at the end of the section.

As usual in chemical dynamics a determination of the rate of the reaction, the reaction products, and the explanation of the reaction in terms of the physics of the reactants and solvent are the primary goals of a study of a Ps reaction with an ion. Traditionally, three types of Ps reactions have been assumed to take place in polar solutions, namely spin conversion, oxidation, and bound state formation. However, just as is the case for nonpolar reactants, discussed in Sect. 8.3, the two last mentioned reactions seem not to have been clearly separated for aqueous solutions. Other scientists have claimed the opposite. Also the spin conversion reactions seem not to be clearly separable from the other reactions in some cases. This is one of the main problems in investigations of Ps reactions, and hence, it is useful to discuss some typical cases of Ps reactions with oxidizing ions in water to elucidate the difficulties involved.

Lifetime measurements give good information on the rates of Ps reactions. In practice, little or no information on the product of the reactions can be obtained by LT methods. Typical examples of LT studies, namely the investigations of Ps reactions with MnO_4^-, $Cr_2O_7^{--}$, IO_4^-, and Cu^{--} in water, were discussed in details in Sect. 7.5 and illustrated in Fig's. 7.7–9 [7.10]. Because the o Ps lifetime is short and its intensity is fairly low in water, great care is necessary in the computer analyses of LT spectra for aqueous solutions, as discussed in further details below. The remarks on the nonlinearity of λ_3 versus concentration of free radical in Sect. 8.4 are very relevant for paramagnetic ions, too. This nonlinearity is also too small for a clear separation between spin conversion and the other two reaction mechanisms for ions.

The important conclusion is that the product of the Ps ion reactions can only be identified by AC and DB measurements. DB studies are useful in the separation of spin conversion and the two other reactions. However, because of the poorer resolution of the DB detectors, detailed studies of spin conversion reactions must also be performed on AC equipments. For the elucidation of the reaction products in the cases of oxidation and/or bound state formation, AC studies are absolutely necessary, and even the very best AC measurements and analyses of the AC spectra have only, as yet, resulted in very uncertain identifications of the reaction products. Apart from the work of the Moscow and Risoe groups, AC studies of Ps reactions with ions in solutions have not been performed recently. The main attempts to identify the reaction mechanism have been done by the Moscow group ([8.39,43] and ref's therein). However, their analysis is probably too simple, and their assignment of a bound state formation with TMOPO in methanol [8.39] is incorrect, as discussed in Sect. 8.4.

The best possible illustration of the problems encountered in the assignment of the reaction mechanism for a Ps reaction with diamagnetic ions, is a discussion of the interpretation of some older Risoe results in terms of various theories. In Chap. 5 and Sect. 6.1 the separation of the water AC curve into its three components due to "free" (i.e., hydrated) positron, pick off, and intrinsic p Ps annihilation was discussed. Two sets of components given by sums of gaussians are shown in Table 5.3. As shortly discussed in [6.4], the components marked by 1 were used to analyze the AC spectra for MnO_4^-, $Cr_2O_7^{--}$, NO_3^-, $MnCl_2$, and TMOPO aqueous solutions, apart from the very extensive use of this set of components in positron reaction studies, as discussed in detail in Chap. 6. All these analyses form the most comprehensive attempt to fit the AC spectra of aqueous solutions in terms of three components. The main results of the analyses are the following.

First, if the spectra contained components of shapes significantly different from those of the water components, the fits were unacceptable, either because the component intensities became unphysical (e.g., negative), or the variances of the fits were much too large. AC spectra for solutions of Cl^-, Br^-, I^-, CN^-, S^{--}, and SCN^-, including $MnCl_2$, gave that result at middle and higher concentrations because positron bound states are formed (see Chap. 6). However, TMOPO in water above 0.2–0.5 M gave unacceptable fits, too. The reason is probably that the positron in Ps annihilates on TMOPO after Ps has reacted with TMOPO, although we cannot totally exclude that the hydrated positron reacts with TMOPO, too. More detailed AC measurements are necessary to determine the Ps-TMOPO reaction product more precisely.

Second, fits in terms of the three components gave very uncertain component intensities for aqueous solutions of MnO_4^-, $Cr_2O_7^{--}$, and NO_3^-, very probably because the shape of the pick off component of set 1 was very close to that of pure water. Hence, the analyses did not give clear information on the Ps reaction products for $Cr_2O_7^{--}$ and MnO_4^-.

Third, the two Ps components were added to give a total Ps component of the same shape as for pure water, and the spectra were fitted by a sum of the hydrated positron and total Ps components. The results were fully acceptable for the inhibitor, NO_3^-, giving an inhibition constant in agreement with LT measurements. Furthermore, adding the positron-halide bound state component resulted in acceptable three term fits of very many halide solution spectra (see Chap. 6). Surprisingly, the two component fits of the MnO_4^- and $Cr_2O_7^{--}$ spectra gave almost acceptable fits, too, which shows that such analyses might sometimes give uncertain answers.

Let us now discuss the validity of such analyses of AC spectra. First, the very recently discovered second o Ps component of the LT spectra, namely the distribution of shorter lifetimes due to secondary o Ps reactions with spur species, must be taken into account. These reactions of o Ps occur probably partly with the spur electrons, giving mainly a narrow component, and with the positive ions resulting in oxidation of Ps. As discussed in Chap. 5, this component has an intensity of 3–10% and an average lifetime of 0.8 ± 0.5 ns in the liquids studied until now, i.e., hydrocarbons, diethylether, methanol, water, CS_2, and some mixtures. For water the component is rather uncertain because the normal o Ps lifetime is short and its intensity fairly low. However, the high quality LT results show clearly that a fit by a sum of only three exponentials is somewhat wrong for water. At Risoe we realized already 10–15 years ago that the aqueous solution spectra contained more than what could be fitted well by three exponentials only. However, we had to wait for the high resolution measurements to get a reasonably explanation. The conclusion is, therefore, that the three water components used in the analyses of the AC spectra discussed above, are only approximately correct.

Let us try to select better AC component shapes. First, the arguments for selecting the shape of the free (i.e., hydrated) AC component are still valid, and hence, the shape of the 1.74 M PPS spectra is still our best choice. Our previous choice of the intensity of o Ps in the LT spectrum for pure water of 27.1% [6.4] was influenced by the second o Ps component. Free three exponential fits of water tend to give 1.95 ns (25%) and $\tau_2 = 480$ ps in stead of 395 ps, which is the τ_2-value obtained if Ps is totally missing. It is reasonable to assumed that the second and third exponentials both contain parts of the second o Ps distribution, resulting in the $\tau_2 = 480$ ps instead of 395 ps. The choice of the total yield of o Ps, I_o, is difficult, but unavoidable, if we wish to derive AC component shapes.

Let us choose $I_o = 30\%$, of which 25% annihilates by pick off giving a lifetime of 1.95 ps. The rest, 5%, is assumed to give 1.45% pick off, 1.77% narrow, and 1.78% free component AC shapes. The p Ps yield of 10% results in 0.60% pick off and 9.4% narrow component shapes. These assumptions result in an AC curve of pure water consisting of a narrow peak of 11.17%, a pick off component of 27.05%, and a free component of 61.78%. A constrained PAACFIT analysis then gives the shapes given as curves 2,H_2O in Table 5.3, assuming four gaussians for the total Ps component and constraining one gaussian to give the narrow peak intensity. A third assumption, namely that the o Ps shorter-lifetime distribution is due to oxidation only, apart from pick off, gives other shapes than 2,H_2O in Table 5.3. On comparing the two different sets of AC component shapes in Table 5.3, we see that the older pick off component contains a small intensity of a fairly narrow gaussian, the second pick off component contains a negative intensity gaussian to make it somewhat flat in the top, while the third pick off component is in between the two other components. Strictly speaking we do not know which of the sets of components, or other similarly chosen sets, is the correct one. The new sets of components have not, as yet, been used in detailed analyses of the available AC spectra.

It is important to realize that a change in the assumed yields of the two o Ps components influences mainly the shape of the pick off and narrow AC components. The shape of the total Ps component changes only little if the yield of free component is changed by some percents only. Hence, the computer analyses of AC spectra in terms of total Ps component and other curves, such as those used in the hydrated positron works, are influenced little by the changes in assumptions.

Of course, other assumptions than the ones mentioned until now are used, if aqueous solution AC spectra are analyzed in terms of the chosen shapes. For example, it is assumed that the AC component shapes and the pick off lifetimes are independent of the concentration of the solutes. In particular, the pick off AC component shape might very well change with the solute concentration at higher concentrations, because the structure of the Ps "bubble" is expected to be influenced by water properties which depend on the presence of the solute at high concentration. Furthermore, if the variances of the fits are acceptable, and if the yields of the different components are reasonably well determined and do not have unphysical values, it is reasonable to interpret the derived yields in terms of the kinetic equations of Sect. 8.1, which, however, are approximations, too, as the second o Ps distribution is not included in the equations. Another way of interpretation is to assumed that the kinetic equations are valid, and that the shape of the free AC component is equal to that of pure water, and then to derived the shape of the Ps components as functions of solute concentration. These shapes can then be compared to the shapes obtained for pure water, and a conclusion concerning the product of the Ps reaction might be reached. Such procedure was attempted in [8.39]. The Strasbourg group has used DB component shapes given by single gaussians only to fit many DB spectra. This procedure has mainly been used with good results in Ps inhibition studies, but it resulted in incorrect results for TMOPO/methanol solutions [8.34].

If the ion is diamagnetic a fit of (8.1) to the measured long lifetimes versus solute concentration can be used to decide whether a reaction with a constant rate constant is reasonable model. Figure 7.7 illustrates that the quality of such fit is good for MnO_4^-, $Cr_2O_7^{--}$, and IO_4^-, resulting in the well defined rate constants shown in Table 7.2. The assumptions that the diffusion coefficients, D, of Ps and the ion are equal to that of water of 2.5×10^{-5} cm^2s^{-1} and a reaction radius, R, of 0.7 nm give $k = 2.6 \times 10^{10}$ $M^{-1}s^{-1}$ by use of (6.14), as measured for $MnO_4{}^-$ and $IO_4{}^-$. Because Ps is light its D value might be expected to be higher than that of water, which means that R is correspondingly lower than 0.7 nm. Hence, the largest rate constants correspond to diffusion controlled reactions, as for nonpolar solvents. For the $CuCl_2$ solutions (8.1) does not give an acceptable fit because $CuCl^+$ and $CuCl_2$ reacting more strongly with Ps than Cu^{++} are present in the solution at higher concentrations.

The next step in the interpretation of the Ps reaction with ions is to explain the LT and AC spectra in terms of the properties of the ion and solvent. Only paramagnetic ions can react with Ps by spin conversion, but paramagnetic ions can also oxidize or form bound states with Ps if they are strong electron acceptors. For all ions it is important to know the structure of the ion. Ps reactions are studied at concentrations above about 0.002 M and in many cases up to 1 M. Hence, complex formation might be important, as measured for the $CuCl_2$ solutions.

The physics of the spin conversion reaction is different from that of the other Ps reactions. Hence, we shall at first discuss the Ps reactions resulting in oxidation of Ps and/or a positron bound state formation. For such reactions the change in the Gibbs free energy on a Ps ion reaction is roughly speaking influenced by three effects. First, the ion must have an electron affinity, i.e., it must provide a potential well for accommodation of the electron. Second, the positron must be accommodated either outside the ion in a hydrated state, on a positron acceptor, or on the ion together with the electron. Third, the Gibbs free energy of the initial Ps bubble state is determined mainly by the solvent properties.

Let us first discuss the effects of the electron affinity of the ion. In Sect. 7.5 we discussed the correlation between Ps quenching and inhibition by an ion and the electron accepting property of the ion. It was concluded that the standard reduction potential of an ion could not be used to explain its Ps inhibition properties. A specially calculated electrode potential for single electron reactions had to be used to explain the change from Ps inhibition to no inhibition by ions in water. Table 7.3 illustrates that point very clearly. Similarly, such special potentials have been used in the interpretation of hydrated electron reactions [4.2,7.61]. Of course, the same special electrode potentials are also the best ones to use in the correlation of Ps rate constants with the electron accepting strength of the ions (see [1.3], p. 104). Such correlation has not been published except by *Byakov* et al. [8.44], who, however, do not explain how their electrode potentials are calculated. If we accept their special potentials, experimental facts seem to show that a necessary (but not sufficient) condition for a Ps oxidation or bound state formation reaction with the ion is that the special electrode potential of the ion is above about 0.0 V. For certain ions the standard reduction potential of roughly 0.0 V seems also to be the limit for those reactions. This is to be compared with a special potential above -2.7 V for hydrated electron reactions, and above about −2.0 V for full (i.e., not partial) Ps inhibition. As mentioned in Sect. 7.5, spin conversion reactions of Ps occur with ions having much lower special potentials. For example, Ps reacts with Co^{++} having a special potential of about −3.1 V, which seems to be the lowest value for an ion shown to react with Ps.

If the positron state after the reaction is the hydrated positron, the reaction is a simple Ps oxidation. An expected positron hydration energy of about 2 eV, as for the electron, compared to a fraction of an eV in nonpolar solvents, leads us to expect that oxidation of Ps is more probable in water than in nonpolar liquids. The fact that MnO_4^-, $Cr_2O_7^{--}$, and PPS give almost identical AC spectra at high concentration supports this expectation, as discussed in Chap. 6.

If one of the ligands in the first solvation sphere of the ion is a positron acceptor, i.e., if that ligand reacts with the hydrated positron in water, the positron is expected to be trapped on the positron acceptor during the reaction, and the positron-ligand bound state will probably split off from the ion. Hence, the product is an electron on the ion and a positron on the free positron acceptor. Examples of such ligands are, of course, Cl^-, Br^-, I^-, CN^-, S^{--}, and SCN^-. Such a process makes the Ps reaction more exothermic by the difference between the energies of the hydrated and the bound positron states. If the reaction without the positron ligand reaction is strongly exothermic, this positron bound state formation might not influence the rate significantly. However, if the reaction is barely exothermic or endothermic without the positron ligand reaction, the rate constant might be strongly increased if the ligand is added to the solvation sphere. Maybe the increase of the rate constant on adding $Cl^-$to the solvation sphere of Cu^{++}, shown clearly in Fig. 7.7, is partly caused by such effect. It is important to note that the formation of such positron bound states in Ps reactions will give strong effects in the measured AC spectra, while it is probably not detectable by LT measurements except for the possible change in the o Ps lifetime. Such effects seem not to have been proposed or reported in the literature. Further AC studies would be very useful.

The last possible outcome of a Ps solute reaction is an electron and positron both bound on the solute. Of course, both the electron and positron will influence, and be influenced by, the solvation sphere, if they are bound to the ion, whether alone or together. Hence, strictly speaking they are bound both to the ion and the ligands in such states. We need not discuss this case in detail, as the discussion of the Ps reactions with solutes in nonpolar solvents in Sect. 8.3 applies also for Ps ion reactions in polar liquids. In particular, the special states giving anomalous magnetic quenching of Ps might occur also for Ps ion reactions in water. Such case has not been published yet. Furthermore, a positron-electron-ion state with no measurable Ps attributes should not be called a Ps bound state.

The energy of the initial Ps bubble state in hydrogen bonded solvents is determined by the dispersion, and maybe polarization, components of the surface tension. The hydrogen bond component does not influence the size of the Ps bubble, as discussed in Chap. 5. We expect that the discussion of the influence of the properties of the Ps bubble on the Ps reaction in Sect. 8.3 applies as well for reactions with ions. In particular, an increase in temperature is expected to decrease the energy of the Ps bubble compared to that of the Ps reaction product. A maximum in the rate constant versus temperature has not been found for the Ps ion reactions in water. The rate constants seem to increase with temperature, normally as expected from the viscosity variations, in the reported works.

Of course, energy conditions tell us only whether a Ps ion reaction is possible, in principle. Other conditions determine the actually measured rate constants, too. This is illustrated by the lower rate constant of $Cr_2O_7^{--}$ than that of MnO_4^-, although these two ions are so strong oxidizers that their reactions with Ps are expected to be strongly exothermic for both. Properties of the reaction, such as energy barriers and tunnelling through the barriers, steric hindrances, available states on the ion for the electron in Ps, rates of dissipation of the reaction energy, etc., might also influence the rate constant. The hydrated electron reactions with ions might serve as good examples to be used in the interpretation of Ps ion reactions [4.2]. For the hydrated electron the rates are not determined solely by the gain in Gibbs free energy between the initial and final products. More important is the availability of a vacant orbital at a low energy level in the ion. The capacity of ligands to "conduct" an electron to the central atom seems to influence strongly the rates of the hydrated electron reactions with ion complexes. Similar effects are expected in the Ps reactions with ions. However, too few rate constants of Ps reactions with ions are available, and a detailed interpretation is, therefore, almost impossible at present.

The rates of the spin conversion reactions of Ps with ions are determined by ion and solvent properties which are different from those influencing the two other Ps reactions. The discussion in Sect. 8.4 applies also for Ps ion reactions. The spin conversion rates are mainly influenced by the overlap of the wave function of the electron in Ps and those of the unpaired electrons on a solute. This overlap occurs during an encounter of Ps and the ion, i.e., when the ion is part of the surface of the Ps bubble. Hence the ion needs not be a strong electron acceptor to react with Ps by spin conversion. The largest rate constants of Ps spin conversion reactions of ions, and other solutes as well, have been measured to be 8–10 times lower than diffusion controlled rate constants, in agreement with theory (see Sect. 8.4). Hence, the largest spin conversion rate constants for aqueous solutions are about 3×10^9 $M^{-1}s^{-1}$. Experiments seem to indicate [8.45,1.3] that for an ion complex, the rate constant seems to be determined by the concentration and accessibility of unpaired electrons in the outer coordination sphere, and not just by the number of unpaired electrons belonging to the paramagnetic ion itself. Consequently, the rate constant seems to be largest for those ion complexes where the ligands "conduct" most efficiently the unpaired electrons of the central atom to the outer parts of the complex.

After thus having discussed the basic principles of the Ps ion reactions in polar liquids, we shall end this section by a description of selected cases.

As usual we cannot review all the papers on our topic because of lack of space. The criteria of the selection are that the recent trends of the field are well described, and that older results illustrating the basic principles are included.

The "classical" studies of Ps reactions with ions by use of LT measurements alone are well illustrated by the recent papers of Lazzarini's group, [8.46–49] and ref's therein. Their LT equipment had a resolution of about 330 ps. They analyzed in terms of three non-constraint exponentials by use of POSITRONFIT. No o Ps yields, I_3, were given, in agreement with the fact that I_3 is very uncertain if the second lifetime is not fixed in cases of Ps reactions. The type of reaction, spin conversion or not, was decided by reference to AC results of *Trumpy* [8.50] or estimated by reference to the standard reduction potential of the paramagnetic ions. The main experimental results were the rate constants versus temperature in the interval $274\ K < T < 313\ K$ [8.46,47], versus viscosity in H_2O/sucrose [8.48,49], and in various solvents [8.48].

Selected rate constants at room temperature are shown in Table 8.3. Other rate constants are given in Table 7.2. Except for $[Ni(NH_3)_6]Cl_2$, all the spin conversion rate constants are below $2.6 \times 10^9\ M^{-1}s^{-1}$, while the highest rate constant is $28.3 \times 10^9\ M^{-1}s^{-1}$ for MnO_4^-. Maybe $[Ni(NH_3)_6]Cl_2$ is partly a Ps oxidizer. The activation energy of the rate constant in water was roughly equal to that of the viscosity for both spin conversion and oxidation (bound state) reactions [8.46–48]. The rate constants were interpreted in terms of the Smoluchowski equation (6.14).

The Strasbourg group have also studied Ps ion reactions recently. In particular, they use the DB technique to identify the spin conversion reactions. In [7.65] they studied aqueous solutions of $Co(ClO_4)_2$. The anion was chosen to avoid ion association effects, and because it influences the o Ps lifetime and intensity very weakly (Table 7.2). They calculated the DB results by use of the LT data and the shapes of the water components given by only one gaussian each, as discussed above. By fitting the calculated to measured LT and DB results a spin conversion rate constant of $1.7 \times 10^9\ M^{-1}s^{-1}$, a bound state rate constant of $0.43 \times 10^9\ M^{-1}s^{-1}$, and a bound state fwhm = 2.36 keV (9.24 mrad) could be derived. However, the use of only one gaussian for each component is probably a fairly bad approximation, and it may lead to incorrect conclusions, as discussed in the case of TMOPO/methanol solutions in Sect. 8.4. Hence, the results should not be considered as the final answer, yet. On the other hand, this seems to be the only recent attempt to determine whether spin conversion occurs together with another reaction for a Ps ion reaction. A detailed analysis of LT and AC spectra, as the one performed for the TMOPO/methanol case in Sect. 8.4, has not been performed, as yet, because only very few AC spectra for spin conversion reaction of ions are available. Apparently, no clear case of a mixed reaction has been identified for any ion, and even the TMOPO/methanol case in Sect. 8.4 should be considered to be fairly undecided at present. These facts show, in particular, that very little has been done yet, i.e., AC spectra for the study of Ps reactions are strongly needed.

Table 8.3. Rate constants of selected ions in dilute aqueous solution at 20°C. Ox, oxidation; SC, spin conversion. Consult the original literature for solution details, since extra solutes, such as HCl to prevent hydrolysis, might have been added (see also Table 7.2)

Compound	k [$M^{-1}(ns)^{-1}$]	Ref.	Type
$(NH_4)_2[Ce(NO_3)_6]$	16.0	[8.46]	Ox
$GdCl_3 \cdot 6H_2O$	0.25	[8.46]	SC
$CeCl_3 \cdot 6H_2O$	0.05	[8.46]	SC
K_2CrO_4	14.1	[8.46]	Ox
$NH_4Fe(SO_4)_2 \cdot 12H_2O$	8.95	[8.46]	Ox
$K_2Cr(SO_4)_2 \cdot 12H_2O$	0.87	[8.46]	SC
$(NH_4)_2Fe(SO_4)_2 \cdot 6H_2O$	2.03	[8.46]	SC
$KMnO_4$	28.3	[8.47]	Ox
$K_3[Fe(CN)_6]$	18.0	[8.47]	Ox
$NiCl_2 \cdot 6H_2O$	2.29	[8.47]	SC
$[Ni(NH_3)_6]Cl_2$	4.84	[8.47]	SC
$[Co(NH_3)_5H_2O]Cl_3$	5.42	[8.48]	Ox
$K_4[Fe(CN)_6]$	0.05	[7.68]	Ox
Fe^{3+}	9.7	[8.47]	Ox
Hg^{2+}	6.7	[8.47]	Ox
Cu^{2+}	2.9	[8.47]	Ox
Fe^{2+}	2.3	[8.47]	SC
Cr^{3+}	1.0	[8.47]	SC
Ni^{2+}	2.6	[8.47]	SC
Co^{2+}	2.1	[7.65]	SC+Ox
$Co(CN)_6{}^{3-}$	0.185	[7.81]	Ox
$Co(NH_3)_6{}^{3+}$	2.65	[7.81]	Ox
$[Co(NH_3)_5F]^{2+}$	1.94	[7.81]	Ox
$[Co(NH_3)_5Cl]^{2+}$	15.7	[7.81]	Ox
$[Co(NH_3)_5Br]^{2+}$	31.5	[7.81]	Ox
$[Co(NH_3)_5H_2O]^{3+}$	5.9	[7.81]	Ox

Ion association effects can be very important for the Ps ion reactions. Examples are solutions of Gd^{3+} [8.46], Ni^{2+} and $SO_4{}^{2-}$ [8.51], Co^{2+} -polyanion systems, and Cu^{2+} (Fig. 7.7) [7.10].

The effect of complex formation on Ps reactions with inorganic ions was studied very early by the Moscow group [8.45,53,1.3]. Ache et al. studied Ps reactions with complexes of Co^{3+} [7.81], Fe^{3+} [7.82], and Hg^{2+} [7.83]. The last six diamagnetic ions in Table 8.3 illustrate that the rate constants depend very much on the ligands in Co^{3+} complexes. This seems to be general for the complexes which have been studied until now. The very high rate constants for complexes with Cl^- and Br^- ligands might perhaps be caused the formation of positron bond state with these ions on Ps reaction. The much lower rate for the complex with the F^- ligands seems to correlate with the fact that F^- does not react with the positron.

Very strong effects of the ligands on the Ps reactions with Co^{2+} and Fe^{2+} complexes in water propanol solutions (LT studies only) were reported in 1966–1967 by a Russian group [8.45,1.3]. As a function of the propanol content the rate constant for $CoCl_2$ and $FeCl_2$ decreased from about 2.9×10^9 $M^{-1}s^{-1}$ in pure propanol to zero at 0.5–0.6 mole fraction of water. It started to increase again from zero at about 0.8 mole fraction and became about 2.3×10^9 $M^{-1}s^{-1}$ in pure water. For $Co(ClO_4)_2$ only the strong decrease on adding propanol to water was found, i.e., the rate constant was roughly constant, $(0.1–0.2) \times 10^9$ $M^{-1}s^{-1}$, for a water content up to about 0.8 mole fraction in propanol. Similar, but much less pronounced, rate constant effects were found for spin exchange reaction of two free radical with the ion complexes. The strong changes in the Ps rate constants were explained as caused by strong changes in the accessibility of the unpaired electrons in the outer coordination sphere on adding of water to propanol. Direct reaction of Ps with the central paramagnetic ions was assumed not to take place. These old results seem to be the most interesting ligand effects in the Ps reaction with paramagnetic complexes, which have been published, and it is surprising that the studies have not been followed up in other works.

In summary, complicated effects have been found in Ps reactions with ions in water. Only very few Ps ion reactions have been studied in other polar liquids, see e.g., [7.82,83,8.45,48,53]. Judged from the complexity of the Ps ion reactions, too few experimental results are available, in particular concerning the product of the reactions (AC spectra). Hence, the theoretical explanations of the Ps ion reactions, given above, are fairly uncertain, probably a lot more uncertain than the explanations in the Ps inhibition cases. They are based more on theoretical common sense and results of electron reactions than on experimental facts on Ps ion reactions, and should be used with care.

9 Principles of Positron Annihilation in Molecular Solids

Chapters 9 and 10 contain mainly a discussion of positron annihilation in those solids which are normally treated within chemistry. In this chapter we shall discuss the basic principles, which are best illustrated by the results for molecular crystals. In Chap. 10 a general review of positron annihilation in miscellaneous solids and in the liquid crystals will be given. We shall not discuss metals and semiconductors. Ionic crystals will only be treated in the cases where particular effects are well illustrated. Furthermore, rare-gas solids are also considered to be mainly outside the topics of this book.

The behaviour of positrons and Ps in molecular solids depends on the structural defects apart from the dependence on the chemical composition. Hence, positron annihilation in molecular solids is more complicated than in molecular liquids. The repulsion between Ps and molecules which are not strong electron acceptors, causes Ps to become trapped in hole type defects, such as vacancies and vacancy clusters, in the solids. Also the excess electron and the positron may be trapped in structural defects, which may strongly influence the Ps formation, in particular.

In general, Ps formation and the behaviour of Ps and the positron are governed by the same principles in the solid as in the liquid state. For example, the spur model of Ps formation applies as well for solids. On the other hand, several very interesting effects, such as Ps in Bloch function states, very "squeezed" Ps, Ps reactions with defects, and slow positrons impinging on surfaces, can only, or mainly, be studied in the solid state. Historically, these effects have played a great role in the development of the theoretical concepts and ideas of the models used in the explanation of positron annihilation, in general, and hence, a detailed discussion of the effects is appropriate here.

In solids strongly distorted Ps states might exist. It should be emphasized that a state is called a Ps state in this book if it gives at least one of the experimental Ps attributes, i.e., a long lifetime ($\tau > 0.4$–0.5 ns) due to o Ps pick off annihilation and/or a narrow component (fwhm < 4–5 mrad) in the AC spectra due to p Ps intrinsic annihilation. Such Ps can be very different from Ps in vacuum, as will be shown below.

Several good reviews on positron studies of molecular solids have been published by *Eldrup* [4.24,1.41] and *Goworek* [1.42,9.1].

9.1 Positronium Bloch Functions and Ps Trapping in Defects

In solid crystalline compounds the excess electron is normally in the Bloch function state. Because of the strong repulsion between Ps and most atoms, ions, and molecules the Bloch function state for Ps is found only in some very brittle, non-defected solids, as discussed in Sect. 4.5. Evidence for Ps Bloch functions was first measured for quartz [1.39], MnF_2 [9.2], MgF_2 [9.2], and later for H_2O and D_2O ice [1.40], and in several alkali halides (at LN_2 and LHe temperatures in NaF, NaCl, NaBr, NaI, KCl, KBr, KI, and RbCl, but not in LiF, KF, and CsI) [9.3,4]. Hence, the only molecular crystal in which Ps Bloch functions, i.e., Ps delocalized over many molecules, have been found is ice. However, Ps in ice seems to constitute the best illustration of some of the basic characteristics of Ps behaviour in solids, and we shall, therefore, discuss delocalized Ps and Ps trapping in defects in ice in some detail. Furthermore, ice seems to be the molecular crystal investigated in most detail by positron annihilation techniques.

The AC spectra of ice single crystals at low temperatures contain very narrow (fwhm < 1 mrad) peaks caused by intrinsic p Ps annihilation in the Bloch function states [1.40,2.2,9.5]. This is clearly shown in Fig. 3.4, where the ice spectrum at $-1°$C, the H_2O water spectrum at 20°C, and the water components (Table 5.3 and Fig. 6.1) are given too. In Fig. 9.1 the intrinsic p Ps AC components are shown at various temperatures between $-182°$C and the melting point. As the temperature is increased above roughly $-100°$C the intensities of the very narrow central peak and sidepeaks decrease and a middle-broad component of fwhm $= 3.9$ mrad appears. The latter is due to p Ps intrinsic annihilation in localized states on vacancies, the concentration of which seems to increase exponentially with temperature as normally found for solids.

The first term in (2.6) is the intrinsic annihilation probability of p Ps. A reasonable assumption [1.40] is to put $\psi_2(\mathbf{r}_1, \mathbf{r}_p) = X(\mathbf{r}_c) f(\mathbf{R})$, i.e., a product of a function $X(\mathbf{r}_c)$ of the center-of-mass coordinate $\mathbf{r}_c = (\mathbf{r}_1 + \mathbf{r}_p)/2$ and a function $f(\mathbf{R})$ of the relative coordinate $\mathbf{R} = \mathbf{r}_1 - \mathbf{r}_p$, e.g., the s-state wave function. (This assumption is not necessary, i.e., we would also get (9.3) if we only assume $\psi_2(\mathbf{r}_1, \mathbf{r}_1)$ in (2.6) to be a Bloch function). We assume that Ps is delocalized in the center-of-mass motion, i.e., $X(\mathbf{r}_c)$ is a Bloch function in agreement with the translational invariance of the lattice. Hence, we have

$$X(\mathbf{r}_c) \propto \exp(\mathrm{i}\mathbf{p} \cdot \mathbf{r}_c) \sum_{\mathbf{g}} a_{\mathbf{g}} \exp(\mathrm{i}\mathbf{g} \cdot \mathbf{r}_c), \tag{9.1}$$

where $\mathbf{p}$ is the crystal wave vector and $\mathbf{g}$ is a vector in the reciprocal lattice. We then get from the first term in (2.6)

$$dw \propto \sum_{\mathbf{g}} |a_{\mathbf{g}}|^2 \delta(\mathbf{p} - \mathbf{k} + \mathbf{g}) d\mathbf{k} d\Omega. \tag{9.2}$$

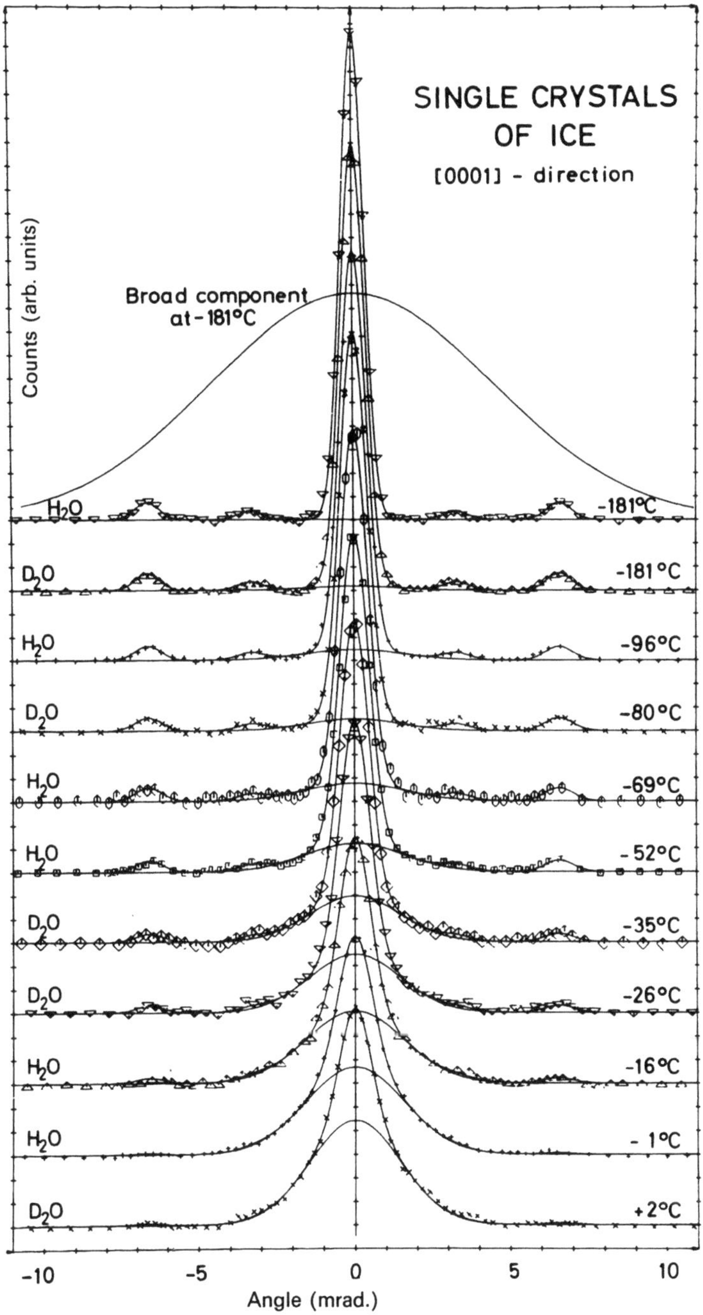

Fig. 9.1. The temperature dependence of the AC spectra for single crystals of H_2O ice oriented along the *a*-axis ([1121]). The measured points and the fitting curves are shown after the subtraction of the broad component from the normalized curves. Only uncertainties greater than the size of the points are shown [2.2,9.5]

The linear slit AC setup determines in a good approximation

$$dw_1 \propto \int_{\Delta\Omega} d\Omega \int_{-\infty}^{\infty} dk_x \int_{-\infty}^{\infty} dk_y \sum_{\mathbf{p}} S\{E(\mathbf{p})\} \sum_{\mathbf{g}} |a_{\mathbf{g}}(\mathbf{p})|^2 \delta(\mathbf{p} - \mathbf{k} + \mathbf{g}) dk_z \tag{9.3}$$

folded with the angular resolution curve. Here $S\{E(\mathbf{p})\}$ is the Maxwellian distribution, and $\Delta\Omega$ is the effective detector area. Peaks may, therefore, occur at angles

$$\theta = 2\pi\hbar g_z/mc, \tag{9.4}$$

where g_z is the projection of $\mathbf{g}$ on the z-direction. The intensity, $I_{\mathbf{g}}$, at the reciprocal point, $\mathbf{g}$, is given by

$$I_{\mathbf{g}} = \sum_{\mathbf{p}} S\{E(\mathbf{p})\}|a_{\mathbf{g}}(\mathbf{p})|^2. \tag{9.5}$$

Only states $\mathbf{p}$ for $|\mathbf{p}| \ll |\mathbf{g}|$, $\mathbf{g} \neq \mathbf{0}$ are occupied, as p Ps is thermalized or nearly thermalized. Hence the shapes of the sidepeaks are nearly identical to that of the central peak and all the peaks are very narrow. Experiments show that $I_{\mathbf{g}}$ is only significantly different from zero for small $\mathbf{g}$, in particular in ice.

This simple theory [1.40] has been extended to take into account the influence of the thermal motion of the atoms on the Ps Bloch function state in a "nearly-free-Ps" approximation [2.2]. As usual in solid state physics, the contributions from the reciprocal lattice points are then given by the numerical value of the square of a "structure factor", which varies with the temperature through the Debye-Waller factors. Also in this, next-higher, approximation very narrow peaks are found at the reciprocal lattice points.

The AC component from p Ps intrinsic annihilation in a localized state was discussed in Sect. 4.4. If we assume that the middle-broad AC component is due to Ps trapped in the vacancy in an harmonic potential, (4.5) gives that $\text{fwhm}_z = 0.27$ nm for $\text{fwhm}_\theta = 3.9$ mrad, which is a fairly reasonable value as the O–O distance in ice is 0.276 nm.

The positions of the peaks, and the relations between the side peak intensities which follow from the Bloch function model, in the measured AC spectra agree well with the theory given above. The measured widths (0.86 mrad at -180°C) are 10–13% larger than what can be explained by the geometrical angular resolution of about 0.65 mrad and the thermal motion of Ps of 0.17 mrad at -180°C, i.e., fwhm $= 0.77$ at -180°C. On the other hand, the temperature dependence of the peak intensities disagrees with that obtained from the "nearly free" Ps model. If we assume that the intensity of the central peak decreases because of Ps trapping in vacancies, the ratio of the intensity of a sidepeak and that of the central peak is expected to be explainable through the temperature dependence of the structure factor as given by the Debye-Waller factors. However, the side peak intensities decrease much more than expected at increasing temperature. The reason for that might be that Ps, like electrons in metals, is not delocalized over the whole crystal as assumed if a Bloch function is used. A more realistic model would be to consider a Ps state delocalized only within a certain region, the size of which is given by a parameter, the coherence length [2.2].

The LT measurements [2.2,9.6] give a long lifetime, $\tau_3 = 0.67$ ns at low temperatures. Above -100°C this lifetime starts to increase and it is roughly 1.0 ns at -5°C, where an extra LT component $\tau_4 = 1.7$ ns (5.1%) has also appeared. The latter is probably due to Ps trapped in divacancies. Low temperature LT measurements on irradiated and fast cooled HF-doped ice give clear evidence of a lifetime of Ps trapped in vacancies of $\tau_3 = 1.2$ ns [9.7,8] (see below). The total o Ps intensity of roughly 54% seems to be fairly independent of temperature. Hence, also the changes in the LT spectra can be explained as caused mainly by Ps trapping in vacancies. However, the simple trapping model cannot account for the observed changes in the combined AC and LT spectra [2.2]. This model predicts that the fraction of trapped o Ps is much larger than that of p Ps at a certain temperature since o Ps lives 5–6 times longer than p Ps in bulk ice, while the measured fractions of trapped o Ps and p Ps are almost identical at all temperatures. A detailed analysis of the data and fairly unsuccessful attempts to use other models of the trapping of Ps can be found in [2.2]. Altogether, the conclusion seems to be that the fundamental concept of a trapping probability per unit time, i.e., a rate constant as discussed in Sect. 6.2, cannot be used; a very interesting result indeed. A similar analysis of both AC and LT results on trapping of Ps in other solids seems not to have been reported.

Positron annihilation in γ-irradiated and HF-doped ice crystals has been studied in details at Risoe [9.7,8]. The results have been of great importance for the understanding of the trapping of Ps in hole type defects in ice. We can only give a short review of the main conclusions here.

The LT results for pure ice γ-irradiated (Co-60) at -196°C and measured for increasing temperature are shown in Fig. 9.2 [9.7,8]. The LT spectra could be resolved into four lifetime components. The most interesting parameters are given in Fig. 9.2. Each point represents about 2.5 days of measuring time. Hence, a series of points, e.g., the circles, is a typical isochronal annealing experiment, where the defects can migrate 2.5 days at each temperature. The onset of significant migration of certain defects introduces changes in the LT spectra. We can here only describe shortly the defects and their influence on the LT spectra. Arguments for the given interpretation are found in the references.

The γ-irradiation causes several changes in the LT spectra. First, the Ps formation is inhibited mainly because interstitial OH radicals, which are created by the irradiation, scavenge the electrons in the positron spur. Second, Ps is trapped in the vacancies left behind when the interstitial OH radicals were injected into the lattice giving rise to the $\tau_3 = 1.2$ ns (I_3) component. Third, a longer lifetime $\tau_4 = 2.3$ ns (I_4) component due to Ps trapping in mainly divacancies appears. The irradiation introduces also some Ps quenching [9.7]. During the annealing the first strong change in the LT spectra occurs at roughly -180°C where the interstitial OH radicals begin to migrate. They disappear by reactions with each other, the vacancies, and divacancies. Consequently, the

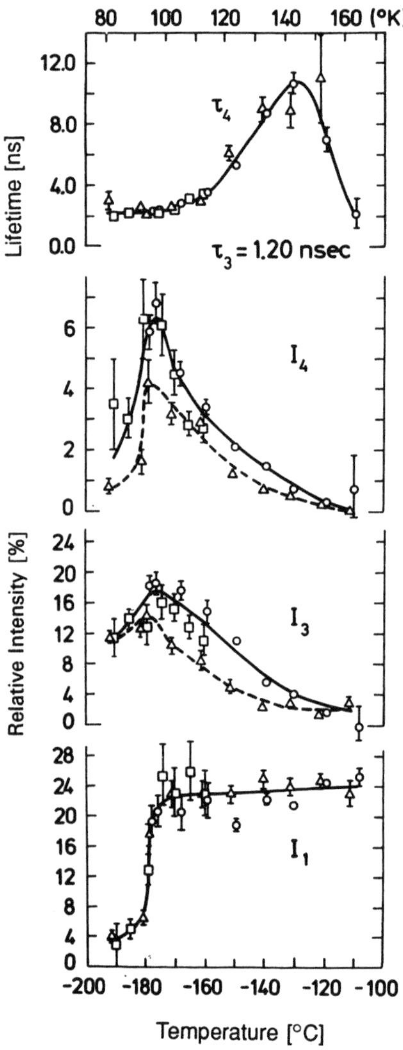

Fig. 9.2. Lifetime results for pure H_2O ice γ-irradiated at -196°C and measured as a function of increasing temperature. (○) polycrystalline samples (11 Mrad), (△) single crystals (4.1 Mrad), and (□) another single crystal (11 Mrad). The full and dashed curves for I_3 and I_4 are drawn for the 11 Mrad and 4.1 Mrad points, respectively [9.7,8]

Ps formation returns to the value found in non-irradiated ice. In particular, I_1 due mainly to p Ps annihilation shows this very clearly. Also the Ps quenching disappears at -180°C.

At temperatures above -175°C a pronounced increase in τ_4 and simultaneous decreases in I_3 and I_4 take place. This is caused by the onset of vacancy, and probably divancancy, migration. Some of the vacancies cluster into divacancies and polyvacancies, and the polyvacancies increase in size as the larger polyvacancies acquire vacancies from the smaller ones having higher surface tensions. The o Ps pick off lifetime in the polyvacancies increases, therefore, up to 11 ns at -130°C, corresponding to a void of about 1.5 nm diameter, i.e., roughly 50 vacancies. At temperatures above -130°C the voids shrink by vacancy "evaporation" from the void surface, and τ_4 decreases again. At still higher temperatures the concentration of voids becomes so low and their sizes so small that the Ps trapping cannot be detected. The LT spectra are then roughly as in the defect-free ice. Apparently the H_2O_2 and H_2 formed as the result of the annealing [9.9] do not appreciable influence the Ps behaviour for the used radiation doses of 4.1 and 11 Mrad.

Interstitial OH radicals created by irradiation have been found by radiation chemists [9.9], whose annealing results indicate that the OH radicals can be expected to migrate at $-180°$C. The vacancy migration and the formation of small polyvacancies have not been detected before for ice.

Measurements of high resolution AC spectra for ice [9.10] showed that Ps is essentially thermalized at 4 K. However, at 4 K the homogeneous radiation damage due to the slowing down of the positrons caused a strong decrease of the intensities of the narrow p Ps peaks and the appearance of a middle-broad (3–4 mrad) component associated with p Ps trapped in vacancies. On heating, the middle-broad component disappeared and the narrow component intensities increased strongly at about $-170°$C. This indicates that the main annealing occurs at about $-170°$C and that the vacancies created at lower temperatures disappear at this stage.

The above given interpretation is strongly supported by the results for HF-doped ice [9.8]. For example, a 25 p.p.m. HF doped crystal was quenched from $-18°$C to liquid-nitrogen temperature (cooled in less than 1 min), and LT measurements were then made as a function of increasing temperature with a measuring time at each temperature of about 3 days. Apparently, quenched HF-doped ice contains a high concentration of vacancies. During the annealing the migration of vacancies to form divacancies caused an increase in I_4 at about $-170°$C for $\tau_4 = 2.3$ ns, which corresponds to Ps trapped on a divacancy. At $-150°$C τ_4 starts to increase and I_4 to decrease as the clustering of the divacancies into larger polyvacancies sets in. A τ_4 value of about 14 ns is obtained at $-120°$C. Above $-120°$C the Ps trapping in the voids becomes insignificant because the concentration of the large voids is too small. Of course, the HF doping does not give OH radicals, and hence, the strong changes in the LT spectra at $-180°$C in Fig. 9.2 are not measured at annealing of the quenched HF-doped ice.

The slow positron studies of ice will be discussed in Sect. 9.5.

In the other crystals in which Ps is in a Bloch function state the trapping of Ps cannot be studied as in ice for several reasons. First, the time dependence of the LT spectra does not show clear effects of trapping in any defects in the temperature interval, where the p Ps peaks disappear [9.11]. Apparently, Ps is self-trapped at higher temperatures resulting in the disappearance of the p Ps peaks in the AC spectra [9.4]. Second, a well defined middle-broad AC component associated with Ps trapped in a certain defect does not appear in the same temperature interval. Third, irradiation and doping do not give defects which trap Ps. In some alkali halides irradiation creates F-centers which trap positrons [9.12].

From defect physics point of view the results of the studies of Ps trapping in hole type defects in ice were very important. It was the first time a vacancy formation energy, a vacancy migration energy, and clustering of vacancies into small polyvacancies have been directly studied for any molecular crystal. The vacancy concentration in ice at the melting point was increased more than a factor of 10^4 as a result of the studies. Ice is the "standard" proton semiconductor. The higher concentration of vacancies implies that vacancies make bound states with an appreciable part of some of the defects, e.g., the D-defects, on the hydrogen bonds responsible for the proton conduction. Hence the "standard" model of proton conduction must be changed considerably [9.5], which probably means that the normally used models of proton conduction in other systems, including, e.g., some biological membranes, must be changed too.

9.2 Positronium Trapping in Vacancies in Plastic Crystals

In plastic crystals Ps trapping in vacancies has also been detected but only for vacancies in thermodynamic equilibrium for temperatures above roughly $0.7T_m$, where T_m is the melting point in K. Clear cases of Ps trapping in vacancies at low temperatures have not been found, mainly because it seems to be unknown how to create such non-equilibrium vacancies. Plastic crystals consist of larger molecules where irradiation splits the molecules into radicals and does not create vacancies, as in most molecular crystals. Furthermore, vacancies created by doping seem not to have been detected in molecular crystals except in ice. The plastic crystal results are good illustrations of some of the basic principles of Ps behaviour in molecular compounds, and hence, they will be discussed in some details here.

The plastic crystals, reviewed in [9.13], are in most cases composed of globular molecules. At temperatures close to the melting point the reorientational barriers between neighboring molecules are reduced, partly due to the globular shapes of the molecules and partly because of the structure of the molecular interaction. This allows a high degree of orientational molecular disordering while the translational order in the lattice structure is maintained. In this high temperature phase the crystals are plastic. All these solids undergo first-order solid-solid phase transitions to a more dense, orientationally ordered, brittle phase with decreasing temperatures. Disorder-disorder phase transitions occur also in some crystals.

The nature and properties of the intrinsic point defects, in particular the vacancies, in the plastic crystals have been studied by use of positron annihilation in a collaboration between Risoe (M. Eldrup) and Glasgow (J.N. Sherwood). It was hoped that Ps trapping in vacancies in these crystals would give as useful information as in ice; a hope which was only partly fulfilled. Nine plastic crystals: succinonitrile, adamantane, camphene, bicyclooctane, pivalic acid, cyclohexane, cyclooctane, hexamethylethane, and norbornylene, and a brittle crystal, anthracene, were investigated. References

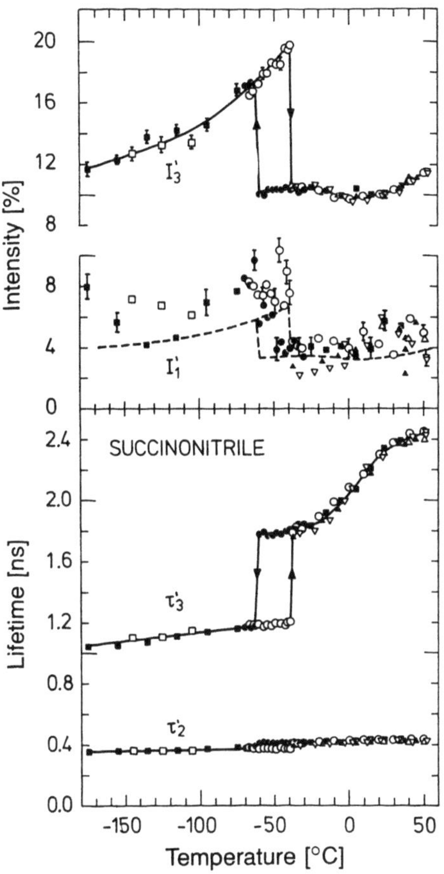

Fig. 9.3. Lifetimes and intensities for succinonitrile as derived from a three-component analysis with $\tau_1 = 0.12$ ns [9.18]

can be found in [4.24], except for [9.14–17]. The hexamethylethane and norbornylene results are unpublished.

The characteristic LT results for a plastic crystal are well illustrated by the lifetimes and intensities for succinonitrile [9.18] shown in Fig. 9.3. Let us discuss the behaviour of the long o Ps lifetime, τ_3. In the brittle, monoclinic phase at lower temperatures τ_3 increases slowly for increasing temperatures. At the solid-solid phase transition of −40°C the lifetime changes abruptly from 1.2 ns to 1.8 ns. This phase transition is shifted to −60°C for decreasing temperatures. In the plastic bodycentered-cubic phase τ_3 shows a sigmoidal behaviour with increasing temperature up to the melting point of 57°C. This characteristic behaviour, found for many plastic crystals, is very probably caused by o Ps trapping into thermally generated vacancies. At the lower temperatures Ps exists in self-trapped "cage" state in the bulk. We repeat that no very narrow AC components caused by p Ps annihilation in a delocalized state have been found for any molecular crystals except for ice. However, with increasing temperature the vacancy concentration increases and Ps begins to get trapped in the vacancies. Probably the trapping sets in at a relative concentration of about 10^{-7}, somewhat dependent on the diffusion constant of Ps. Close to the melting point almost all o Ps is trapped before annihilation and the measured lifetime is close to that of o Ps in a vacancy. The relative vacancy concentration is not well known. It might be 10^{-3} to 10^{-4} at the melting point. In particular, adamantane gives a clear case of such sigmoidal behaviour [9.19]. A similar sigmoidal lifetime variation at increasing temperatures is well known in the studies of positron trapping in thermally generated vacancies in metals. The associated intensity, I_3, increases from 12% to 20% with increasing temperature up to the phase transition. It decreases roughly a factor of two on transition to the plastic phase, where it is fairly constant at 10–12%. The decrease at the transition might be caused by more trapping of the light spur particles in the plastic than in the crystalline phase (see below), or by a much easier loss of CN^- by the succinonitrile anion formed on trapping of the electron in the spur (see Chap. 7) in the plastic phase.

A more detailed analysis of the LT spectra for the plastic phase shows that four lifetime components can be extracted for many plastic crystals and, in particular, for succinonitrile. Furthermore, the variation of the two longest o Ps lifetimes and their intensities are consistent with the trapping model description of o Ps trapping in vacancies. The trapping model for positron trapping was described in Sect. 6.2 ((6.4) and (6.5)). From the trapping model analysis the trapping rate is obtained. In Fig. 9.4 it is shown as a function of inverse temperature for succinonitrile. The slope of this Arrhenius plot is equivalent to an activation energy of 0.36 ± 0.1 eV. If it is assumed that the trapping rate is proportional to the vacancy concentration with a temperature independent proportionality constant (the rate constant in (6.4) and (6.5)) a vacancy formation energy, $E_V^f = 0.36$ eV, is derived for succinonitrile. Similarly, vacancy formation energies have been obtained for adamantane (0.73 ± 0.1 eV), camphene (0.33 ± 0.07 eV), bicyclooctane (0.55 ± 0.1 eV), and cyclooctane (0.16 ± 0.05 eV). Apparently, these values, together with that of ice, seem to be the first directly determined vacancy formation energies for any molecular crystal. A value of 0.10 ± 0.02 eV for methane deduced from o Ps trapping data has recently been reported [9.20].

Some years ago it was often expected that the vacancy formation and migration energies for molecular crystals should be close to the crystal lattice energy, L_s, i.e., roughly the sublimation energy [9.13]. For the less plastic crystals, adamantane and bicyclooctane, $E_V^f \simeq L_s$ is in fact observed, while for the softer materials succinonitrile and camphene E_V^f is about $0.6 \times L_s$. For the soft cyclooctane E_V^f is only about $0.3 \times L_s$. Selfdiffusion in plastic crystals (and ice) takes place by a vacancy mechanism [9.13], and hence, the selfdiffusion activation energy $E_d = E_V^f + E_V^m$, where E_V^m is the vacancy migration energy. For the less plastic of these crystals both tracer diffusion and NMR measurements give activation energies of roughly $2 \times L_s$, while the results for the softer plastic crystals are more uncertain. Hence, also in diffusion experiments the softer crystals give rise to unresolved problems.

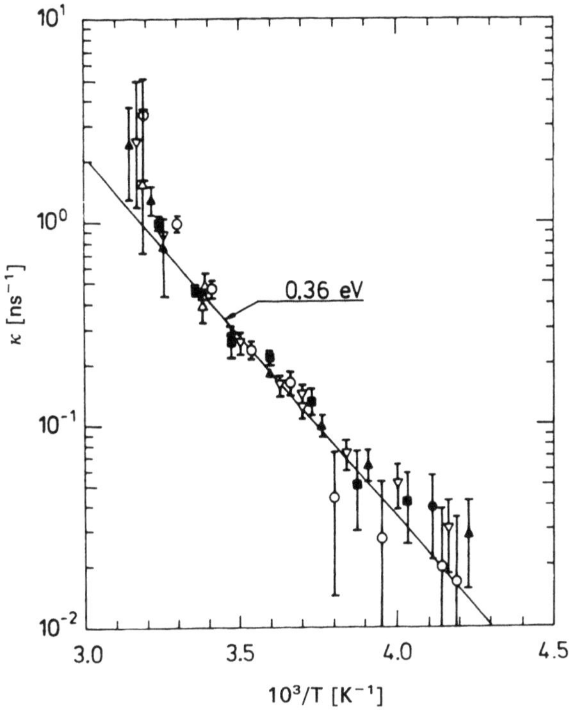

Fig. 9.4. The trapping rate, κ, of o Ps into vacancies in succinonitrile as extracted by use of the trapping model, versus the inverse temperature [9.18]

For hexamethylethane and norbornylene an additional longlived component of about 1 ns has made the detailed analysis in terms of the trapping model impractical, and no vacancy formation energies could be derived. In cyclohexane and pivalic acid no sigmoidal behaviour of the o Ps lifetime could be observed and hence, no vacancy formation energy could be deduced. For pivalic acid the transition to the plastic phase takes place at a very high temperature ($0.9 \times T_m$), which reduces the range of observation to temperatures at which nearly all Ps seems to be trapped [9.21]. Cyclohexane has a phase transition to the plastic phase at −85.6°C ($\sim 0.7 \times T_m$) but the longest lifetime just above the transition is very long (2.26 ns), which indicates that most Ps is trapped in vacancies already at temperatures just above the transition point [9.15]. The reason might be that cyclohexane is a very soft crystal. For both pivalic acid and cyclohexane a second long lifetime component of about 1 ns is found too.

The interpretation of the sigmoidal o Ps lifetime behaviour in the plastic phases as due to Ps trapping in vacancies is further supported by two interesting relationships. The first one is shown in Fig. 4.1 in Sect. 4.4, where the measured o Ps lifetime versus cavity volume is plotted for various solids and liquids. For the molecular crystals (circles) the cavity volume is taken to be the volume per molecule in the crystal. This volume is approximately equal to the volume of a molecular vacancy. The lifetime for the molecular crystals is the measured lifetime in the plastic phase just below the melting point. As expected the lifetime increases with increasing volume of the vacancies. For the three liquids the cavity volume is the Ps bubble volume as calculated for an infinite spherical potential for Ps, as discussed in Sect. 4.4, and the lifetime is the measured lifetime. It must be emphasized that the bubble volumes are not measured but derived by use of the measured lifetime and a simple model of the bubble. As can be seen the open circles are in excellent agreement with the used model. The closed circles are for the crystals: pivalic acid and hexamethylethane, for which selfdiffusion measurements give evidence of an appreciable divacancy concentration close to the melting point. In particular, the lifetimes for cyclohexane, norbornylene, hexamethylethane, and pivalic acid, for which a clear interpretation of the measured data was fairly difficult, agree with the general trend observed for the other crystals. Generally speaking, it may be concluded that the curves in Fig. 4.1 to a good approximation represent the o Ps lifetime in a cavity in an ordinary molecular crystal as a function of the volume of the cavity. In Chap. 10 we shall refer to Fig. 4.1 in the discussion of Ps in polymers and zeolites.

The second relationship correlates the o Ps lifetime in various hydrocarbon crystals (plastic as well as non-plastic) to the packing coefficient of the crystal. The packing coefficient is defined as the volume of the molecule V_M divided by the total volume per molecule Ω. V_M was calculated from Kitaigorodsky's values [9.22] for the volumes of various atom groups, while Ω was determined from available density and X-ray data. The lifetimes are those which can reasonably be associated with o Ps in the bulk. For plastic crystals they are the lifetimes for temperatures just above the phase transition. This correlation is plotted in Fig. 9.5. The fact that a reasonable correlation is obtained for crystals of different structures and molecular shapes seems to indicate that Fig. 9.5 can be used to estimate the o Ps lifetime in the bulk of such crystals, in general. In particular, the measured cyclohexane lifetime is too high compared with the value expected by use of the correlation in Fig. 9.5. This is in agreement with the interpretation given above. In [1.42] new points for other crystals have been added to Fig. 9.5. For some long chain alcohols the points deviate appreciably from the correlation, which seems to indicate that the lifetimes chosen in [1.42] are not those of the brittle phases.

It is important to realize that the fairly good correlation in Fig. 9.5 indicates only that the cage in which the localized o Ps is situated is larger the smaller the packing coefficients are. It does not show that o Ps is mainly situated in the available free volume of the crystal. The cage is probably enlarged appreciably due to the repulsion between Ps and the molecules; how much is difficult to estimate. By use of the correlation shown in Fig. 4.1 the volume of the cage for a lifetime of 1.2 ns can be estimated to be roughly that of the water molecule in ice. Hence, a rough guess of the structure of the Ps cage is an irregularly shaped cavity between the molecules of approximately 0.2–0.3 nm size. Furthermore, Ps is frequently swollen compared to Ps in vacuum as indicated by the low density of the electron at the positron found in certain crystals by magnetic quenching experiments (see Sect. 9.3).

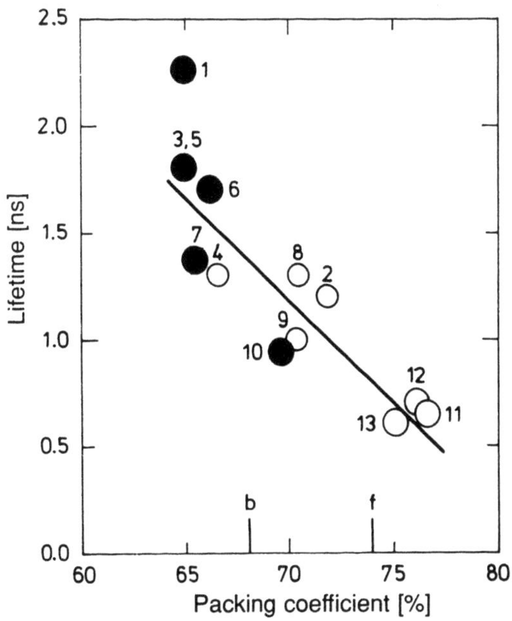

Fig. 9.5. The longest lifetime ascribed to o Ps pick off annihilation in bulk material, plotted versus packing coefficient for a number of hydrocarbon molecular crystals. Closed circles refer to plastic crystals and open circles to brittle crystals. The numbers refer to the following crystals: 1 and 2 cyclohexane; 3 and 4 camphene; 5 norbornylene; 6 hexamethylethane; 7 bicyclooctane; 8 eicosane; 9 naphthalene; 10 and 11 adamantane; 12 pyrene; 13 para-terphenyl. The arrows *b* and *f* show packing coefficients for touching spheres in a bbc and an fcc lattice, respectively [4.24,9.15]

The intensity of the o Ps lifetimes decreases by a factor of about 1.7 on transition from brittle to plastic phase in several crystals (succinonitrile, adamantane, cyclohexane, and cyclooctane). The intensities in molecular crystals will be discussed in Sect. 9.4.

Probably the lifetimes of the "free" positron in the molecular crystals studied at Risoe are amongst the best determined values for the crystals in which Ps is formed because of the quality of the experiments and the fact that four term analyses were applied. Of course, it is expected that secondary Ps reaction with spur species will take place also in the solid compounds. However, because of the short o Ps lifetimes and the fairly low o Ps intensities normally found in these solids, and because of the Ps trapping phenomena, the influence of the secondary Ps reactions on the lifetime spectra is more difficult to determine than in the liquids. Typical well defined values of the second lifetime, τ_2, ascribed to "free" positron annihilation are: benzophenone 331 ps (293°C), no Ps [1.47]; CBr_4 370 ps (−50°C), 380 ps (40°C), 408 ps (48°C), 409 ps (90°C), no Ps [9.23]; CCl_4 385 ps (−75°C), 395 ps (−47°C), 410 ps (−41°C), 424 ps (−23°C), no Ps [9.23]; anthracene 320 ps (20°C), 350 ps (200°C), no Ps below 200°C [9.24]; norbornylene 320 ps (−103°C), 350 ps (42°C) (unpubl.); cyclooctane 320 ps (−153°C), 360 ps (7°C) [9.16]; succinonitrile see Fig. 9.3; pivalic acid 300 ps (−153°C), 360 ps (0°C) [9.21]; and adamantane 380 ps (30–235°C) [9.19]. We note that apart from CCl_4 and CBr_4 τ_2 is always below 400 ps, it can be even as low as 300 ps, and it increases normally with increasing temperature. At the phase transition τ_2 is 10–20 ps higher in the plastic phase. In liquids the "free" positron lifetime, τ_2, is normally roughly 400 ps, with some exceptions, e.g., CCl_4, provided the fourth LT component is taken into account (Table 5.1).

It could be speculated that the equilibrium state of the "free" positron in matter involves some sort of positron solvation by electrons bound in the molecules and that it, therefore, has the lowest energy if a small cavity between the molecules can be formed around the combined positron-electron state. Such cavity can form much easier in the liquids and more difficult in the solids, in particular, the rigid solids at low temperatures. The formation of the cavity might reduce the density of electrons at the positron and, thus, increase the positron lifetime. The larger cavities formed in the liquids and in the solids at high temperatures explain that the longest τ_2 values are found in these compounds. Also in the liquid state τ_2 is larger in CCl_4 (458 ps) than in most other liquids. Generally speaking, we know less about the "free" positron behaviour than that of the two other light particles in solids, just as is the case for liquids.

9.3 Positronium Formation in Solid Aromatic Hydrocarbons

In several aromatic hydrocarbon crystals (biphenyl, p-terphenyl, quarterphenyl, naphthalene, pyrene, and chrysene) a transition from a non-Ps state at low temperatures to Ps localized in a cage at higher temperatures is observed. Figure 9.6 shows that the intensity of the long lived component in the lifetime spectra increases from zero to a finite value in a sigmoidal way for rising temperatures in biphenyl and naphthalene. For both crystals the longest lifetime increased slightly from 0.9 ns to 1.1 ns in the given temperature interval. For biphenyl the middle lifetime was roughly given to be around 300 ps for a fixed shortest lifetime of 123 ps [9.25]. For the shortest lifetime fixed to 124 ps the

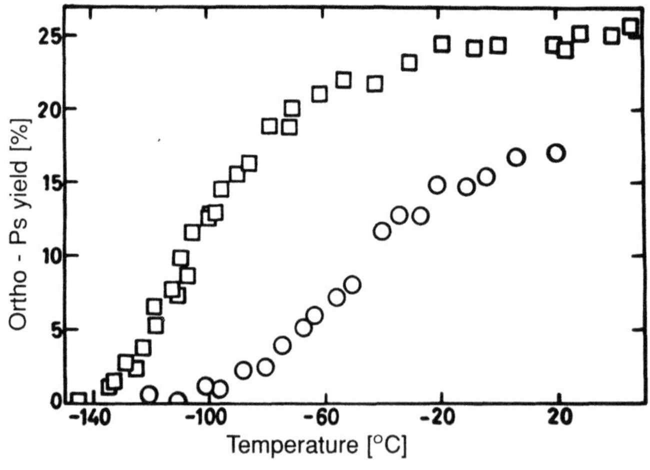

Fig. 9.6. The o Ps yield versus temperature in biphenyl (□) and naphthalene (○) [9.25,26]

middle lifetime increased from about 320 ps at −110°C to 350 ps at 20°C in naphthalene [9.26].

The explanation of these results is the following. At higher temperatures the longlived component is due to o Ps annihilation in a cage in the lattice, just as for the brittle crystals discussed in Sect. 9.2. The longest lifetimes agree with those expected from the correlation to the packing coefficient shown in Fig. 9.5. As the temperature is decreased below 20°C for naphthalene and −60°C for biphenyl the o Ps yield decreases, and it is reduced to zero below −100°C in naphthalene and −140°C in biphenyl. Clearly, the most reasonable explanation of the missing Ps formation is that the energy of Ps in its cage state is higher than that of a non-Ps state at low temperatures. As Ps is formed by a spur reaction between an electron and the positron at higher temperatures, and since the positron spur is formed by processes occurring at energies far above thermal energies, the positron spur is, of course, created also at the temperatures where Ps is not formed. At low temperatures the electron and positron, which would have formed Ps at higher temperatures, are, of course, mutually attracted by the Coulomb field and they will diffuse towards each other. However, they cannot, apparently, perform the final step into the Ps cage state because of lack of energy to create the Ps cage. The dielectric constant for these crystals is fairly low, e.g., 2.54 for naphthalene, and hence, the Coulomb forces are of long range. The Onsager radius (6.19) is large, i.e., above 20 nm for $\varepsilon < 2.89$ at 20°C. Hence, an electron-positron pair at mutual distances much smaller than 20 nm cannot escape the Coulomb attraction. This is the state which was called the "fourth positron state" in Sect. 5.5, where we discussed the transition from Ps to non-Ps state in liquid CS_2 at decreasing temperatures. CS_2 is the only liquid for which such transition has been found. We have, therefore, good reasons to believe that the main number of positrons which at high temperatures would form Ps, are in "fourth positron states" at the low temperatures where the Ps yield is zero, in the aromatic hydrocarbon solids.

There are many experimental facts and theoretical arguments speaking in favour of the given explanation, as discussed in detail in the reviews [1.41,42,4.24,9.1]. For example, AC and 3γ-annihilation studies give essentially results in agreement with the explanation for p-terphenyl, the mostly investigated aromatic hydrocarbon solid. The change from non-Ps to Ps state occurs between 100°C and 160°C in p-terphenyl. Furthermore, if aromatic molecules which are smaller than p-terphenyl are added to p-terphenyl at concentrations above about 0.001 mole fraction longlived components and narrow peaks in the AC spectra are observed at 20°C, indicating the formation of Ps. The smaller molecule, e.g., anthracene, substitutionally incorporated in the p-terphenyl crystal, is surrounded by a small free volume in which Ps can be formed, i.e., Ps can now be formed without the supply of the main part of the energy to form the cage. Hence, Ps formation is more exothermic if preexisting holes are available in the crystal. Furthermore, the Ps formation in anthracene illustrates the main points. In liquid anthracene the o Ps yield is about 40%, while no Ps is formed in the perfect solid, except for temperatures in the small premelting interval [9.24]. If the solid anthracene contains certain defects some Ps is formed. The reason is very probably that Ps cannot be formed in the crystalline, nondefected anthracene because the energy of the Ps cage state is higher than that of the "fourth positron state". The defects which provide small cavities for Ps make the Ps formation just possible.

Qualitatively, the Ps yield versus temperature in Fig. 9.6 resembles the sigmoidal behaviour of the o Ps lifetime versus temperature, discussed in Sect. 9.2. At first sight, one might assume that thermally generated defects give rise to the increase in the Ps yield with temperature, and defect formation energies could then be calculated from a simple model. A detailed analysis [1.42,9.25] shows clearly that vacancies cannot be the defects responsible for the Ps formation. First, the lifetimes above the transition are much too short compared to those expected from the size of the molecules, see Fig. 4.1, but they agree with those expected from the correlation to the packing coefficient (Fig. 9.5). Second, the formation energies of the proposed defects derived from the measured Ps yields are much higher than expected for vacancies implying very low vacancy concentrations. For example, for quarterphenyl $E_{\mathrm{V}}^{\mathrm{f}} = 2.4$ eV is obtained, which by use of generally accepted models would give a relative vacancy concentration of about 10^{-15} at the melting point. Consequently, Ps formation in the aromatic hydrocarbon solids is not associated with vacancies.

The most probable explanation is simply that Ps is selftrapped in many solid, aromatic hydrocarbons, and hence, that the Ps formation sets in at the temperature where the attractive forces between the molecules are weakened so much that the available binding energy of Ps is large enough to permit the formation of the cage around Ps. Consequently, Ps formation is not associated with a particular defect in these solids.

A detailed discussion of the energies of the Ps formation in solids is appropriate here. Only the simplest case of an electron and positron far away from other spur species will be treated. At larger distance between the particles the total energy consists roughly of the sum of the solvation energies of the particles and the Coulomb energy of attraction. The solvation energies are approximately the sum of the polarization energy of a particle in the solid and the particle affinity of the molecules if it is positive. A negative particle affinity means that the particle is situated in between, not on, the molecules. As the distance between the particles is reduced the total energy decreases at first, mainly because the Coulomb energy of attraction increases with decreasing mutual distance. The loss in total energy is dissipated by non-radiative processes etc., as is usually the case in, e.g., spur recombination processes. However, at smaller distances, i.e., below about 2–3 nm, the total energy is influenced by other physical processes than just the Coulomb energy change. This can be qualitatively described in the following way. The polarization energies are reduced if the distance between the particles is decreased so much that important parts of the polarization spheres around the two particles overlap. In the extreme case of Ps formation the polarization energies of the particles have disappeared, except for weak, induced-polarization (van der Waals) energies. Furthermore, the effective dielectric constant decreases with increasing field strength because of saturation effects etc., causing an increase in the Coulomb energy and a decrease in the polarization energies. At close distances, below about 1 nm, the effective dielectric constant approximation might be fairly poor.

At still smaller distances Ps formation may take place. In principle, any state between the fully dissociated particles (e.g., in polar liquids) and a Ps state weakly perturbed compared the vacuum state can be formed as the final product of the reaction. Of course, the final product is statistically distributed according to the principles of statistical mechanics, i.e., the state is perturbed by the thermal motion of the particles and the molecules. Obviously, a Ps state is only formed if its energy is lower than that of the alternative states of the two particles. Apparently, if Ps is formed the total energy is lower if the final product is a Ps atom localized in a cavity in the solid. The cavity is formed by Ps self-trapping, or it is preexisting as a defect in which case it is enlarged due to the presence of Ps. In a few rigid ionic solids the Ps energy inside the crystals have been measured by time-of-flight methods (see Sect. 9.5). Energies between 1.9 and 4.5 eV were obtained for Ps in the self-trapped state, and 3.27 eV for delocalized Ps in quarts. Hence, the energy of Ps is much higher inside the solid. Consequently, if the molecules of the solid have large enough electron and/or positron affinities it is expected that the Ps state has a higher energy than some other non-Ps state, and Ps is not formed. Theoretical considerations tell us that two oppositely charged species in a solid of low dielectric constant cannot escape their mutual Coulomb attraction in a time comparable to the positron lifetime of about 400 ps, if they once have been much closer than the Onsager radius. Hence, if Ps is not formed most of the electron-positron pairs will be in the "fourth positron state" at the time of the positron annihilation, irrespective of the detailed structure of the particles, i.e., the particles might be bound on the molecules for example.

The solid aromatic hydrocarbons showing the special Ps yield behaviour have all a positive electron affinity ranging from 0.15 for naphthalene to 0.59 eV for pyrene [5.34]. Apparently, no Ps is formed if the electron affinity is appreciably higher (e.g., in benzophenone), and Ps is always formed if the electron affinity is significantly lower or negative (e.g., benzene and normal hydrocarbons), other things being equal. However, the addition of naphthalene and biphenyl to hexane increases the Ps yield indicating that the positron is not bound to these molecules in hexane, and hence, that their positron affinities are negative or very slightly positive. The discussion given above and the experimental facts indicate that the Ps state has the lowest energy at high temperatures and the "fourth positron state" the lowest at low temperatures in these solids. The change from non-Ps state to the Ps state is caused by a rather strong decrease in the energy for the formation of the Ps cavity as the temperature is increased. The special values of the electron affinities combined with the rigidity of the lattice are the main properties which make these solids particular with respect to Ps formation [5.34]. It is important to realize that we could expect such solids to exist on the basis of the wide variation in the electron affinities and rigidities of solids. Hence, the special Ps behaviour for these solids speaks strongly in favour of the general explanations used in the rest of the book.

Recently, many new experiments have shown that very strongly perturbed Ps states in condensed matter exist and the strong perturbations are of great importance for the understanding of the obtained results. States which were believed to be only slightly distorted Ps states ten years ago are now realized to be Ps states which are very different from the vacuum state. Furthermore, several clear cases of transition from non-Ps to Ps states have now been experimentally studied, as illustrated by the results for liquid CS_2 and the special solid aromatic hydrocarbons at rising temperatures. Where to placed the boundary between Ps and non-Ps states is solely a nomenclature problem, which should not be over-emphasized.

Positronium in naphthalene at 20°C is a very good illustration of how perturbed the Ps state can be and still be Ps according to our definition in introduction to Chap. 9. Figure 9.7 shows the results of a magnetic quenching investigation of Ps in naphthalene [9.27]. The quenching ratio, R, is defined as the counts in a time interval of the LT spectrum, which contains only o Ps contributions in a good approximation, at a given field divided by the counts in the same interval at zero field [9.27]. The important point is that R versus magnetic field is very different from that in vacuum. In a simple theory the contact density of electrons at the positron for naphthalene has to be $\alpha' = 0.13-0.25$ of that in vacuum in order to fit of the obtained results. Hence, Ps in naphthalene giving a long lifetime of about 1.1 ns is extremely distorted Ps! This is one of the lowest α'-values ever measured. In normal liquids $\alpha' = 0.6-0.9$ are usually found. Similarly, such anomalous magnetic quenching results have been found in solid octadecane and in nitrobenzene/hexane solutions. A detailed explanation was given in [1.44] and in Sect. 8.3. Apparently, anomalous magnetic quenching occurs if the Ps state is just stable, which it is in naphthalene at 20°C as shown in Fig. 9.6.

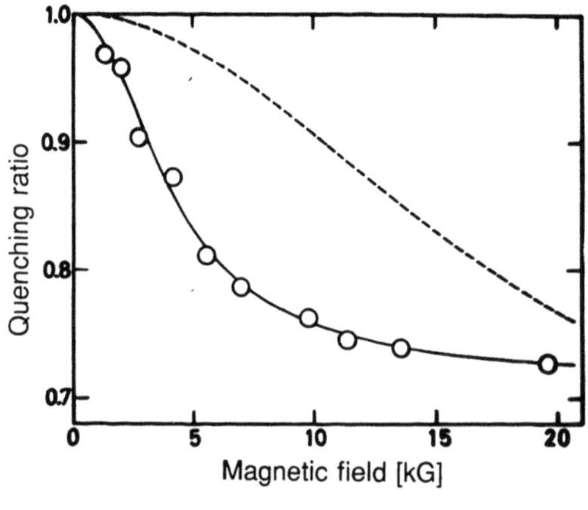

Fig. 9.7. Magnetic quenching of o Ps in naphthalene. The points show the measured quenching ratio (see text). The broken line is the calculated quenching ratio for unperturbed Ps. [9.27]

9.4 Positronium in Some Other Molecular Crystals

We have until now discussed positron annihilation in selected molecular crystals in the previous three sections. This order of presentation was chosen in order that certain fundamental Ps processes could be well treated. However, in most molecular solids the formation and behaviour of Ps are influenced by several effects simultaneously and cannot be described in terms of one dominating process. Consequently, the interpretation of the measured results is very complicated and much more uncertain. Hence, it is important to realize that well accepted explanations of the data for these compounds do not exist, as yet. We can only suggest causes of a given measured effect, and we have to accept that new experiments might very well disprove the suggested explanations. In particular, this is the case for the more complicated molecular compounds such as polymers, which shall be treated in Chap. 10. We shall here discuss Ps in certain classes of molecular crystals.

Levay et al. [9.28] performed LT and DB measurement on two long-chain linear hydrocarbons, n-dotriacontane (n-$C_{32}H_{66}$) and n-tritriacontane (n-$C_{33}H_{68}$), between 80 K and their melting points. Their results illustrate some of the complex phenomena responsible for the formation and lifetime of Ps in molecular compounds, in general. Both crystals gave qualitatively similar results. Hence, we shall mainly discuss the C_{33} crystal. The o Ps lifetime was about 1.05 ns at 80 K. It increased almost linearly to 1.4 ns at 310 K. These lifetimes seem to indicate that Ps is self-trapped in a cage between the molecules since they are very similar to the lifetimes shown in Fig. 9.5 for brittle crystals. At about 310 K the slope of τ_3 versus T increased, which correlates to the onset of molecular oscillations and rotations according to NMR investigations. At three phase transitions (328, 339, and 342 K) the τ_3 versus T slope increased further. The lifetime reached almost 3.0 ns at the melting point of 345 K, which is not much lower than that expected for the liquid phase. The phase from 342–345 K is a so-called "rotator" phase, in which rotational and longitudinal disorder and interchain kink defects are present. Hence, the o Ps lifetime simply reflects the fact that the crystal becomes softer as the temperature is increased, i.e., Ps is self-trapped in larger holes the softer the crystal is. For these long-chain hydrocarbons the lifetime just below the melting point is almost as large as that of the liquid, and the Ps cage is, therefore, nearly as large as the Ps bubble in the liquid. The reason seems partly to be that a "rotator" phase exists just below the melting point.

The intensity of the long lifetime component decreased strongly from 31% (C_{32}) and 25% (C_{33}) to 15% (both crystals) as the temperature was increased from 80 K to about 220 K. This has been correlated to CH_3 group rotation [9.28], which has been used to explain changes in NMR spectra in the same temperature interval. The "free-volume" Ps-formation model used in [9.28] seems to be incorrect (see below). Here we shall discuss another explanation. It seems reasonable to assume that the low Ps formation at temperatures above 220 K is caused by the trapping of the electrons and the positron in the positron spur. The trapping reduces drastically the mobility of the particles. They are thereby partly prevented from forming Ps within the lifetime of the positron of about 400 ps except by encounter pair effect and/or non-solvated particle reactions. Hence, Ps formation is limited in a way which is similar to the cases of polar liquids discussed in Chap. 5. In that way the strong reduction in the Ps yield in the interval 80–220 K is simply explained by the increasing trapping of the positron and electrons in the positron spur made possible by the increase in the importance of the rotations of the CH_3 groups in the same temperature interval.

In the spur model of Ps formation there are essentially three mechanisms which can reduce the Ps formation strongly, namely scavenging of the particles in the positron spur, a strong increase of the spur size so that it becomes much larger than the Onsager radius, and the reduction of the spur particle mobilities and diffusion rates so that they cannot react in time before the positron decays. Clearly, in the cases discussed here only the last mechanism is reasonable. Hence, by exclusion we have to accept the above given explanation. Furthermore, we shall see below and in Chap. 10 that the Ps yield is limited by the finite positron lifetime in very many molecular compounds where particle trapping is reasonably expected and has been found in many cases. On the other hand, in the brittle crystals where no significant trapping of the positron and/or electrons takes place the o Ps yield is high, e.g., 54% in ice and 48% in crystalline methanol, while it is low in amorphous compounds, as shall be discussed in details below. This is one of the most important principles in the understanding of Ps formation in solids.

The Ps yield increased again above about 320 K and reached roughly 31% at the melting point in both crystals. The most reasonable explanation is that the particle detrapping due to thermal motion of the molecules and softening of the crystals starts to become important at about 320 K. The particles can then move faster and more reactions into Ps can take place before the positron decays. Apparently the slope of the Ps intensity versus T shows also changes at the phase transitions mentioned above. This interpretation of the rise in the Ps yield in terms of particle detrapping is somewhat uncertain.

In many solids characteristic effects in the positron annihilation parameters caused by particular treatments of the samples before the measurements have been found. In [9.28] such effects in the Ps yield only were reported. For example, the C_{32} sample was rapidly cooled (4 K/min) from 293 K to 80 K, resulting in an o Ps yield of 21% in stead of 31% at 80 K. After three days of relaxation at 80 K the o Ps yield reached 31% again. Other similar Ps yield effects were reported in [9.28]. The positron studies of the plastic crystals, cyclohexane [9.15] and adamantane [9.29], gave results depending on the number of times the samples were taken through the solid-solid phase transition and on the impurity content. *Walker* [9.30] showed that the o Ps lifetime and yield decreased over times of 50–100 days at 21°C in certain binary mixtures of n-eicosane and n-docosane. In general, one should always be aware of that the history of the thermal and mechanical treatments of solid molecular samples before the measurements might show up in the positron annihilation parameters. Furthermore, very pure samples of molecular compounds are difficult to obtain. Many results reported in the literature seem to be influenced by effects due to impurities and to the fact that the samples contain defects caused, e.g., by the thermal treatment prior to the measurements [4.24].

In [9.31] LT results were reported for n-tetradecane, n-hexadecane, and n-octadecane in the temperature interval −80°C to 100°C, i.e., around the melting points. The o Ps lifetime increased from about 1.5 ns to 3.1 ns at the melting points, while the o Ps yield was higher in the solid than in the liquid phases just above the melting point. These results differ appreciably from the results for the longer chain hydrocarbons [9.28], discussed above.

Another problem concerning Ps formation in hydrocarbons was first discovered by *DeBlonde* et al. [9.32]. They found that the intensities of the narrow components in the AC spectra were almost zero although the long lifetime intensities were large in solid methane (33% at −187°C), cyclohexane (21% at −70°C), and benzene (21% at −10°C). The reason seems to be that the intrinsic annihilation of p Ps gives fairly broad AC components (fwhm = 4–5 mrad), which were mistakenly included in the broad components in [9.32]. This problem is not fully sorted out yet. In this context, results on magnetic quenching of Ps in solid octadecane at 20°C [9.33] are interesting. The quenching ratio decreased strongly at a magnetic field which was about four times lower than for Ps in vacuum, while the result for liquid octadecane was normal for liquids ($\alpha' = 0.62$). Furthermore, the results could not be well fitted by variation of α'. Hence, the magnetic quenching was fairly similar to that found for solid naphthalene, discussed in Sect. 9.3. Consequently, Ps is very distorted in solid octadecane, which might be the case in most solid hydrocarbons. The very broad p Ps AC components might be caused by such distortion of Ps. As naphthalene has a positive, and octadecane is expected to have a large negative, electron affinity the distortion of Ps might be very different in the two solids.

Next, let us discuss positron annihilation in polar molecular crystals exemplified by alcohols. In [9.34,31] the long lifetime, τ_3, and its intensity, I_3, in the temperature interval −140°C to −70°C for CH_3OH and CD_3OD were reported. We discuss the normal methanol case as the results for CD_3OD were very similar. At the melting point of −97.8°C τ_3 increased from 1.48 ns in the solid to 2.8 ns in the liquid phase, while I_3 decreased from 48% in the solid to 21% in the liquid. This behaviour of the o Ps yield is very similar to that of ice, discussed in Sect. 9.1. In Sect. 5.5 the low Ps yield in polar liquids was explained. The main reason was that most of the positron-electron pairs could not form Ps before the decay of the positron having a lifetime about 400 ps. Ps was mainly formed only if the particles were in close encounter position after full solvation or if it could be formed before both particles were fully solvated. In crystalline methanol the particles are not solvated and the dielectric constant is very low. The particles have, therefore, fairly high mobilities and diffusion constants, and they can form Ps in times which are small compared to 400 ps. Apparently, the spur particles in crystalline short chain alcohols and ice are not trapped in defects in the crystalline state. Hence, the Ps yields in these solids are comparable to those found in liquid hydrocarbons of fairly high electron mobility. A solid-solid phase transition at $-110°C$ in methanol gave small changes in τ_3 and I_3.

The Ps yields in amorphous short-chain alcohols were reported in [9.35]. They found Ps yields comparable to those in the liquid state, while crystalline alcohols gave much higher yields. The low yields are caused by the trapping of the positron and electrons in the positron spur in the defects in the amorphous solid. The Ps formation processes are then similar to those found in polar liquids and in the long chain solid hydrocarbons at intermediate temperatures discussed above. The low Ps yield in glassy compared to crystalline alcohols speaks strongly in favour of the spur model of Ps formation. An alternative model, used in [9.28], assumes that Ps is only formed if large enough "free volume" is available in the solid. According to the latter model one would expect that the Ps yield is much higher in the amorphous alcohol containing many preexisting holes than in the crystalline alcohol, in clear disagreement with experimental facts. Cavities in solids are mainly of importance for the mere existence of Ps only in those solids, e.g., some aromatic hydrocarbons, where Ps formation without the cavity to reduce the Ps zero-point energy cannot occur because of lack of energy, as discussed in Sect. 9.3.

Lenc [9.36] measured LT spectra for the alcohols: n-dodecanol-1 (C12), n-tetradecanol-1 (C14), n-hexadecanol-1 (C16), n-eicosanol-1 (C20), and n-docosanol-1 (C22) in large temperature intervals. The alcohols contained between 100 and 900 ppm of organic impurities, and the water content was not determined. C14 and C16 were also studied after degassing. No effects of the content of oxygen were found in the solids. The other alcohols were only studied without degassing. The long lifetime, τ_3, increased roughly linearly with temperature from 1.0 ns at $-160°C$ to 1.4–1.5 ns at the solid-solid phase transition a few degrees below the melting points for C14, C16, C20, and C22. For C12 similar results were found but only temperatures above $-120°C$ were studied. These τ_3 values indicate that Ps is self-trapped in the crystals. Above the melting points τ_3 values typical for liquid alcohols were obtained, i.e., about 2.7–3.0 ns for non-degassed alcohols. I_3 increased approximately linearly with temperature in the main part of the brittle solid phases. The values were between 16% and 26%. Small differences in the I_3 value for decreasing and increasing temperatures were found in the brittle phase. In a small temperature interval 3-5 degree wide just below the melting points phenomena due to the existence of a "rotator" ("waxy") phase were observed. For decreasing, but not for increasing, temperatures in C20 I_3 was about 6% lower in the rotator phase (60–64°C) than in the brittle phase, while τ_3 was 2.2 ns in the rotator phase compared to 1.4 ns in the brittle phase and 2.6 ns in the non-degassed liquid phase. The other alcohols gave less pronounced effects in the "rotator" phases. In n-octadecanol-1 (C18) [9.37] effects similar to those of C20 were found. However, effects similar to those of C20 measured for C16 in [9.36] were not found in [9.38]. The higher τ_3 value in the "rotator" phase is in agreement with a hypothesis of

Ps trapping in a vacancy but might also be caused by a bubble like Ps state in the soft crystal. Effects of fast cooling for C12 and C20 were also reported.

Positronium inhibition and quenching, i.e., the addition of molecules and ions and its influence on the Ps yield and o Ps lifetime, in molecular compounds have been investigated very little. The reasons are, of course, partly identical to those which cause the main differences between normal chemistry in liquids and solids. In most solids the solubility of foreign molecules or ions in one homogeneous phase is very low. Clusters of a second phase of different concentration of the additive are formed in many cases even at low concentration. Furthermore, interaction of additives with defects may occur, and/or small clusters of additives may be formed, even if the presence of a second phase is not established. The additive might also change the concentration and distribution of defects. This might influence the Ps yields and lifetimes so that a determination of the inhibition and/or quenching by the additive becomes very difficult. Great care should be exercised in the interpretation of such experiments!

The AC results for frozen aqueous solutions [9.39] are good illustrations of the very strong differences between positron annihilation in some solid and liquid mixtures. The solutions were fast frozen and cooled to $-140°C$, the temperature of the measurements, in order to reduce segregation of the ions during the freezing and cooling. The salts $MnSO_4$, $KMnO_4$, and $K_2Cr_2O_7$ did not change the AC spectra at concentration of 4–5% weight by weight. In the liquid aqueous solutions Mn^{++} would increase, and $Cr_2O_7^{--}$ decrease, the p Ps narrow peak by a factor of two at these concentrations. The acids H_2SO_4 and HNO_3 and the bases LiOH and NaOH gave no appreciable effects. Only the additives containing halide ions strongly influenced the spectra. These additives would not influence the liquid solution spectra at the concentration used. Very few, scattered results of frozen solutions of polar liquids have been reported [9.40]. Ps chemistry studies in frozen nonpolar solutions are also very rare.

The interpretations of the results are fairly uncertain. Very few additives can be dissolved in ice. 100 ppm of HF in ice has been reported but it is not completely clear whether HF is actually homogeneously distributed at these high concentrations. The halide ions influence probably the AC and LT spectra because they cause vacancies to be build into the ice during the freezing, and the vacancies trap Ps, as discussed in Sect. 9.1. Apparently, the other solutes mentioned above cannot be dissolved into ice even at ppm concentrations. They are included into a second hydrate-like phase, where they seem not to act as converters, oxidizers, inhibitors, etc., as they do in the liquid solutions. This is the best explanation we can offer.

Apparently, additives influencing positron annihilation can be more easily build into the molecular compounds by irradiation, which seems to give homogeneous solid solutions, although the homogeneity has normally not been tested. The few reported results for molecular crystals seem to indicate that both the o Ps lifetime, τ_3, and intensity, I_3, can change on γ-irradiation. For example, τ_3 increased by 17% at a dose of 200 Mrad Cs-137 irradiation (1.6–1.9 Mrad/h) in adamantane at room temperature, while I_3 was roughly independent of the irradiation [9.41]. In acetyl methionine γ-irradiation of 30 Mrad (Co-60, 1.77 Mrad/h, 20°C) decreased I_3 from 44% to 25%, while the lifetimes were unchanged, at 20°C [9.42]. The concentration of EPR centers created by the irradiation and the change in the I_3 value disappeared mainly during annealing at 20°C over 50 days. No simple correlation could be established between the number of EPR centers and the decrease in I_3. In eicosane both τ_3 and I_3, measured at −78°C and 20°C, decreased on γ-irradiation at −78°C [9.43]. Correlation to EPR center concentration was discussed and annealing results up to 400 hours were reported.

It is fairly easy to find possible explanations of these effects of irradiation. The main problem is to know which of the possible explanations are the correct ones. For example, an increase in τ_3 could be due to trapping of Ps in irradiation created hole-type defects. More far fetched, τ_3 might be reduced by reactions with dissolved O_2 in the pure solid, and the irradiation might remove this O_2 because of a reaction of O_2 with free radicals created by the irradiation. A decrease in τ_3 could be caused by Ps reactions with the irradiation created free radicals. I_3 might decrease because the electrons and positron in the positron spur are trapped in irradiation created defects which prevents some of them from forming Ps before the positron decays. The free radicals might also react with the spur particles thereby reducing the Ps yield.

Concerning irradiation created damage in molecular compounds, it must be realized that defects giving no EPR signal or light absorption are very rarely studied in other experiments than positron annihilation (see Sect. 4.6). Hence, information on, e.g., hole type defects in molecular compounds might be totally absent, or the given discussions might be incorrect, in the available literature. For example, a correct description of the Ps behaviour in ice was only found after it had been realized that the concentration of vacancies given in almost all the literature was too low by a factor of about 50,000. Similarly, the prevailing ideas amongst specialists concerning the formation of small vacancy clusters in metals during irradiation and annealing had to be partly discounted before positron trapping in these clusters could be correctly interpreted.

Many problems of molecular compounds have never been studied or have been studied very little by positron annihilation. Even at room temperature only fairly few molecular crystals have been examined. In [9.44] about 3000 structures are given in the index, of which 200–300 have been studied by positron annihilation. For example, no inorganic hydrate crystals seem to have been investigated. Similarly very few of the structures given in [9.22] have been examined. Many special crystal structures, such as host/guest (clathrate) compounds and layer structures, are uninvestigated except for very few cases (see Chap. 10). Ice [9.10] and methane [9.20] seem to be the only molecular crystals investigated at temperatures below LN_2 temperatures. Apparently, very few compounds have been studied at high pressures [1.45,46,9.45]. The mean o Ps lifetime decreased exponentially with the change in volume due to the increasing pressure [9.45]. Electric field effects have been studied mainly in polymers (see Chap. 10) but not in molecular crystals except probably in eicosane [9.46].

As for liquids very little is known about the behaviour of positrons in molecular solids where Ps is formed.

9.5 Slow-Positron Beam Studies of Solids

The principles of slow-positron beam measurements were discussed in Sect. 3.4. Slow positrons of energies 1 to 10,000 eV hit the surface of a solid, which must have low vapour pressure to permit a high vacuum at the positron beam. The number of o Ps atoms decaying by 3γ-annihilation after having escaped through the surface, the energy of the escaped Ps, and/or the width of the 2γ-annihilation peak in the γ-spectrum may be measured versus energy of the incident positrons. We shall here discuss selected results for ice, which seems to be the only molecular compound studied, and some ionic crystals.

Eldrup et al. [9.47–49] studied crystalline and amorphous ice using positrons of energies in the range 0.0–4.5 keV. In Fig. 9.8 two typical curves of the 3γ-annihilation fraction, $F_{3\gamma}$, versus the logarithm of incident positron energy, E_p, for crystalline ice at 44 and 150 K are shown. Some of the positrons in ice form Ps, which then diffuses to the surface and escapes into vacuum, where o Ps decays by 3γ-annihilation. The fraction, $F_{3\gamma}$, can be determined from the shape of the γ-energy spectrum, measured with a semiconductor detector (Sect. 3.4). After a small correction for the 3γ-events inside the ice, $F_{3\gamma}$ gives the amount of o Ps escaping the sample. Because of the repulsive interaction between Ps and ice molecules Ps has a negative affinity to the crystal,

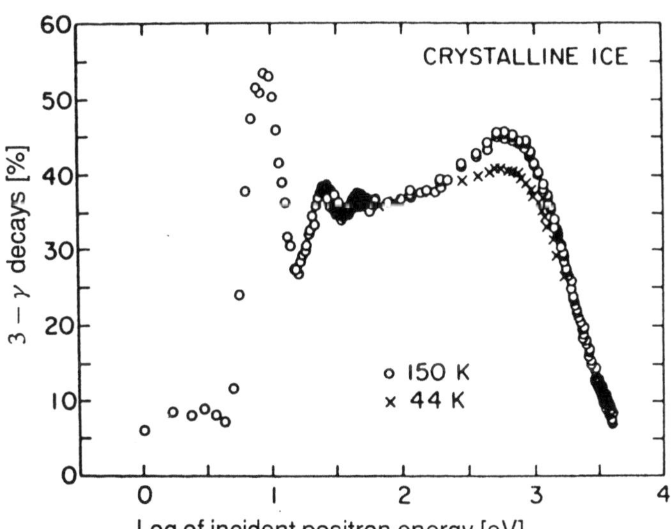

Fig. 9.8. Typical curves for the fraction of 3γ-annihilation of o Ps versus energy of incident positrons in crystalline ice [9.47]

which could be estimated to be (-2 ± 1) eV. Hence, all Ps atoms reaching the surface are expected to escape into vacuum. The positron work function was measured to be positive and thus, thermalized positrons do not escape from the sample.

Figure 9.8 shows three characteristic energy regions. At low energies $F_{3\gamma}$ varies strongly and exhibits sharp maxima and minima, in the intermediate energy interval, 50 eV $< E_p <$ 700 eV, $F_{3\gamma}$ has a slow, gradual increase, while $F_{3\gamma}$ shows a steady decrease at higher energies, $E_p >$ 700 eV. Let us start the discussion of $F_{3\gamma}$ versus E_p in Fig. 9.8 at high E_p values. At high positron energies the positrons penetrate rather far into the crystal. Very rough guesses are that the secondary electron and positron ranges are 20–30 nm in ice, and the mean penetration depth of positrons is 20–30 nm at $E_p = 1$ keV. Hence, above roughly $E_p = 1$ kev very few of the spur electrons and positrons are lost through the ice surface, i.e., the positron spur is not "cut" by the surface. The o Ps yield is, thus, expected to be about 54% as in the bulk ice. However, some Ps does not diffuse to the surface before annihilation, and the measured $F_{3\gamma}$ is, therefore, somewhat below 54% in Fig. 9.8. At increasing E_p above 1 keV the $F_{3\gamma}$ value decreases because more Ps annihilates before reaching the surface by diffusion at increasing penetration depth. A diffusion coefficient, $D = 0.17 \pm 0.09$ cm^2/s, for Ps could be determined.

As the positron energy decreases below 1 keV the Ps yield is influenced by three effects. First the penetration depth decreases and it becomes easier for Ps to diffuse to the surface, which increases the Ps yield. Second, more spur particles can escape from the surface as the penetration depth decreases. This effect decreases the Ps yield. Third, less spur electrons are created in the positron spur as E_p decreases, which reduces the Ps yield. These effects lead to a maximum in $F_{3\gamma}$ at about E_p = 700 eV followed by a slow decrease in $F_{3\gamma}$ below 700 eV.

As E_p is reduced to about 50 eV there is only energy for the creation of one or a few electrons in the positron spur, and energy conservation gives rise to strong variations in $F_{3\gamma}$ versus E_p. Below the lower boundary of the Ore gap, E_l, (see Sect. 7.7) Ps formation is impossible for energy reasons. The energies determining the Ore gap and the Ps yield inside ice are, of course, different from those in vacuum. The Ore gap limits are estimated to be $E_l = 5 \pm 1$ eV and $E_u = 10 \pm 1$ eV in [9.47] by use of the slow positron results. The $F_{3\gamma}$ value begins to be different from zero at E_l. It then increases strongly up to a maximum at about E_p = 8 eV in Fig. 9.8, followed by a drastic decrease above 10 eV. This decrease at E_u is explained in the following way [9.50]. On increasing E_p from just below E_u to just above E_u, a new process becomes possible. Above E_u the electron can escape the positive ion without necessarily being bound to the positron. In the bulk of ice most electrons and positrons will slow down at a distance small compared to the range of the Coulomb force anyway, and a strong change in the Ps yield at E_u is not expected. However, in ice close to the surface a large number of the electrons will be ejected through the surface, and many electrons which just below E_u would have formed Ps can escape the positron just above E_u. This means that Ps formation must compete with a new process, and hence, the Ps yield decreases strongly just above E_u. In the same way the maximum in $F_{3\gamma}$ around 25 eV and the small minimum at roughly 35 eV can be explained as caused by the Ore gap associated with the electron band at around 30 eV. Curves of $F_{3\gamma}$ versus E_p similar to those shown in Fig. 9.8 have been found in mica [3.14] and a semiconductor, GaAs [9.51].

The discussions of the slow positron data for ice in [9.47–49] have resulted in wrong statements in several papers that the spur model of Ps formation is not valid at positron energies below about 50 eV. An answer has been given in [9.50]. In this context it must be realized that energy conservation applies in the spur model as well as in the Ore model. Hence, the necessity to use energy conservation to explain the $F_{3\gamma}$ values does not, of course, disprove the spur model. The main difference between the spur and Ore models in the cases where only one spur electron can be created during the slowing down of the positron are the following. The spur model describes the Ps formation as a result of many intermediate processes, such as slowing down, diffusion, trapping, and reaction of the light particles and other spur species. In the Ore model the Ps formation is a single-step process, and hence, the spur processes are disregarded, except as expressed in the total probabilities of Ps formation and positron escape from the Ore gap. It must also be realized that even if the Ore model can describe the Ps yield at low energies in ice by variation of several unknown parameters, it cannot explain the Ps formation in many other systems, as discussed in Sect. 7.7.

The results of the slow positron experiments on amorphous ice illustrate that point. In amorphous ice at 43 K the maxima and minima in $F_{3\gamma}$ at $E_p < 50$ eV are missing, and the o Ps yield is lower than that of crystalline ice [7.47–49]. An o Ps yield of 34% was found in amorphous ice in agreement with a measured yield of 34 ± 2% for 10 M NaOH glassy ice at 85 K [9.52]. It is important to realize that this drastic change in the slow positron results on going from crystalline to amorphous ice could not be explained in terms of the Ore model, because the Ps formation is assumed to take place at positron energies far above thermal energies in that model. However, the amorphous ice results seem to be well explained in terms of the spur model. In the amorphous ice the trapping of the spur particles in preexisting defects reduces the Ps yield and the possible escape of the electrons through the surface which caused the maxima and minima in the Ps yield, according to the discussion given above.

The Ps diffusion coefficient for amorphous ice was determined to be about $D = 0.001$ cm²/s at 43 K. The much lower D value is very probably caused by Ps trapping in defects in amorphous ice. Another important result was that $F_{3\gamma}$ did not approach a very low level at high positron energies. In stead it reached a level of about 12%. The reason is that o Ps annihilates by 3γ-decay in large cavities (> 1.7 nm diam.) which exist in the as-grown amorphous ice. The voids disappeared after annealing to about 100 K. In general, these results indicate that the formation and annealing of amorphous solids could be a very fruitful field of research in positron annihilation in the future.

The Ps energy compared to vacuum of the lowest state in some ionic crystals, i.e., the Ps affinity, A_{Ps}, has been measured in quartz [9.53], some ionic crystals [9.54], and rare gas solids [9.55] by time-of-flight techniques. The following values were obtained: Quartz (327°C) −3.27 eV, LiF (350°C) −4.5 eV, NaF (310°C) −4.3 eV, KCl (370°C) −2.4 eV, KBr (370°C) −2.2 eV, KI (370°C) −1.9 eV, solid Ar (−238°C) −2.6 eV, solid Kr (−238°C) −2.4 eV, and solid Xe (−238°C) −2.3 eV. Hence all the Ps affinities were negative, as expected. For quartz, where Ps is delocalized at 327°C [9.56], the temperature dependence of A_{Ps} was not reported. For the other ionic crystals, where Ps is localized at the given temperatures [9.4], the effect of increasing temperature is to shift the Ps affinity to higher values. The average temperature dependence was 1.5 ± 0.3, 1.9 ± 0.5, 1.6 ± 0.3, and 1.6 ± 0.1 meV/K for NaF, KCl, KBr, and KI, respectively. Hence, the A_{Ps} values were less negative at the higher temperatures, which can be explained by a deeper trap of localized Ps at higher temperatures in agreement with the expectations. It is remarkable that Ps is emitted with a rather well-defined kinetic energy at an incident positron energy of up to 4 keV [9.54] even if Ps is selftrapped in a localized state. In the rare gas solids, where Ps always seems to be localized, the distribution of kinetic energy of the ejected Ps was fairly broad.

In [9.54,55] the spur model was claimed to be wrong and the Ore model was used. However, the spur model was misunderstood in [9.54,55], where it was incorrectly assumed that the intrinsic binding energy of Ps, E_b, i.e., 6.8 eV in vacuum, contributes to the kinetic energy of Ps outside the crystals. Actually, E_b is dissipated by non-radiative processes during Ps formation in the positron spur, as is the case for all electron-ion recombination processes in normal spurs. A broader peak of the number of emitted Ps versus Ps energy at incident positron energies below 500–1000 eV [9.54] seems to be caused by the effects of "cutting" of the positron spur by the surface.

Very interesting results of slow positron research of rare gas solids at low energies of the incident positrons have been reported in [9.57,58].

Actually, slow positron beam studies of chemical problems have been performed mainly by physicists until now, and very few results have been reported. It can be expected that such studies will give very interesting information in positron and Ps chemistry in the future, when chemists begin to be interested in applying the technique.

10 Polymers and Miscellaneous Molecular Compounds

In the previous chapter positron annihilation in the simpler molecular solids was discussed. In particular, the first three sections dealt with molecular crystals in which one simple effect dominated the Ps behaviour and/or Ps formation, while Sect. 9.4 treated the more complex results obtained for most normal, molecular solids. In this chapter a discussion of positron annihilation in several molecular compounds which have very complicated structures is given.

As an introduction to amorphous structures the formation and behaviour of Ps in certain molecular systems during the transition from amorphous to normal liquid state is discussed. The main part of this chapter concerns positron annihilation in polymers. Especially, some electric field and charging (electret) effects will be treated. A discussion of liquid crystals and micelle systems then follows, and the last part of the chapter is a section on certain selected solids such as zeolites, oxide powders, clathrates, and biological systems.

10.1 Transition from Glassy to Normal Liquid State

Positron annihilation LT measurements of the transition from glassy to normal liquid state were first performed on two phenyl ethers: m-phenoxy-phenyl-m-(m-phenoxyphenoxy)-phenyl ether (5PO) and bis [m-(m-phenoxyphenoxy)phenyl] ether (6PO) [10.1]. A similar LT study of ortho-terphenyl was reported in [10.2], followed by a report on AC and DB measurements on 6PO [10.3]. All three liquids exhibit super cooling and form glassy solids at low temperatures.

The LT and AC results for 6PO given in Fig. 10.1 illustrate the most important information obtained. At first, let us discuss the LT studies. The LT results for 5PO and 6PO were very similar, and hence, we shall mainly discuss only the 6PO data. The LT spectra for 6PO were measured at temperatures between 213 and 373 K and analyzed into three terms. The longest lifetime, τ_3, caused by o Ps pick off annihilation, is shown in Fig. 10.1. Clearly, τ_3 shows a sigmoidal dependence on temperature. Its intensity, I_3, varied roughly linearly with temperature from 33% at 213 K to 42% at 373 K. At 213 K 6PO is a glassy solid of a shear viscosity of $\eta > 10^{14}$ P, while it is a normal liquid of $\eta \simeq 0.1$ P at 373 K. Above a particular glass transition temperature, $T_g = 223$ K, τ_3 increases markedly with raising the temperature.

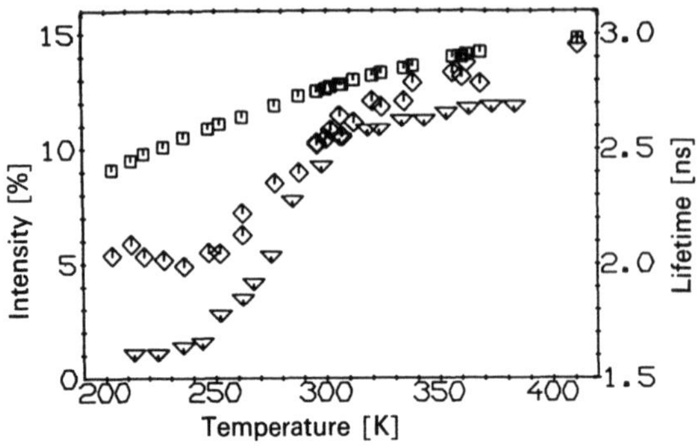

Fig. 10.1. Lifetime and AC results for 6PO versus temperature. (▽) the o Ps lifetime, (◇) the intensity of the narrowest gaussian, I_1, in a three gaussian fit to the AC spectra, and (□) the p Ps intrinsic annihilation intensity calculated from the LT results [10.3]

At temperatures below T_g the super cooled 6PO has a rigid amorphous structure. During the slowing down of the electron and positron in the positron spur, and as thermalized, the particles move in an irregular potential due to the molecules. The particles might get trapped in the preexisting potential wells and thereby be prevented from forming Ps before the positron decays, as discussed for polar liquids in Sect. 5.5. During its lifetime Ps will diffuse by delocalization and/or tunnelling effects in the repulsive potential of the molecules. It tends to get trapped, or formed, in the preexisting cavities of the structure, from where it might move to a larger neighbor cavity with a lower potential for Ps. This diffusion from trap to trap will stop when Ps decays or if the trap is so deep that Ps cannot move within its lifetime. Clearly, p Ps will have a tendency to annihilate in shallower traps than o Ps due to its shorter lifetime. The cavities will be only slightly enlarged due to the presence of Ps. Hence, Ps is mainly a "seeker" below T_g.

The glass transition temperature, T_g, determined by the o Ps lifetime is somewhat lower than that obtained from viscoelastic measurements, $(T_g)_V$. The reasons might be that T_g is a local glass transition temperature of the average cavity in which Ps is trapped, and/or that the Ps formation process raises the local temperature somewhat.

At rising temperatures above T_g the structure gets increasingly softer and the repulsion between the molecules and Ps causes the preexisting cavities in which Ps is trapped to be appreciably enlarged. The higher the temperature is the more the cavity is enlarged. Consequently, Ps is both a "seeker" and a "digger" above T_g.

Above a still higher temperature, $T_r = 303$ K, the structure has become so soft that o Ps can make the Ps bubble in thermal equilibrium in a time which is short compared to the o Ps lifetime in the bubble. The lifetime of o Ps in the bubble state is now obtained, and the slope of τ_3 versus T is much smaller than at temperatures below T_r. Hence, τ_3 is approximately the lifetime expected from the correlation (4.7) between surface tension and o Ps lifetime. Consequently, Ps is mainly a "digger" at temperatures above T_r. As in normal liquids, Ps probably begins to get self trapped in the bubble during the Ps formation process. In [10.2] T_r is correlated to other properties of the liquid and another interpretation is given.

Angular correlation spectra were measured for 6PO at various temperatures [10.3]. They were fitted by a sum of three gaussians. The intensity of the narrowest gaussian, I_1, versus temperature is shown in Fig. 10.1. The associated fwhm ($\sim$ 3.23 mrad) did not show systematic variation with the temperature below about 350 K. The p Ps intrinsic annihilation intensity, I_p, calculated from the LT results is also given in Fig. 10.1. Clearly, I_1 is much smaller than I_p at low temperatures. As the temperature increases above 250 K I_1 approaches I_p from below. This was the most important result of the AC measurements.

The reason is very probably that the intrinsic annihilation of p Ps results in an AC component consisting of a narrower gaussian of fwhm = 3.23 mrad and a somewhat broader gaussian of fwhm = 4.5–5.0 mrad. In the three gaussian analysis the latter is included into the broad component given by the two broader gaussians. In [10.3] it was shown that such hypothesis did not give a broad AC component with an unacceptable shape (e.g., a too flat top). The situation is similar to that found for liquid CS_2 discussed in Sect. 5.5.

These results illustrate clearly that great care should be exercised in the assignment of the narrowest gaussian to p Ps intrinsic annihilation. In general, in the larger holes (bubble, cavity, etc.) the annihilation of Ps gives results which are mainly dependent on the size of the hole. However in the smaller holes the Ps wave function spills over into the surrounding matter and the simple model of Ps in a hole giving LT and AC components depending on the hole size is a bad approximation.

The LT results for ortho-terphenyl [10.2] in the super cooled, glassy state were similar to those of 6PO and 5PO. T_g and T_r were measured to be 240 K and 305 K, respectively compared to a melting point, T_m, of 331 K. Typical τ_3 (I_3) values were 1.55 ns (39%) below T_g, 2.55 ns (40%) just above T_r, and 2.7 ns (42%) in the liquid at 367 K. At temperatures between 300 K and 331 K crystallization occurred in the glassy state typically after four to six days. In the crystalline state τ_3 (I_3) were 1.1 ns (18%) and 1.2 ns (20%) at 300 K and 331 K, respectively.

10.2 Positron Annihilation in Polymers

Polymers have been studied a great deal by positron annihilation. In particular, LT results were reported in many papers. The behaviour of Ps and positrons depends on many details of the polymer structure, which have been investigated very little by other methods. Furthermore, several physical effects seem to influence simultaneously the shapes of the measured spectra. Consequently, many positron annihilation results for polymers can only be explained in a fairly rough way, at present. It is, therefore, rather difficult to give a good description of the available data, and hence, the given discussion will necessarily be qualitative.

On the other hand, the presently available experimental results on positrons in polymers can, apparently, be accounted for by use of the basic ideas discussed in the previous sections and chapters. Positronium is formed by the spur process in polymers, too. Being a "seeker" and a "digger" Ps is trapped in preexisting holes, which are enlarged somewhat after the trapping, it can be transferred from one hole to another, deeper hole within its lifetime, and it can selftrap if no preexisting holes are available. We expect that the secondary spur reactions of Ps take place also in polymers, where they, however, cannot be detected because the LT spectra are very complicated.

The preparation and characterization of polymer samples are difficult tasks. In most papers, normally available polymer samples obtained from commercial firms seem to have been used. Apparently, such samples contain additives (antioxidants, lubricants, etc.) of concentrations in the order of one per cent weight in many cases. Furthermore, the properties of a polymer detected by positron annihilation may depend on density, degree of crystallization, number-average and weight – average molecular weights, and thermal and mechanical treatment before the measurements. These properties are frequently not reported. Taking into account the uncertainties in the characterization of the samples it is surprising that a rough agreement between the results of several groups seems to have been obtained.

Recent reviews on positron annihilation in polymers are [10.4–6]. Many papers on polymers were published in the two latest proceedings [1.7,14]. Because of lack of space we cannot here discuss all the literature on positrons in polymers. A discussion of certain experimental results selected to illustrate, in particular, the problems of interpretation of the data, will be given.

Kindl et al. have measured LT spectra for several polymers in large temperatures intervals: high- and low-crystalline polytetrafluoroethylene (PTFE), −200–420°C [10.7,8]; linear and branched polyethylene (PE), −190–180°C [10.9]; and polymethylmethacrylate (PMMA), −190–250°C [10.10]. They used commercial samples. The spectra were analyzed into four lifetime components, τ_1 (I_1) to τ_4 (I_4), of which the two longest lifetimes were associated with o Ps pick off (see below), the second lifetime with positron, and the first lifetime with p Ps annihilation. The fitting parameters for lowcrystalline PTFE are shown in Figs. 10.2,3. In [10.8] the same parameters for high-crystalline PTFE are given.

The analyses of polymer lifetime spectra into four free lifetime components are difficult. Apparently, the iteration procedure does not converge, and/or unphysical results appear, in most cases. It must be realized that the PTFE results, shown in Figs. 10.2,3, are rare exceptions, mainly because the long-lived components are fairly well separated in the spectra. In [10.10] two constraints, namely $I_1 = (I_3 + I_4)/3$ and $\tau_1 = 0.125$ ns, were used in the PMMA case, while only the former constraint was used for PE in [10.9]. *Serna* et al. [10.11] constrained the first three lifetimes to fixed values in four term analyses of PE spectra. Three component analyses of polymer lifetime spectra seem normally to converge and give well defined parameters if long-lived components are present. However, the variances of the three term fits are too large for high

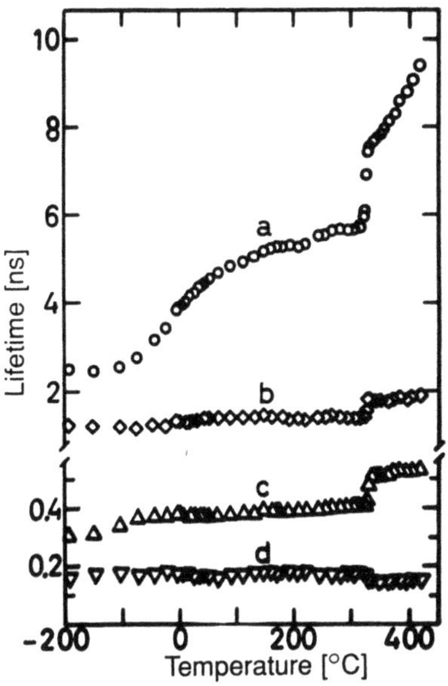

Fig. 10.2. Positron lifetimes versus temperature for polytetrafluoroethylene (PTFE). a, τ_4; b, τ_3; c, τ_2; d, τ_1 [10.7]

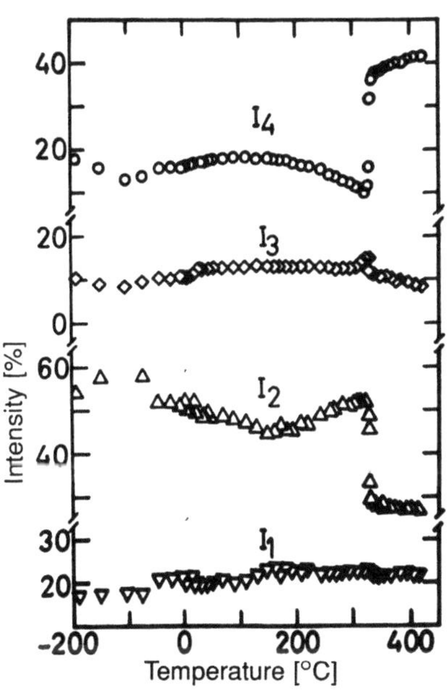

Fig. 10.3. Relative lifetime intensities versus temperature for polytetrafluoroethylene (PTFE) [10.7]

quality spectra. Consequently, great care should be exercised in the assignment of physical processes to the fitting parameters both in three and four component analyses of polymer spectra.

The assignment of the two long-lived components in four term analyses to o Ps decay in the polymer seems now to be well accounted for. The nature of the third lifetime component has been questioned mainly because *Kerr* [10.12] found a low narrow peak intensity, I_N, in AC spectra for PTFE and PE, which agreed with the hypothesis that only the longest lifetime was caused by o Ps annihilation. However, the p Ps intrinsic annihilation might give an AC component which has a larger intensity than that extracted in a simple analysis if the o Ps lifetime is short, as discussed for 6PO in Sect. 10.1 and CS_2 in Sect. 5.5. Hence, Kerr's I_N values are very probably not the p Ps yields. Recently, results of high field magnetic quenching LT studies indicated that the third lifetime component for polypropylene (PP), PMMA, PTFE, and PE had to be ascribed to a Ps-like system, which is different from that giving the fourth component [10.13]. Furthermore, 3γ-annihilation results for γ-irradiated PTFE agree with the assignment of both long-lived components to o Ps decay [10.14].

In principle, the detailed interpretation of the LT spectra must be based on what is known about the physical structure of polymer. However, the properties which can be associated with the o Ps lifetimes, such as the density and size of the hole type defects and the local viscosity, are fairly unknown. In most positron papers the simple two-phase approximation is used, i.e., the polymer is assumed to consist of small crystalline regions, the crystallites, and the amorphous regions. Actually, the structures of most polymers are very probably more complicated. The longest lifetime component is assumed to be the average pick off lifetime of o Ps in the cavities in the amorphous regions, where the cavities are enlarged due to the presence of Ps, as discussed in Sect. 10.1 for 6PO. Furthermore, the third lifetime is assumed to be an average lifetime of o Ps self-trapped in small cavities in the crystalline regions similarly to the Ps lifetimes found for the bulk of molecular crystals. We recall that delocalized Ps has never been found for molecular compounds except for ice. Additionally, it has been assumed that Ps in the crystalline regions can diffuse rapidly enough to have a significant chance of getting trapped in the amorphous regions. The second and first lifetimes are then associated with annihilation of "free" positrons and p Ps, respectively.

Actually, these rough assumptions on the four term fitting parameters seem to give a reasonable, approximate explanation of the LT spectra. The o Ps lifetimes, τ_4 and τ_3, in Fig. 10.2 are roughly the values expected from the correlations shown in Fig. 4.1 for holes in the amorphous parts and in Fig. 9.5 for the crystalline parts, respectively. In Fig. 10.2 τ_4 versus temperature shows a change in the slope at the glass transition temperature for the amorphous region of about $T_g = -110°C$. This change in slope is more pronounced in low- than in high-crystalline PTFE, as expected. The increase in τ_4 above T_g can be explained similarly to the way we explained the 6PO long lifetime. The reason is mainly a decrease in the local viscosity of the amorphous parts for increasing temperature. In the temperature interval around T_m, where the crystallites are melting, τ_4 increases abruptly to a remarkably high value of 7.6 ns in melted PTFE. Hence, the Ps bubble in melted PTFE is larger than the enlarged average void which trap Ps in the amorphous regions just below T_m.

The small change in I_3 at about 20°C was ascribed to two phase transitions in the crystalline phase of PTFE [10.7,8].

The second lifetime, τ_2, is close to that expected for the "free" positron except that its value of 300 ps at low temperatures seems to be somewhat too short. As the polymers are partly amorphous we expect intuitively that the positron lifetimes are larger than those found for brittle molecular crystals, such as anthracene, for example. Unrealistic τ_2 values shorter than 300 ps below -175°C were reported for PE in [10.9]. The reason for the too short τ_2 might be that it is correlated to the longer lifetimes, which are short at lower temperatures, in the computer analyses. Hence, the short τ_2 might be unphysical. Furthermore, I_1 in Fig. 10.3 is too large to be the p Ps yield. For four term fits of PE spectra at 20°C τ_2 values of 300 ps were reported in [5.16] in rough agreement with [10.9], while [10.11] reports $\tau_2 = 370$ ps.

The best explanation of the Ps yield for polymers not reacting with the electron and/or positron seems to be that it is determined mainly by the effect of the trapping of the spur particles in the defects in the structure. The mobility and diffusion constant of the trapped particles will thereby be fairly low so that some of the pairs cannot react to form Ps in time before the positron decays. Hence, the Ps yield is expected to be limited by the finite positron lifetime similarly to the yields in polar liquids, and in, e.g., the solid, long chain, linear hydrocarbons between 220 and 320 K discussed in Sect. 9.4. As in the latter solids the Ps yields might increase at higher temperatures because of particle detrapping due to thermal motion of segments of the molecules. In particular, this will happen in the liquid polymer if the particles are not strongly solvated. This effect might explain the strong increase in the total o Ps yield from 25% just below T_m to about 50% in liquid PTFE. A fairly high mobility of the electron and/or positron in liquid PTFE is, therefore, predicted. In general, the o Ps yields are between about 20%, as measured for alcohols, and about the 40%, obtained for liquid, linear hydrocarbons, for most non-reacting polymers, in agreement with the given explanation. The exceptions seem to be the high yields found in the very flexible poly(dimethylsiloxane) [10.15] and in low temperature (i.e., very rigid) PE [10.9]. Furthermore, the Ps yields seem to depend rather strongly on the thermal history and crystallinity of the polymer [10.8–11], as expected.

The only measured positron mobility for a polymer is for PE, where $\mu_+ = 10$ cm^2/Vs was found in [10.16], while [10.17] gives $\mu_+ = 28$ cm^2/Vs. At first sight, these large values speak against the given interpretation of the Ps yield, as they would give very short spur reaction times (5.8) of some picoseconds. However, within the framework of the spur model the above given explanation seems to be the most reasonable one. Furthermore, we have good reasons to believe that the particle mobilities in the crystalline are much higher than in the amorphous regions of the polymer, where the particles are trapped. The measured μ_+ values are average mobilities for all positrons, as the used DB measurements cannot determine separately the contributions from the positrons in different parts of the polymer. In fact, even the Ps contribution cannot be separated from that of the "free" positrons. The given explanation of the Ps yield can be maintained if we assume that μ_+ for the crystalline regions of PE is so high that the measured values are obtained even if a larger part of the positrons are trapped in the amorphous regions. In [10.16] a PE sample with a fairly high amorphous content of 40% was used, which might explain the measured lower μ_+ value compared with that given in [10.17]. It must be emphasized that the given explanation of the Ps yield is preliminary and it must be tested in detail in the future.

The o Ps yield in polyvinylchloride (PVC) is only about 6% [10.6]. This low value is very probably caused by the spur electron reaction with the molecules followed by a splitting off of Cl^-, which reacts with the positron (see Chap. 7). The only clear case of no Ps formation in a polymer at 20°C seems to be polyimide (Kapton) [4.26], which, thus, probably reacts with the electron and/or positron. Poly(pyrrole tosylate) and poly(pyrrole fluoride) seem to have no Ps formation at 10–100 K [10.18].

In [10.11] LT results for 9 PE samples of various crystallinities between 31% and 80% were reported. For the three shortest lifetimes fixed to $\tau_1 = 0.12$ ns, $\tau_2 = 0.37$ ns, and $\tau_3 = 1.1$ ns, they found that τ_4 and I_4 decreased, and I_3 increased with the crystallinity. They interpreted the data in terms of the model described above, and a simple assumption that the Ps formed in the crystallites could diffuse to the amorphous phase with a constant transition rate, K. They derived the size of holes trapping Ps in PE by use of Figs. 4.1 and 9.5, and extracted K values; density of voids; and Ps diffusion constant and length. Of course, this interpretation can be questioned. On the other hand, it seems to be one of the most detailed discussions of positron annihilation in well characterized polymer samples. In general, the higher τ_4 and I_4 values are found in the less crystalline polymers, as shown for PTFE in [10.8], as might be expected.

Several positron annihilation studies of glass transition points in polymers have been published. Recent works are the following. In [10.19] the glass rubber transition in four poly(alkylmethacrylates) was investigated. *Jean* et al. [10.19] found two clear transition points in the o Ps lifetime versus temperature in three amine-cured epoxy polymers. The highest transition point was the glass transition temperature. Isothermal relaxation and thermal response for atactic polystyrene were studied in [10.20] (see also [10.6]).

Recent papers on irradiated polymers are on γ-irradiated PTFE [10.21], γ-irradiated PE [10.22], and electron-irradiated PE [10.23]. A much stronger decrease in the longest lifetime intensity, I_4, in low crystalline than in high crystalline PTFE was explained mainly in terms of the increase in crystallinity of about 30% at a γ-dose of 20 Mrad for low crystalline PTFE. Also the longest lifetimes, τ_4 and τ_3, decreased due to the irradiation of PTFE. For PE the irradiation caused mainly a fairly small decrease in the Ps yield. Older works on irradiated polymers are reviewed in [10.4,6].

The polymerization of epoxy was studied in [10.24].

Finally, let us discuss briefly the so called "free volume model" [1.38,10.25], which has been used for the discussion of polymer results in many papers. It is assumed that the molecules prevent Ps from penetrating the "excluded volume" which is defined as the volume occupied by the van der Waals spheres around each atom in the molecules. Hence, Ps moves in the "free volume" between the molecules. The o Ps pick off lifetime is correlated to the average free volume per molecule, or molecule segment, in a polymer. This is incorrect, except in the rare cases where Ps is delocalized, such as in ice. As discussed extensively above, Ps is a "seeker", i.e., it tends to become trapped in hole type defects in solids, and a "digger" because it enlarges the preexisting holes and becomes selftrapped in cavities in softer solids and in the bubble state in liquids. The o Ps lifetime depends on the hole sizes, which are normally not correlated with the average free volume at all. Figure 8 in [10.25] showing the o Ps lifetime in three term fits, τ_3, versus "reduced free volume" for PTFE (80–640 K) has been used several times in arguments for the "free volume model". At increasing temperature (80–640 K) a pronounced maximum at (480 K, 4.3 ns) and a minimum at (600 K, 3.6 ns) in τ_3 were found and "explained" in terms of the "free volume model". However, the maximum and minimum have not been measured by others, see Fig. 10.2 and [10.8]!

10.3 Electric Field Dependence of the Ps Yield

It is reasonable to devote a full section to the discussion of the electric field dependence of the positron annihilation spectra as very interesting results have been obtained recently in particular for polymers. An external electric field influences the yield of Ps, while the o Ps lifetime is independent of the field.

In [9.46] the electric field dependencies of LT spectra for pentane (65 kV/cm), hexane (60 kV/cm), nonane (50 kV/cm), tetradecane (95 kV/cm), heptadecane (65 kV/cm), liquid (no field) and solid (15 kV/cm) eicosane, five PE samples (150–250 kV/cm) of crystallinity between 46% and 80%, two PTFE samples (175–185 kV/cm), and one PMMA sample (225 kV/cm) were studied. The highest field is shown in the parentheses. In three term fits the lifetimes remained constant for all electric fields. However, the intensities for the liquids depended strongly on the electric field. A fairly strong decrease in the Ps yield for increasing field up to 50 kV/cm followed by a slower decrease at higher fields were found. For the typical linear hydrocarbon, hexane, the intensity of the longest lifetime due to o Ps pick off annihilation decreased with an initial slope of 3.0 cm/MV from 0.37 at zero field to 0.27 at 60 kV/cm.

The spur model predicts [7.9] that the presence of an electric field decreases the Ps yield simply because some of the positrons and excess electrons which otherwise would have formed Ps are pulled out of the positron spur by the field. In [7.9] it is shown that the relative electron positive-ion recombination in the electron spurs is roughly 50% of the relative Ps formation as a function of the external electric field in hexane. This is a satisfactory agreement because the positron spurs are expected to be somewhat different from the electron spurs, as discussed in Sects 5.4 and 7.2.

In [9.46] the electric field effect was interpreted in terms of the Ore model of Ps formation. It was assumed that the field enhances the scattering of positrons out of the Ore gap, thereby decreasing the Ps yield. It was also assumed that the field raises the effective temperature of the positrons so that some positrons reached energies above the lower limit of the Ore gap; this increases the Ps yield. The former effect dominated at low field, the latter at high field. Strong arguments against this explanation were given in [7.9].

The PE samples gave Ps yields versus electric field curves similar to those of the liquid hydrocarbons, however with initial slopes 25 to 50% of those of the liquids. PTFE showed still smaller effects, while PMMA gave no electric field effect on the Ps yield up to 230 kV/cm [9.46]. In a polar polymer, such as PMMA, the positron spurs are expected to be much smaller than those of a nonpolar polymer, e.g., PE. The effect of the external electric field is, therefore, expected to be much smaller than that of the field due to the charged spur species. Furthermore, the spur particles are expected to become trapped in deeper potential wells in the polar polymers. Hence, the low or missing electric field effect on the Ps yield at lower fields in polar polymers is well explained in the spur model. In another polar polymer, cellulose acetate (CA) [10.26], no change in the Ps yield for fields up to 400 kV/cm was found, while the decrease in the Ps yield at lower electric fields has been measured for all the studied nonpolar polymers (PE, PTFE, and PP) [9.46,10.26]. Polyethyleneterephalate (PET, mylar) gave no field effects up to 250 kV/cm [10.26].

A delayed electric response of the Ps yield in PE was reported in [10.27]. A field of 100 kV/cm decreased the relative Ps yield from 100% to 59%. However, the relative Ps yield, *R*, increased roughly linearly with time up a saturation value of about 71% within 30 hours, with the field on and the positron source present. After 100 hours with the field on the field was removed. The *R* value then increased within about 100 hours to roughly 100% under short-circuit condition. The delay-effect seemed to be roughly independent of the source strength, and a sample maintained in a field of 100 kV/cm for a few hours only, recovered its unperturbed Ps yield immediately. At 50 kV/cm the delayed electric response was much slower. The most likely explanation [10.27] is that the Ps yield is determined by a local electric field which is the sum of the external field and the field due to the polarization charges of the sample. The delay is then caused by the relaxation times for the build up and the removal of the polarization charges. Such charge-storage and charge-decay processes in insulators are discussed in books on electret research, see, e.g., [10.28] in which charging of many polymers is treated. Much faster delayed electric response in low-density PE at high fields up to 1.25 MV/cm has been reported in [5.16]. Only PE has been investigated for this delay-of-charging effect on the Ps yield.

The next important discovery was made by *Bisi* et al. [10.26,29] who first measured the LT spectra for very large external electric fields. For PE *R* decreased at first to a minimum of 0.4 at 0.3 MV/cm. However, at 0.5 MeV *R* started to increase strongly, and it reached about 1.9 at the maximum field of 1.7 MV/cm [10.29,26]. As an example Fig. 10.4 shows *R* versus field for

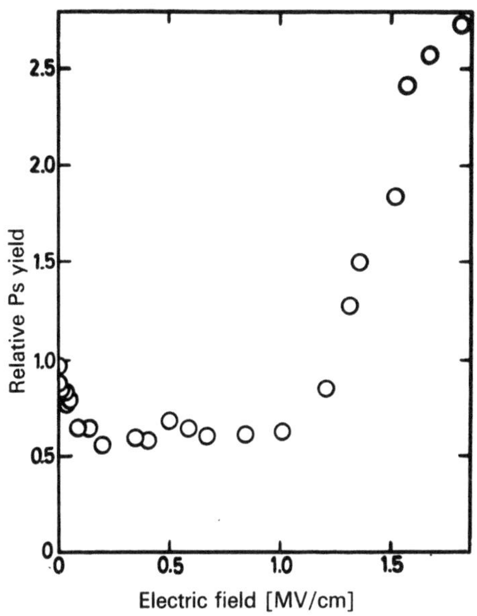

Fig. 10.4. Relative Ps yield in polypropylene versus external electric field [10.26]

PP. For PET *R* remained constant up to 0.25 MV/cm, decreased to 0.25 at 1.2 MV/cm, and increased again to 2.1 above 1.8 MV/cm. PTFE gave only a decrease in *R* up to 0.8 MV/cm the maximum field possible without breakdown, and *R* for CA was independent of the field up to 0.4 MV/cm at which breakdown occurred [10.26]. Recently, *Cornaz* et al. [10.30,5.16] measured LT and AC spectra for PE in greater details confirming several of the effects found by Bisi et al. and discovering new phenomena.

The high field results for PE [10.29] triggered *Brandt* [7.93] to claim that they "decide in favor of the preponderance of the Ore model mechanism of Ps formation". He showed that his model of the electric field dependence of the Ps yield, discussed above, could explain fairly well the measured PE results. A discussion of Brandt's arguments shows that many other effects may lead to the rise in the Ps yield [5.15]. *Bisi* et al. [10.26] realized that the model of Brandt could not explain well the results for PP shown in Fig. 10.4. Hence, they explained the decrease in the Ps yields at low fields for PE, PP, and PTFE in terms of the spur model [10.26] and the strong increase at high fields as caused by the heating up of the positrons to energies above the lower limit of the Ore gap by the electric field. A similar explanation was used by *Cornaz* et al. [10.30,5.16].

A detailed discussion of the electric field effect on the Ps yield is appropriate here. At first it is very important to realize that the electric field can heat up the positrons into the Ore gap, or prevent them from thermalizing out of the Ore gap, in the Ore as well as the spur model. Energy conservation is, of course, also valid in the spur model. Hence, if the high field effects are really caused by this mechanism this fact does not disprove the spur model. However, other effects might explain the high field results, too [5.15]. First, the polymers can become strongly charged in the high fields. For example, at an electrode-polymer distance of 10^{-3} cm, breakdown occurs at a very low voltage of about 400 V in air (Fig. 2.2 in [10.28]). The polymer will, therefore, be surface and/or bulk charged so that there is a maximum of 400 V across the air gap at higher electric fields, at those parts of the sample where the distance is 10^{-3} cm. In general, only if there are no air gaps between the electrodes and the sample, which must be ensured by use of elaborate sample preparation techniques and/or very special electrode-sample configurations, the charging can be prevented for fields up to the intrinsic electric field strength of around 6.5 MV/cm for PE. For instance, for PET at 4 MV/cm a surface charge of 10^{-6} C/cm^2 and a bulk density of 10^{17} electrons per cm^3 have been obtained (p. 32 in [10.28]), and they were detected after the field was switched off. These densities correspond to an average distance between the electrons of 4 nm and 21.4 nm, respectively. It seems reasonable to assume that similar high densities occur also in other polymer samples, when the electric field is applied during the entire measuring period and irradiation due to positron slowing down takes place.

The electrons in the bulk are trapped in defects in the polymers, and, of course, their density increases with increasing electric field. Electron densities on the order of 10^{16}–10^{17} cm^{-3} are probably comparable to the excess electron density in the positron spur in polymers. A reasonable guess of the spur size is roughly 10 nm. Consequently, the positron can form Ps also with the injected electrons, and hence higher Ps yields are expected at higher electric field strengths. The effects of the surface charge is more difficult to estimate, as it will depend on the details of the source-sample configuration [5.15]. Charge injection effects on the Ps yields in low density PE was found in [5.16], although very carefully prepared electrode-sample configurations were used in the AC measurements.

Furthermore, in high electric fields the recombinations of the electron positive-ion pairs created by the slowing down processes of the positron are very different from those at low fields. An external electric field of 2 MV/cm gives a field of 1 MV/cm inside the sample at dielectric constant of about two typical of a polymer. This is larger than the field of a unit charge at distances larger than about 2–3 nm. Most of the electron positive-ion pairs are, thus, separated, i.e., the electrons are pulled out of the tracks to become trapped in defects, recombined with the positive ions of other tracks in homogeneous reactions, trapped in surface layers, and/or transferred to the positive electrodes. What actually happens to the electrons depends on the structure of the solid and on the previous irradiation dose. The positive ions are probably bound to the molecules and will contribute to the bulk charge density.

The positron will be pulled out of many positron spurs, too. However, it is important to realize that some spurs are oriented so that the field will pull the positrons into the spur, or into other parts of the track formed by the positron itself. Hence, it is expected that the electric field does not reduce the Ps yield to zero, which is in agreement with experiments. In [10.26,30,5.16] it is incorrectly assumed that only that part of the Ps yield which can be removed by the electric field at the minimum of the Ps yield versus field is formed by the spur model process. The spur model is also used incorrectly in [10.26], where it is assumed that the positron-electron pairs are outside the range of the Coulomb attraction and repulsion of the other charged spur species. Actually, the positron will be thermalized at a distance smaller than the Onsager radius from several electrons and positive ions in the positron spurs of polymers.

A positron having a mobility of 10 cm^2/Vs will move 0.04 mm in 400 ps in a field of 1 MV/cm. Actually, the mobility will very probably be reduced drastically if the positron is heated up in the field and more vibrational modes can be excited in collisions. The positron may also get trapped before annihilation. However, it is expected that the positrons can move appreciable distances during their lifetime in the high electric fields. In that motion they may encounter electrons trapped in the defects, on the surfaces, etc. and form Ps. This effect might also contribute to the larger Ps yields at higher electric fields.

Additionally, detrapping of electrons and positrons in the potential wells of the polymers will take place by effect of the temperature and electric field. The motion of the particles will, therefore, depend on temperature and electric field strength. As discussed in Sect. 10.2, the Ps yield is probably limited by the inability of the particle to form Ps before the positron decays because of particle trapping in many polymers. Hence, the detrapping might also increase the Ps yield. This makes the situation even more complicated.

In [5.15] it was argued that radiation chemistry data seems to indicate that electrons cannot be heated up in polymers at the used electric field strengths. However, according to [10.30] heating up electrons in PE to the energies of the lower limit of the Ore gap has been observed, which, of course, speaks in favour of the heating up of positrons, too.

In summary, the electric field effect on the Ps yield in nonpolar polymers can be well explained in terms of the spur model. The strong increase in the Ps yield at higher fields may be caused by the heating up of positrons above the lower limit of the Ore gap, by one or more of the other processes discussed above, or by a combination of several of these mechanisms. Further experiments are necessary to solve that problem. Concerning the application of the Ore model to explain the electric field effects in polymers, one should also remember that a model of Ps formation must explain the Ps yield in all compounds. The Ore model without the "hot Ps" reactions included is simply inapplicable in many cases, as discussed in Sect. 7.7.

It is appropriate here to discuss more results of the detailed studies of PE by *Cornaz* et al. [5.16,10.30]. They used ^{64}Cu as positron source in the AC measurements. Because two electrons are emitted for each positron, a very high irradiation dose of up to 200 Mrad/day was given to the samples resulting in creation of up to 1 M of electron-ion pairs, of which a greater number were separated in the high electric fields. Very strong relaxation effects depending on radiation dose and electric field were found, in particular at temperatures below −30°C, but also at higher temperatures. The AC spectra were fitted by a sum of only two gaussians, of which the narrower one was assigned to p Ps intrinsic annihilation. By that procedure they found 100% Ps formation at electric fields above 1.0 MV/cm. As discussed several times in this book, this is a very rough way of determining the Ps yield. Apparently, they got too high Ps yield at high fields by that method. The used four component fits of the LT spectra fixing the second lifetime to 300 ps tends to give too high Ps yields, too. 300 ps seems to be too low a lifetime for free positrons at room temperature in PE.

Recently, PMMA which was charged to become an electret (PMMA-electret) was studied by LT technique in the temperature interval −190–140°C [10.31]. Differences between the Ps yield for PMMA-electret and PMMA were found. Their low temperature results for PMMA seem to be very different from those in [10.10]. A very long relaxation time of 1000 (400) hours was needed before the Ps yield of PMMA-electret (PMMA) reached its equilibrium value at −190°C.

The effects of an electric field on the positron annihilation in polymers are very interesting indeed. In particular, the electret phenomena deserve more attention. This is a very open field, where many new unexpected results might be obtained in the future. Clearly, these investigations ought to be extended to other molecular compounds.

10.4 Liquid Crystals and Miscellaneous Liquids

Liquid crystals are organic compounds in transition states between solid and liquid forms. They are viscous, jelly-like materials which exhibit rheological properties of liquids (viscosity) as well as anisotropic physical properties typical of crystalline solids (light scattering and reflection). Perhaps the best known examples are various esters of cholesterol. Liquid crystals are technologically important.

About 35 papers on positron annihilation in liquid crystals are available. Reviews with lists of literature have been published by *Jain* [10.32,33]. The work was started by *Walker* et al. [10.34–37]. Four other groups have been involved [10.32,33,38–41] in the work. Around 30 different liquid crystals have been investigated by use of positrons.

Positron annihilation has mainly been used to determine the phase transition temperatures of the liquid crystals. Various phases, such as crystalline solid, nematic, cholesteric, smectic, and isotropic liquid phases, may occur in a liquid crystal compound. Normally, the long lifetime due to o Ps decay and/or its intensity change at the transition temperatures. In some materials such studies have resulted in the discovery of new phases not revealed by other conventional techniques. Because of lack of space we cannot here give a full account of all the measured results.

The most important point to discuss is the interpretation of the o Ps lifetimes and yields because it has been claimed in several recent reviews that no good explanations of the measured data are available. Actually, the properties of Ps in liquid crystals seem to be very similar to those obtained for other molecular liquids and solids. However, a detailed discussion of the liquid crystal results in terms of the models described above has not been published. We shall, therefore, discuss the properties of Ps in some typical liquid crystals to illustrate the basic principles.

The mostly investigated liquid crystals have structures

$$R - C_6H_4 - B - C_6H_4 - A, \qquad (10.1)$$

where R is an alkyl (C_nH_{2n+1}) or alkoxy (OC_nH_{2n+1}) group; C_6H_4 the benzene ring; B is CH=N, COO, NO=N, N=N, or nothing; and A is an alkyl, alkoxy, CN, or NO_2 group. As discussed in Chaps. 7,8, NO_2 attached to a benzene ring reacts with Ps and inhibits Ps formation. The best example is nitrobenzene. Hence, for A equal to NO_2 a low Ps yield and a short o Ps lifetime are expected. *Jain* et al. [10.32,33] have studied five such cases. They found a Ps yield between 2% and 6%, and in four of the samples long lifetimes of 0.5–1.0 ns. Small decreases in the lifetimes at transitions from less to more disordered liquid phases were found. This can be explained as caused by a stronger Ps reaction with the NO_2 group at lower viscosity. In the fifth liquid the lifetimes were somewhat longer, but maybe also uncertain because of a low intensity of 2% in the liquid crystal phases. On taking into account A) that the NO_2 group are more diluted in the liquid crystals than in nitrobenzene and B) that the reactivity of NO_2 benzene-ring group towards Ps depends on the other groups attached to the benzene-ring (e.g., NH_2 in the para position in nitrobenzene reduces the reactivity of Ps), the results agree well with the expectations. Compared to the NO_2 group the other groups on the molecules (or dissolved O_2) are not expected to influence strongly the o Ps yield and lifetime.

In seven other liquid crystals A is equal to CN [10.32,33,41]. A benzene-ring with a CN group attached inhibits normally the Ps yield, but it does not react with Ps. For example, the o Ps yield (lifetime) for benzonitrile is 7.3% (2.87 ns), chlorobenzene 8.5% (2.83 ns), and bromobenzene 3.4% (3.24 ns) [1.34]. In the liquid phases for A equal to CN, o Ps yields between 9.1% and 14.5% and lifetimes between 2.0 ns and 2.4 ns were found, except for one case, namely for R equal to the ethoxy and B the CH=N groups, where the lifetimes were 1.16 ns and 1.33 ns in the nematic and isotropic liquids, respectively. Both the Ps lifetime and yield changed fairly little at the transition from one liquid phase to another. Hence, these results are also in rough agreement with the expectations.

It must be realized that most liquid crystals studied by positrons have not been degassed before the measurements. For example, *Jain* et al. [10.42] did not degas the samples except in the works reported in [10.41]. The samples used in [10.39] were degassed, while Walker et al. [10.34–37] do not mention degassing. Oxygen in air at 1 bar above the liquid reduces the o Ps lifetime in hexane by 40% and in ethanol by 17% [1.33]. The o Ps yield in hexane is not inhibited by O_2. It is expected that the solubility of O_2 in liquid crystals in their liquid phases depends on the temperature. Furthermore, because of the high viscosities the equilibrium concentration is obtained after very long time (days) in unstirred liquids. The effect of O_2 on the Ps parameters in part of the published work is, therefore, difficult to estimate. The o Ps lifetimes in the liquids containing CN groups might be influenced by the dissolved O_2.

All the other studied liquid crystals containing no groups which strongly influence on the Ps parameters have o Ps lifetimes above 2.0 ns and intensities above roughly 17%. The lifetimes are probably influenced by dissolved O_2 in nondegassed liquids. Again the changes in lifetime and yield at the transitions from one liquid phase to another are rather small. In these liquid phases the Ps yield is probably limited by the finite lifetime of the positron, as, e.g., in polar liquids (see Sect. 5.5). Consequently, o Ps yields comparable to those of the alcohols and/or less polar liquids such as diethylether are anticipated. Hence also for these liquids the expected results are found.

At the transitions from one liquid phase to another, less ordered phase at higher temperature, the o Ps lifetime is expected to increase due to the lower effective surface tension (see Sect. 4.4) in the less ordered phase. At high viscosity the lowering of the viscosity at the transition might cause an increase in the lifetime, too. However, if some components of the liquid, e.g., dissolved O_2 or NO_2 groups on benzene rings, react with Ps, a decrease in the lifetime caused by a higher reaction rate at lower viscosity is expected. Experiments show an increase in the o Ps lifetime at the liquid-liquid transition to the less ordered phases in most, but a decrease in several other, liquid crystals [10.33]. Similarly, the Ps yield is expected to increase at the transition from more to less ordered liquid phases due to a decrease in the viscosity if the yield is limited by the finite positron lifetime. On the other hand, the Ps yield might decrease at the transition if a reaction in the spur, e.g., of the electron with O_2, or CN followed by CN^- formation, is more effective in the less ordered phase. Experimentally, most Ps yields increase with decreasing degree of molecular order [10.32,33], but a decrease is found in few cases, too [10.41].

For most of the compounds the changes in o Ps lifetime and yield at the transition from the solid to the first liquid crystal phase are fairly large except for the NO_2 containing molecules. Apparently these changes can be roughly explained in the same way as discussed for molecular crystals in Chap. 9. Surprisingly large changes were reported for degassed p-n-hexyloxy cyanobiphenyl [10.41].

In summary, the o Ps lifetimes and yields in the liquid crystal compounds are well explained in terms of the models discussed in the preceding chapters. It would be interesting to study the Ps inhibition and quenching by additives in the liquid crystal phases. This seems not to have been done except for the uncertain results for dissolved O_2. The samples ought to be degassed.

Results of positron annihilation experiments on micelles and other aggregated systems have been reported in 20–30 papers. These studies were started by *Ache* et al. [10.43–49] and continued in [10.50–54].

Micelles are aggregates of large molecules, such as surfactants, formed in aqueous solutions. Two important parameters characterize the aqueous micelles, namely the critical micelle concentration (CMC) and the aggregation number. The concentration above which the micelles are formed is the CMC. Below CMC only monomers are present, while above CMC micelles with a characteristic aggregation number (independent of concentration) are in equilibrium with a monomer concentration identical to the CMC. Aggregation numbers are generally within 10 to 100 monomers per micelle. A typical surfactant, e.g., sodium dodecylsulfate (NaLS), consists a hydrophilic head group at one end of the molecule and a hydrophobic alkyl chain at the other end. The precise structure of an aqueous micelle seems to be somewhat uncertain. The classical model is a spherical structure with the head groups at the surface and the alkyl chain pointing into the center of the sphere. The surfactant NaLS dissociates into Na^+ ions and a negative head group in water. About 70–80% of the Na^+ ions are associated with the head groups in the so called Stern layer, while the rest of the Na^+ ions are located outside the micelle in the diffuse double layer (GoüyChapman layer). Micelles can be anionic (such as NaLS), cationic, zwitterionic, or nonionic.

Reverse micelles are aggregates of large molecules, such as surfactants, in nonpolar solvents. Apparently, the structures of reverse micelles are more uncertain than those of the micelles. Surfactants in nonpolar solvents seem often to exhibit sequential selfassociation instead of the abrupt aggregation into a micelle of a particular aggregation number found in aqueous systems. A schematic model of a reverse micelle consists of sphere of aggregated molecules having the head groups in the center and the alkyl chains pointing towards the nonpolar solvent. Addition of water into the solution of a surfactant in nonpolar solvents affects the aggregation strongly by increasing the aggregation number. Large amounts of water can be solubilized in the center of the reverse micelles ("water-in-oil" solutions).

The reported positron annihilation results for micelle and reverse micelle systems cover only LT data. Results of AC and DB measurements are not available. Normally only two term analyses of the LT spectra have been used, and only the longest lifetime and its intensity due to o Ps decay are interpreted. The LT spectra have mainly been measured under variation of 1) the surfactant concentration, 2) the amount of water added to reverse micelle solutions, and 3) the concentration of added solutes reacting with Ps. As usual we can only discuss typical results here.

In [10.48] LT spectra were measured versus concentration of the surfactant, NaLS, in water. Apparently, only the o Ps yield, I_2, is strongly influenced by the added NaLS, while the o Ps lifetime is roughly independent of the added surfactant. Below the CMC of about 10 mM NaLS I_2 remained constant, while I_2 versus NaLS concentration decreased above the CMC until it saturated at about 68% of the value in pure water above roughly 0.3 M. At 90°C I_2 behaved similarly, but it saturated at about 62% of the pure water value above 0.15 M. These results are easily rationalized in terms of the spur model. The hydrated electron reacts strongly with electron scavengers solubilized in cationic micelles but normally not with the surfactant monomers. Rate constants of 10^{12} $M^{-1}s^{-1}$

have been measured and explained in terms of a reaction radius equal to the large Onsager radius of the positively charged micelles. Similarly, we expect that the positron is very effectively trapped in the negatively charged NaLS micelles, but that it does not react with the NaLS monomer. The trapping of the positron will reduce the Ps yield compared to that of pure water as found experimentally. In [10.48] it was shown that (7.29) gave a poor fit of I_2 versus NaLS concentration. However, (7.29) is shown empirically to fit well the Ps inhibition only when the Ps yield is inhibited to zero. This need not be the case for the NaLS micelles as the electrons might also be thermalized or trapped in the micelles resulting in Ps formation inside the micelles. Furthermore, the concentration used in (7.29) ought to be the micelle concentration not that of NaLS. Hence, the poor fit in [10.48] does not speak against the spur model.

Reverse micelles formed by bis(2-ethylhexyl)sodium sulfosuccinate (AOT) in benzene and isooctane and dodecylammonium propionate (DAP) in benzene and cyclohexane were studied in [10.44]. The o Ps lifetimes changed very little with surfactant concentration. However, the o Ps yield, I_2, versus surfactant concentration showed strong decreases in surfactant concentration intervals close to certain "operational" CMC's in all solutions. For example, in the AOT isooctane solution a drop in I_2 from 40% to 30% occurred at the very low concentration interval of 0.5–1.2 mM AOT. This Ps inhibition by AOT in isooctane is comparable in strength to that of SF_6 in neopentane at 1 mM SF_6, the strongest inhibition case for single molecules. For isooctane, rate constants per AOT molecule for excess electron attachment to "dry" micelles of 4×10^{11} $M^{-1}s^{-1}$, and of up to 10^{15} $M^{-1}s^{-1}$ for electron trapping per AOT micelle containing water pools, have been reported [10.57]. Hence, the very strong Ps inhibition by AOT micelles in isooctane can be well explained in terms of trapping of electrons and maybe also positrons in the positron spur. As found for alcohol clusters in nonpolar liquids (Sect. 7.6), the trapping of the light particles reduces their mobility drastically resulting in a Ps yield limited by the finite lifetime of the positron as in polar liquids, and hence, the Ps yield is strongly reduced compared to that of nonpolar liquids. The LT spectra for the AOT/isooctane system were remeasured in [10.54]. A more gradual decrease in I_2 from about 34 to 31% in the concentration interval 1 to 3 mM AOT was found. However, the reported I_2 value for pure isooctane is much too low in [10.54] (see Table 5.1).

The o Ps yield, I_2, versus water contents in 67 mM and 400 mM AOT/isooctane solutions was reported in [10.47]. At 67 mM AOT a strong decrease in I_2 from about 28% to 21% occurred on addition of roughly 1 vol.% of H_2O (about 8–9 H_2O molecule per AOT), followed by an I_2 saturation at 19% up to 10 vol.% H_2O, where I_2 started to increase strongly at increasing water content. At 50 vol.% H_2O $I_2 = 29\%$ was found. A 67 mM AOT/heptane solution gave qualitatively similar results. In 400 mM AOT/isooctane a decrease in I_2 from 21% to a saturation value of 17% at 5 vol.% H_2O followed by a saturation at about 16% above 20 vol.% H_2O were obtained, i.e., no increase in I_2 was found. The strong extra drop in I_2 at low water content, apart from the

I_2 decrease due to the presence of the "dry" micelles, can be well explained by the very large excess electron rate constant of AOT micelles containing water pools [10.55]. Probably the positron is trapped in the water pools, too. The I_2 increase above 10 vol.% H_2O is caused by a phase transition from spherical aggregates to a lamellae structure in the AOT/isooctane/water system.

Fairly small temperature dependencies of I_2 in AOT/isooctane and DAP/benzene solutions were reported in [10.45]. Furthermore, I_2 versus surfactant concentration for micelle solutions of the Na^+, NH_4^+, $N(CH_4)_4^+$, and $N(C_2H_5)_4^+$ salts of bis(2-ethylhexyl) sulfosuccinate in benzene was obtained. Only the Na^+ salt gave an "operational" CMC, while I_2 decreased at very low concentrations in the other solutions.

The Ps reactions with nitrobenzene and $CuCl_2$ in various aqueous micelle solutions were studied in [10.43]. The rate constants of the Ps reactions with nitrobenzene and $CuCl_2$ decreased in most solutions on addition of the surfactants; in some cases to 40% of the value in pure water. In [10.44] the rate constants of Ps reactions with nitrobenzene, tetracyanoethylene, and nitrophenol in reverse micelle solutions of benzene, cyclohexane, and isooctane were reported. The interpretation of the results in [10.43,44] in terms of three rate constants (expressions (1) in both papers) seems to be questionable.

Vesicles (liposomes) are bilayer systems which have more complex structures than solutions of micelles and reverse micelles. In the simplest terms these aggregates are closed bilayer structures. The shape, size, and number of bilayers depend on the components and method of preparation. Stable bilayer vescicles of dioctadecyldimethylammonium chloride (DODAC)/water were studied in [10.49]. The rate of o Ps decay versus temperature was given. The reactivity of o Ps with p-benzoquinone in bilayer systems (water-DL-α-dipalmitoylphosphatidylcholine) was studied in [10.52].

10.5 Miscellaneous Solids

In this section we shall discuss positron annihilation in several types of solids of chemical interest, which have not been treated in Chap. 9 and Sects. 10.1–3. The solids with larger holes, such as zeolites, silica gels, and small grain oxide powders, are particularly important as Ps is trapped and/or formed in the holes. Also biological samples will be discussed.

Positron annihilation results for zeolites have been reported in about 20 papers (see [4.28,10.56–60]). According to orthodox views zeolites are crystalline aluminosilicates, whose structures includes large natural cages. Recently, other structural variants of zeolite-type frameworks have been included in the class of zeolites [10.61].

Let us discuss the structure of zeolite 13X as an example. The unit cell composition of 13X is $Na_{86}[(AlO_2)_{86}(SiO_2)_{106}]^-$, xH_2O. The basic structural units are tetrahedral groups of SiO_4 and AlO_4, which are joined to form the larger sodalite units (SU) containing the β-cages of diameter 0.66 nm and entrance diameter of 0.22 nm. Furthermore, each SU is bound tetrahedrally to three other SU. The SU arrangement forms large cavities, the α-cages, having a diameter of about 1.3 nm and an entrance diameter of 0.8 nm. The zeolite water is mainly situated in the α-cages and has a density in fully hydrated 13X similar to that of normal water. This water can be removed by dehydration at temperatures above about 300°C, whereby the main structure remains unchanged. The tetrahedral group of AlO_4 has a negative unit charge which attracts a Na^+ ion. The Na^+ ions are fairly loosely bound and can be exchanged with other positive ions.

The zeolites have mainly been studied by use of the LT method. It is generally agreed that the LT spectra consist of at least four terms. Magnetic quenching effects show that the two longest LT terms are due to o Ps annihilation [10.57]. However, the results differ very much from one paper to another. There are several reasons for this. First, *Ito* et al. [10.57,58] realized that a 50 ns window in the TAC of the lifetime set-up was too short as the spectra contained long lifetime component of 20–80 ns. A window of at least 800 ns had to be used to determine these components (see also [4.28,10.56]). The other lifetime components depended strongly on the TAC window used. Some papers give only results for short TAC windows. Second, the third lifetime, which seems to be due to o Ps annihilation in the largest cages of the zeolites, changed strongly with the temperature of dehydration [4.28]. For 13X it reached 8 ns after evacuation at 400°C but decreased to 5.5 ns after evacuation at 600°C. Most groups have evacuated at about 200–300°C. Third, the LT spectra consist, apparently, of a distribution of many lifetime components, which makes the analyses fairly uncertain.

Of course, these facts make a description of the until now published results fairly difficult. Hence, the following remarks are rather preliminary. The latest works seem to indicate that the third lifetime (3–12 ns) is caused by o Ps annihilation in the largest natural cages of the zeolites, i.e., the α-cages for 13X. In [10.57] six zeolites were studied. The third lifetimes correlated well with the volume of the largest cage using the curve found for molecular compounds (Fig. 4.1) for four zeolites. Two zeolites gave about 40% shorter lifetimes than expected. In [4.28] two out of four zeolites gave the expected third lifetimes after evacuation to 400°C, while the two other lifetimes were too short. Apparently, o Ps in the largest cages can react weakly with "centers" on the surface of the cages resulting in the shorter third lifetimes at evacuation to temperatures above 400°C. These "centers" seem to become active if all the water molecules are removed. Maybe the "centers" are inactive because of associated water even after dehydration to 400°C [4.28]. In [10.57,58] evacuation at 350°C was used. The third component intensities are normally 9–14%.

The most reasonable explanation of the longest, fourth lifetime of 20–80 ns is that it is caused by o Ps annihilation in cavities (2–5 nm diameter) which are much larger than the largest natural cages in the zeolites. In many zeolites, e.g., 13X, the large entrance windows of the largest cages will probably permit Ps to diffuse rapidly from one cage to another. Taking into account the long third lifetime it is expected that o Ps can get trapped in the cavities even if the cavity concentration is very low, e.g., 10^{-7} cavities per large cage. Furthermore, it is expected that the cavities have different sizes, and the structure and concentration of the cavities will depend on the preparation of the samples. This explains why many different fourth lifetimes and intensities (5–14%) have been reported for most zeolites. The cavities, which, apparently, are not just voids between the grains of the zeolites, can probably not be studied by use of other experimental methods. In the papers where the second lifetime is given it seems to be very long; for 13X 0.6–0.8 ns. Apparently, the free positron lifetime on the surface of ionic crystals containing oxides is unknown. Maybe the second lifetime is partly caused by o Ps annihilation in the smaller cages, too, apart from free positron annihilation on the surfaces of the cages. The first lifetimes are given only in few cases. Their values and intensities are too large to be due to p Ps annihilation.

The few AC [10.58] and DB results [4.28,10.59] seem to agree with the given interpretation.

The addition of liquids and gases to the zeolite changes the LT spectra in most cases. As expected, the fully hydrated zeolites seem to give a longest lifetime roughly equal to that of water for 13X and 4A [10.55,60]. Oxygen at 1 bar removes the fourth LT component for 4A, 5A, 13X, and mordenite [10.57,58]. For 5A this was shown to be caused by both Ps inhibition and quenching. In [10.56] He, N_2, O_2, Cl_2, and CH_4 at 1 bar pressure were introduced into the zeolites 4A and 13X. In particular, O_2 and Cl_2 gave strong effects, because they react with spur electrons and Ps.

The next class of solids to discuss is the small-grain oxide powders with large specific surfaces. Since the first study of Al_2O_3, MgO, and SiO_2 [10.62], about 50 papers on fine oxide powders have been published. In particular, a discussion of positron annihilation in silica (SiO_2) powder (silicagel) will be given, as this is the mostly studied case. Typically, two long LT components of $\tau_4 = 30–140$ ns (10–25%) and $\tau_3 = 2–3$ ns (3–7%) have been measured for SiO_2 powders. The shortest lifetimes are normally not given because only few points at short times in the LT spectra were measured. τ_4 and τ_3 are the o Ps lifetimes in the space between the grains and in smaller defects in the bulk SiO_2, respectively.

The AC and LT spectra show that Ps between the grains is not completely thermalized before the annihilation. *Chang* [10.63] measured the LT spectra in silica aerogel of 0.1 g/cm^3 density and an average grain size of about 5 nm, which means a fractional volume of roughly 5% SiO_2 and an effective "void" size of about 70 nm. The longest lifetime $\tau_4 = 137$ ns is due to o Ps decaying essentially by 3γ-annihilation, as the vacuum lifetime of o Ps is 141.9 ns. A second long lifetime $\tau_3 = 2.6$ ns was found. The intensities were not given. By use of a time selecting energy spectrometer the DB spectra were measured simultaneously for eight time windows. When a magnetic field of 0.4 T was applied the longest lifetime component splits into two components, $\tau_4' = 33$ ns and $\tau_4'' = 134$ ns. 67% of the 33 ns component decays by 2γ-annihilation. On adjusting the time windows to select the different sections of τ_4', which correspond to different ages of o Ps, the widths of the time dependent DB spectra could be obtained, and hence, the kinetic energy of o Ps versus o Ps age could be determined. At an o Ps lifetime of 0.0 ns the energy was about 0.68 ± 0.12 eV. It dropped to 0.27 ± 0.02 eV at 5 ns, 0.1 eV at 30 ns and about 0.05 eV at 60 ns the longest lifetime given. A roughly similar energy versus age curve was obtained by converting o Ps to p Ps by use of 0.1 bar of oxygen in the sample.

Hyodo et al. [10.64] measured the AC spectra for SiO_2 aerogel (0.1 g/cm^3, 5 nm particle diameter), and powders of densities 0.3 and 1.0 g/cm^3 and average particle diameter of 7 nm. They determined roughly the p Ps components by ascribing the broad component to a gaussian fitted to the AC spectrum at angles larger than 5 mrad. The fwhm of the p Ps component was 4.8, 4.0, and 3.1 mrad for 0.1, 0.3, and 1.0 g/cm^3, respectively. An external magnetic field gave additionally an AC component of fwhm = 1.4 mrad due to 2γ-annihilation of the o Ps, $m = 0$, state. The fwhm = 4.8 mrad corresponds to a p Ps energy of about 0.74 eV, which is fairly similar to the 0.68 eV for just formed o Ps given by Chang, while the magnetically quenched o Ps gives a much narrower AC component. *Mills* et al. [10.65] measured by a time-of-flight method the energy distribution of Ps emitted from SiO_2 powder of density 0.18 g/cm^3 and particle diameter 3.5 nm after the powder was irradiated by a 50 ns wide pulse of slow positrons of energy E_+. Even at $E_+ = 20$ keV giving a penetration depth of 0.002 cm, most of the emitted Ps atoms were not thermalized. Hence, several independent experiments show that p Ps and partly o Ps annihilate before thermalization in small-grain SiO_2 powder. Furthermore, theory seems also to indicate a long thermalization time (30–140 ns) for Ps in SiO_2 powders [10.65,66].

In all papers on fine oxide powders it is assumed that Ps is formed within the powder grains, and that some Ps atoms diffuse to the grain surface and escape to the voids between the grains [10.67]. However, except for very special conditions Ps might be formed by other processes, too. According to the spur model some excess electrons and/or the positron might escape from the grains, and Ps can form in the voids by escaped particles, or by a reaction on the grain surface between one escaped particle and the oppositely charged particle thermalized in the grain. We expect that the spur size is about 10–20 nm in amorphous SiO_2, and hence, it is larger than the grains in very fine powders. Furthermore, the Onsager radius is about 60 nm in vacuum, and only a little smaller in fine powders, at 20°C. Consequently, the oppositely charged spur species attract each other strongly at distances comparable to or larger than the grain sizes. Only in the unlikely situation where both the electrons and the positron in the voids are always absorbed immediately whenever they hit the surface of the grains, Ps formation inside the grains will be the only significant process. The particles in the voids are probably reflected many times at the grain surface even if they probably have a lower energy inside the grains. Unavoidable impurities on part of the surface might influence the reflection of the particles.

Some of the measured results are interesting in that context. The small p Ps energy of 0.74 eV at annihilation is difficult to explain in terms of Ps emitted from the grains as the energy of Ps emitted from crystalline quarts (SiO_2) was found to be 3.27 eV [9.53]. Furthermore, the lower energy of p Ps at annihilation in powders with larger grains [10.64], and hence presumably larger voids, is also not expected if Ps is emitted with a certain energy independent of grain size. At present the problem of Ps formation and thermalization in SiO_2 powders seems, therefore, not to be solved.

As found for zeolites, the o Ps lifetime is shorter if the SiO_2 powder is degassed at a high temperature of 600°C instead of the normally used lower temperature of, e.g., 200°C [10.68]. The reason is probably that Ps reacts with "centers" on the void surface, which are "covered", e.g., by water molecules, even after degassing at 200°C, but are removed at 600°C.

Positronium inhibition and quenching by added compounds in the voids of oxide powders have been investigated by many groups. Chuang and Tao studied the effects Br_2, O_2, water, n-hexane, nitromethane, I_2, and CCl_4 added to silicagel (see [10.68] and refs therein). For example, I_2 inhibits and quenches Ps while CCl_4 mainly inhibits Ps, as in liquids. A general discussion of the Ps-O_2 reaction in silicagel is found in [10.69], while low-density O_2 gas in SiO_2 aerogel was studied in [10.70], which refers to many older papers.

Small metal particles may be conveniently prepared by dispersing the metal on a high surface area oxide support. Only very few positron annihilation studies of such systems have performed (see refs. in [10.71]). In [10.71] silica (Cab-o-Sil H5) supported iron microcrystals containing 10% Fe (by weight) were studied by use of LT and DB measurements. The samples were carefully prepared. Apparently, the preparation of oxide supported metal microcrystals is a fairly difficult task. Small iron particles have recently attracted much attention as model catalyst for ammonia and hydrocarbon synthesis. The LT spectra (200 ps resolution) were measured at two different time-per-channel settings, and the analyses combined information obtained with both settings. The LT spectrum of the silica support gave the following lifetimes (intensities): 0.169 ns (11.5%), 0.552 ns (58.9%), 2.06 ns (6.2%), 11.21 ns (5.0%), and 71.85 ns (18.3%). The DB spectrum (not deconvonluted for the resolution of fwhm = 5.19 mrad) could be fitted by a sum of gaussians: 5.60 mrad (16.4%), 10.33 mrad (69.7%), and 18.08 mrad (13.9%). This example shows that the LT spectrum for a typical silica sample contains several o Ps contributions and a large narrow p Ps DB component of about 16%. The addition of iron microcrystals reduced the long-lived o Ps and the narrow p Ps component intensities to zero. Chemisorbed, but not physisorbed, hydrogen on the surface of the iron microcrystals gave o Ps components with short lifetimes (1–4 ns, about 18%) and a narrow p Ps DB component of 5.46 mrad (8.9%). When this sample was further slowly oxidized (passivated) only small changes in the LT and DB spectra were measured.

Clearly, Ps in the voids reacts strongly with the clean iron surface. Probably it is physisorbed to the iron surface or it splits up in collision with the iron, leaving the positron in a surface state. If conversion occurs it is completely masked by "chemical" processes as no narrow component is found. On the other hand, hydrogen on the iron surface seems to give o Ps to p Ps conversion because the o Ps lifetimes are strongly reduced compared to the pure silica case, and a rather large narrow DB component is obtained. However, also for the pure silica the narrow DB component is larger than expected from the o Ps LT components. In summary, the positron studies of this type are at a very preliminary stage at present. It seems to be a very fruitful field for future investigations.

A LT study of dodecane/urea clathrate has been performed at Risoe [5.21]. At 21°C 0.136 ns (25.0%), 0.395 ns (24.0%), and 2.12 ns (51.0%) were obtained for the clathrate, while pure urea gave 0.123 ns (20.6%), 0.364 ns (26.7%), and 0.984 ns (52.7%). At increasing temperature the LT spectrum of the clathrate changed to that of pure urea at about 83°C because dodecane escapes from the urea "host". It is remarkable that Fig. 4.1 and the long o Ps lifetime of 2.12 ns with its high intensity of 51.0% indicate that Ps is localized in the channels of the urea lattice between the dodecane molecules. Apparently, o Ps can push the dodecane molecules apart in the urea channels. Other clathrate results seem not to have been reported.

Apart from the data on micelle and vesicle systems discussed in Sect. 10.4 few positron annihilation results for systems of direct biological interest have been reported. The subject has been reviewed in [10.72,73].

Only very few studies of samples cut out of living systems have been performed. In [10.74] AC spectra of rat muscle tissue and water at 4°C and −4°C were compared. At 4°C the muscle spectrum was a little lower in the top than that of water, while the narrow component in the muscle spectrum was roughly half as big as that of water at −4°C. The LT spectra for pork meat and liver were almost identical [5.21]. For liver and meat three term fits of 0.103 ns (12.2%), 0.393 ns (66.0%), and 2.01 ns (21.8%); and 0.189 ns (18.7%), 0.423 ns (59.4%), and 2.02 ns (21.9%), respectively, were obtained.

On consulting treatises of the composition of living systems [10.75] and biochemistry we realize immediately that very few of the atoms, ions, molecules, and/or polymers constituting living matter inhibit Ps formation and/or react with Ps at the concentrations found. The most obvious exception seems to be Cl^- ions. In pork liver and human muscle the Cl^- concentrations are roughly 0.028 M and 0.022 M, respectively [10.75]. In [6.4] Fig. 4 0.025 M Cl^- is shown to reduce the o Ps yield from 26.5% to about 23.3% in pure water. Hence, the measured o Ps yields in liver and meat are roughly explained by the Ps inhibition caused by positron reactions with Cl^-, taking into account that about 70% of the samples is water. The o Ps lifetime in the samples is very little longer than that of water of 1.89 ns (Table 5.1), maybe because the presence of organic molecules allows Ps to blow a slightly larger bubble. In [10.76] the LT data for various parts of animal eye lenses are presented. For most samples the results were fairly similar to those of water. Interesting exceptions are the very low o Ps yields for cow ($I_3 = 7.1\%$) and calf ($I_3 = 5.4\%$) cortex samples. The mentioned results seem to be the only ones for samples directly cut out of animal tissue.

In [10.77] AC spectra for Agave leaves in darkness and under white and red light are given. The light reduces the counts in the top of the AC spectra.

The LT spectra for hydrated DNA at various water contents, and as a function of temperature (20 to −180°C) for 10% and 50% wt of water, are presented in [10.78]. The results are reasonably interpreted with reference to the pure water, ice (Sect. 9.1) and amorphous structure (Sect. 10.1) data.

11 Conclusion

The main purpose of writing this book was to discuss selected experimental results which are typical for positron annihilation in chemistry, and, in particular, to explain these results. It was not intended to give an exhaustive discussion of all the published results. Experimental results and various theoretical explanations believed to be erratic and misleading have mainly been omitted.

As discussed in the book, there are many fields of physics and chemistry where positron annihilation has not been applied, or where it has been used only in very few cases. To illustrate this point let us look at the isothermals in a pressure-volume diagram for any pure compound in a textbook. Clearly, many interesting positron processes, e.g., positron and Ps reactions and Ps inhibition, can be studied at any point in such diagram. In the future a greater part of the positron work ought to be directed to these fields. Furthermore, many unsolved theoretical problems have been pointed out. Obviously, they ought to be solved. It seems to be appropriate to mention here that complicated quantum mechanical modelling cannot be expected to be a realistic way of solving most of these problems. As is normally the case in condensed matter physics and chemistry, the important issue is to find a qualitative or semiquantitative model which explains approximately the basic processes occurring. Apparently, progress in the understanding of positron annihilation in complicated systems has been delayed in the past because many scientists dislike studying phenomena which cannot be described by quantum mechanics or other complex mathematical models. A more realistic, chemistry-oriented attitude is desirable.

Clearly, it would be important for positron annihilation research in chemistry to use the slow-positron techniques for the study of selected phenomena. However, many problems cannot be expected to be solved in that way. For example, very little is known about the state and reactions of the positron in molecular condensed matter. This applies both for the positron not forming Ps and for the positron in the "product" of a Ps reaction. The most profitable way of investigating these positrons seems to be the AC technique, which, however, is used by very few scientists in the field. Other examples are the positron and Ps reactions in high density gases and in fluids at or above the critical point. They must mainly be studied by use of the "classical" LT and AC (DB) techniques.

The discussions in the book show very clearly that numerous phenomena in the field are not, or only partly, explained today. Furthermore, many areas have not been studied yet. Hence, the progress in the field is, certainly, not limited by the lack of unsolved problems. Probably, it will mainly be limited by the science sociology conditions. The amount of funding and number of jobs available are important factors, as is the ever increasing bureaucracy of science. The rapid exchanges of the fashionable new topics of science (examples are anomalous water, cold fusion, and high T_c superconductors) might strongly influence the progress in the field, too. Hence, as the general history of science in the last decades has clearly illustrated, the future scientific output in our field will be determined, in particular, by the science sociology effects influencing the science administrators and the scientists.

References

1.1 D.M. Schrader, Y.C. Jean (eds.): *Positron and Positronium Chemistry* (Elsevier, Amsterdam, 1988)

1.2 H.J. Ache (ed.): *Positronium and Muonium Chemistry*, Adv. in Chem. Series **175**, 1–376 (1979)

1.3 V.I. Goldanskii: Atomic Energy Rev. **6**, 3–148 (1968)

1.4 J. Green, J. Lee: *Positronium Chemistry* (Academic Press, New York, 1964)

1.5 W. Brandt, A. Dupasquier (eds.): *Positron Solid State Physics*, Proc. Int. School of Physics E. Fermi, Varenna, Italy, 1981 (North-Holland, Amsterdam, 1983)

1.6 Y.C. Jean, R.M. Lambrecht, D. Horvath: *Positrons and Positronium a Bibliography 1930–1984* (Elsevier, Amsterdam, 1988)

1.7 L. Dorikens-Vanpraet, M. Dorikens, D. Segers (eds.): *Positron Annihilation*, Proc. 8th Int. Conf. Positron Annihilation, Gent, Belgium, 1988 (World Scientific, Singapore, 1989)

1.8 P.C. Jain, R.M. Singru, K.P. Gopinathan (eds.): *Positron Annihilation*, Proc. 7th Int. Conf. Positron Annihilation, Delhi, India, 1985 (World Scientific, Singapore, 1985)

1.9 P.G. Coleman, S.C. Sharma, L.M. Diana (eds.): *Positron Annihilation*, Proc. 6th Int. Conf. Positron Annihilation, Arlington, Texas, 1982 (North-Holland, Amsterdam, 1982)

1.10 R.R. Hasiguti, K. Fujiwara (eds.): *Positron Annihilation*, Proc. 5th Int. Conf. Positron Annihilation, Lake Yamanaka, Japan, 1979 (The Japan Institute of Metals, Sendai, Japan)

1.11 Collection of large abstracts of 4th International Conference on Positron Annihilation, Helsingoer, Denmark, 1976 (unpublished)

1.12 A.T. Stewart, L.O. Roellig (eds.): *Positron Annihilation*, Proc. 1st Int. Conf. Positron Annihilation, Detroit, USA, 1965 (Academic Press, New York, 1967)

1.13 S.C. Sharma (ed.): *Positron Annihilation Studies of Fluids*, Proc. Int. Symposium, Arlington, Texas, 1987 (World Scientific, Singapore, 1988)

1.14 Y.C. Jean (ed.): *Positron and Positronium Chemistry*, Proc. 3rd Workshop on Positron and Positronium Chemistry, Milwaukee, USA, 1990 (World Scientific, Singapore, 1990)

1.15 R.J. Drachman (ed.): *Annihilation in Gases and Galaxies*, Proc. 5th Workshop on Positrons in Gases, Greenbelt, USA, 1989 (NASA Conf. Publication 3058, 1990)

1.16 J.W. Humberston, E.A.G. Armour (eds.): *Atomic Physics with Positrons*, Proc. of a Nato Adv. Res. Workshop, London, UK, July 1987 (Plenum, New York, 1987)

1.17 P.J. Schultz, G.R. Massoumi, P.J. Simpson (eds.): *Positron Beams for Solids and Surfaces*, Proc. Conf., London, Ontario, Canada, 1990 (AIP Conf. Proc. 218, AIP, New York, 1990)

1.18 V. Devanathan, K.P. Gopinathan (eds.): *Positron Annihilation Spectroscopy*, J. Madras Univ. **B45** (1982)

1.19 P.A.M. Dirac: Proc. Camb. Phil. Soc. **26**, 361 (1930)

1.20 A.I. Akhiezer, V.B. Berestetskii: *Quantum Electrodynamics* (Interscience, J. Wiley, New York, 1965)

1.21 P.A.M. Dirac: Sov. Phys. Usp. **22**, 648–653 (1979)

1.22 C.D. Andersen: Phys. Rev. **43**, 491–494 (1933)

1.23 S. Mohorovicic: Astr. Nachr. **253**, 93–108 (1934)

1.24 M. Deutsch: Phys. Rev. **82**, 455–456 (1951)

1.25 A. Rich: Rev. Mod. Phys. **53**, 127–165 (1981)

1.26 T.A. Pond: Phys. Rev. **93**, 478–479 (1954)
1.27 S. Berko, A.J. Zuchelli: Phys. Rev. **102**, 724–728 (1956)
1.28 A.T. Stewart: Phys. Rev. **99**, 594–595 (1955)
1.29 R.L. De Zafra: Phys. Rev. **113**, 1547–1555 (1959)
1.30 A. Ore: Univ. of Bergen Aarb. Naturvit. rekke **9** (1949)
1.31 J.E. Jackson, J.D. McGervey: J. Chem. Phys. **18**, 300–303 (1963)
1.32 R.A. Ferrell: Phys. Rev. **110**, 1355–1357 (1958)
1.33 J. Lee, G.J. Celitans: J. Chem. Phys. **44**, 2506–2511 (1966)
1.34 P.R. Gray, C.F. Cook, G.P. Sturm: J. Chem. Phys. **48**, 1145–1157 (1968)
1.35 O.E. Mogensen: J. Chem. Phys. **60**, 998–1004 (1974)
1.36 P.W. Percival, E. Roduner, H. Fischer: Chem. Phys. **32**, 353–367 (1978)
1.37 O.E. Mogensen, V.P. Shantarovich: Chem. Phys. **6**, 100–108 (1974)
1.38 W. Brandt, S. Berko, W.W. Walker: Phys. Rev. **120**, 1289–1295 (1960)
1.39 W. Brandt, G. Coussot, R. Paulin: Phys. Rev. Lett. **23**, 522–524 (1969)
1.40 O.E. Mogensen, G. Kvajic, M. Eldrup, M. Milosevic-Kvajic: Phys. Rev. **B4**, 71–73 (1971)
1.41 M. Eldrup: in [1.5], 644–658 (1983)
1.42 T. Goworek: Phys. Stat. Sol. **a102**, 511–526 (1987)
1.43 S. Rochanakij, D.M. Schrader: in [1.8], 193–194 (1985)
1.44 O.E. Mogensen: Chem. Phys. Lett. **163**, 145–150 (1989)
1.45 Y. Kobayashi: Chem. Phys. Lett. **172**, 307–311 (1990)
1.46 Y. Kobayashi: Chem. Phys. **150**, 453–460 (1991)
1.47 P. Kirkegaard, M. Eldrup, O.E. Mogensen, N.J. Pedersen: Comp. Phys. Comm. **23**, 307–335 (1981)
1.48 P.J. Schultz, K.G. Lynn: Rev. Mod. Phys. **60**, 701–779 (1988)
1.49 T.C. Griffith, G.R. Heyland: Phys. Rep. **39**, 169–277 (1978)
1.50 M. Charlton: Rep. Prog. Phys. **48**, 737–793 (1985)
1.51 T.C. Griffith: Adv. Atom. Mol. Phys. **22**, 37–75 (1986)
1.52 M. Charlton, G. Laricchia: J. Phys. **B23**, 1045–1078 (1990)
1.53 J.D. McNutt, W.W. Kinnison, A.D. Ray: J. Chem. Phys. **60**, 4730–4733 (1974)
1.54 M. Charlton: J. Phys. **B18**, L667–L671 (1985)
1.55 Science **251**, 166 (1991)
1.56 G.L. Brownell, T.F. Budinger, P.C. Lauterbur, P.L. McGeer: Science **215**, 619–626 (1982)

2.1 O.E. Mogensen: Thesis, DTH, Lyngby, Denmark (1968)
2.2 O.E. Mogensen, M. Eldrup: Risoe report 366 (1977)
2.3 J.P. Carbotte: in [1.5], 33–63 (1983)
2.4 O.E. Mogensen: in [1.11], G13–G18 (1976)
2.5 K. Siegbahn (ed.): *Alpha-, Beta- and Gamma-Ray Spectroscopy* (North-Holland, Amsterdam, 1968)
2.6 A. Ore, J.L. Powell: Phys. Rev. **75**, 1696–1699 (1949)
2.7 R.M. Drisko: Phys. Rev. **102**, 1542–1544 (1956)
2.8 V.G. Baryshevsky, O.N. Metelitsa, V.V. Tikhomirov: J. Phys. **B22**, 2835–2847 (1989)
2.9 R. Ramaty, J.M. McKinley, F.C. Jones: Astrophys. J. **256**, 238–246 (1982)
2.10 J.S. Nico, D.W. Gidley, A. Rich: Phys. Rev. Lett. **65**, 1344–1347 (1990)
2.11 M.W. Ritter, P.O. Egan, V.W. Hughes, K.A. Woodle: Phys. Rev. **A30**, 1331–1338 (1984)
2.12 Y.K. Ho: Phys. Lett. **A144**, 237–239 (1990)
2.13 A.P. Mills: Phys. Rev. Lett. **50**, 671–674 (1983)
2.14 L.A. Page, M. Heinberg: Phys. Rev. **106**, 1220–1224 (1957)
2.15 A. Greenberger, A.P. Mills, A. Thompson, S. Berko: Phys. Lett. **32A**, 72–73 (1970)
2.16 G. Gerber, D. Newman, A. Rich, E. Sweetman: Phys. Rev. **D15**, 1189–1193 (1977)
2.17 A. Rich, H.R. Crane: in [1.12], 321–327 (1967)
2.18 V.G. Baryshevsky, O.N. Metelitsa.V.V. Tikhomirov, S.K. Andrukhovich, A.V. Berestov, B.A. Martsinkevich, E.A. Rudak: Phys. Lett. **A136**, 428–432 (1989)
2.19 A.Dupasquier: in [1.5], 510–564 (1983)

3.1 G.F. Knoll: *Radiation Detection and Measurement* (J. Wiley, New York, 1979)
3.2 Y.C. Jean, D.M. Schrader: in [1.1], 91–119 (1988)
3.3 L.M. Smedskjær, M.J. Fluss: Meth. Exper. Phys. **21**, 77–145 (Academic Press, 1983)
3.4 I.K. MacKenzie: in [1.5], 196–264 (1983)
3.5 M. Eldrup, I.K. MacKenzie, B.T.A. McKee, D. Segers (eds.): in [1.7], 216–226 (1989)
3.6 EG&G Ortec: *Detectors and Instruments for Nuclear Spectroscopy* (1991/92)
3.7 G. Wikander, O.E. Mogensen: Chem. Phys. **72**, 407–423 (1982)
3.8 J. de Vries, A. Zecca, R.S. Brusa, R.G. Grisenti, S. Oss, A. van Veen: in [1.7], 654–656 (1989)
3.9 P. Kirkegaard, N.J. Pedersen, M. Eldrup: PATFIT-88, Risoe-M-2740, 1–132 (1989)
3.10 O.E. Mogensen: Nucl. Instr. Meth. **84**, 293–296 (1970)
3.11 J.T. Grissom, D.R. Koehler: Am.J. Phys. **35**, 753–760 (1967)
3.12 P. Kirkegaard, N.J. Pedersen, O.E. Mogensen: unpublished (1986)
3.13 C. Dauwe: Appl. Phys. **24**, 279–285 (1981)
3.14 A.P. Mills: in [1.5], 432–509 (1983)
3.15 K.G. Lynn: in [1.5] 609–643 (1983)

4.1 J.M. Warman: "The Dynamics of Electrons and Ions in Non-Polar Liquids" in *The Study of Fast Processes and Transient Species by Electron Pulse Radiolysis*, ed. by J.E. Baxendale and F. Busi, (Reidel, Dordrecht,1982) pp. 433–533
4.2 E.J. Hart, M. Anbar: *The Hydrated Electron* (J. Wiley, New York, 1970)
4.3 A.O. Allen: NSRDS-NBS-57 (1976)
4.4 W.F. Schmidt: HMI-B-156 (1974)
4.5 J.P. Dodelet, K. Shinsaka, G.R. Freeman: J. Chem. Phys. **59**, 1293–1297 (1973)
4.6 J.P. Dodelet, G.R. Freeman: Can. J. Chem. **53**, 1263–1274 (1975)
4.7 J. Belloni, J.L. Marignier: Radiat. Phys. Chem. **34**, 157–171 (1989)
4.8 C.D. Jonah, D.M. Bartels, A.C. Chernovitz: Radiat. Phys. Chem. **34**, 557–565 (1989)
4.9 M.A. Lewis, C.D. Jonah: J. Phys. Chem. **90**, 5367–5372 (1986)
4.10 Y. Gauduel, S. Pommeret, A. Migues, A. Antonetti: Radiat. Phys. Chem. **34**, 5–14 (1989)
4.11 M. Anbar, M. Bambenek, A.B. Ross: NSRDS-NBS 43 (1973), supplement (1975)
4.12 W.P. Helman, G.L. Hug, I. Carmichael, A.B. Ross: Radiat. Phys. Chem. **32**, 89–93 (1988)
4.13 A.O. Allen, T.E. Gangwer, R.A. Holroyd: J. Phys. Chem. **79**, 25–31 (1975)
4.14 A.O. Allen, R.A. Holroyd: J. Phys. Chem. **78**, 796–803 (1974)
4.15 T. Wada, K. Shinsaka, H. Namba, Y. Hatano: Can. J. Chem. **55**, 2144–2155 (1977)
4.16 M.H. Cohen: Can. J. Chem. **55**, 1906–1914 (1977)
4.17 J.L. Magee: Can. J. Chem. **55**, 1847–1859 (1977)
4.18 S. Linderoth, I.K. MacKenzie, S. Tanigawa: Phys. Lett. **107A**, 409–413 (1985)
4.19 F. Heinrich, A. Schiltz: in [1.9], 705–707 (1982)
4.20 R.A. Ferrell: Phys. Rev. **108**, 167–168 (1957)
4.21 A.P. Buchikhin, V.I. Goldanskii, A.O. Tatur, V.P. Shantarovich: Zh. Eksp. Teor. Fiz. **60**, 1136–1140 (1971)
4.22 Y. Ujihira, T. Ryuo, Y. Kobayashi, T. Nomizu: Appl. Phys. **16**, 71–74 (1978)
4.23 J.J. Jasper: J. Phys. Chem. Ref. Data **1**, 841–980 (1972)
4.24 M. Eldrup: in [1.9], 753–762 (1982)
4.25 S.J. Tao: J. Chem. Phys. **56**, 5499–5510 (1972)
4.26 F.M. Jacobsen: thesis, Risoe-R-433 (1981)
4.27 F.M. Jacobsen, M. Eldrup, O.E. Mogensen: Chem. Phys. **50**, 393–403 (1980)
4.28 H. Nakanishi, Y. Ujihira: J. Phys. Chem. **86**, 4446–4450 (1982)
4.29 F.M. Jacobsen, O.E. Mogensen, G. Trumpy: Chem. Phys. **69**, 71–80 (1982)
4.30 B.S. Jensen: in [1.11], paper G16 (1976)
4.31 A.F.M. Barton: *Handbook of Solubility Parameters and Other Cohesion Parameters* (CRC Press, Boca Raton, Florida, 1983)
4.32 A.F.M. Barton: Chem. Rev. **75**, 731–753 (1975)
4.33 C.M. Hansen: J. Paint Technol. **39**, 104–117 (1967)

4.34 O.E. Mogensen: Electrochim. Acta **33**, 1203-1210 (1988)
4.35 J.M. Ziman: *Electrons and Phonons* (Clarendon Press, Oxford, 1960)
4.36 N.G. Gee, G.R. Freeman: Can J. Chem. **64**, 1810-1816 (1986)
4.37 N.G. Gee, G.R. Freeman: Chem. Phys. Lett. **60**, 439-440 (1979)

5.1 A. Mozumder: Adv. Rad. Chem. **1**, 1-102 (1969)
5.2 K. Siegbahn (ed.): *Alpha-, Beta- and Gamma- Ray Spectroscopy*, vol. I, 1-78 (North Holland, Amsterdam,1968)
5.3 Farhataziz, M.A.J. Rodgers (eds.): *Radiation Chemistry. Principles and Applications*, (VCH Publ. Inc., Weinheim, Germany, 1987)
5.4 A. Mozumder (ed.): *Early Events in Radiation Chemistry*, Radiat. Phys. Chem. **34**, 1-191 (1989)
5.5 J.W. Hunt: Adv. Rad. Chem. **5**, 185-315 (1976)
5.6 A. Hummel: Adv. Rad. Chem. **4**, 1-102 (1974)
5.7 K. Funabashi: Adv. Rad. Chem. **4**, 103-180 (1974)
5.8 "Glossary", Can. J. Chem. **55**, 1797-1799 (1977)
5.9 J.E. Turner, R.N. Hamm, H.A. Wright, R.H. Ritchie, J.L. Magee, A. Chatterjee, W.E. Bolch: Radiat. Phys. Chem. **32**, 503-510 (1988)
5.10 N. Itoh: J. de Phys. Coll. C7 **37**, 27-37 (1976)
5.11 N. Itoh, T. Eshita, R.T. Williams: Phys. Rev. **34**, 4230-4234 (1986)
5.12 O.E. Mogensen, N.J. Pedersen, F.M. Jacobsen: in [1.8], 205-207 (1985)
5.13 S. Dannefær, L. Smedskjaer: J. Phys. **F6**, 3536-3544 (1973)
5.14 R.J. Douglas, M. Eldrup, L. Lupton, A.T. Stewart: in [1.10], 621-624 (1979)
5.15 O.E. Mogensen: Phys. Lett. **96A**, 250-254 (1983)
5.16 A.G. Cornaz: Thesis, Diss. ETH Nr. 8902 (ETH Zürich, 1989)
5.17 P.G. Varlashkin, A.T. Stewart: Phys. Rev. **148**, 459-462 (1966)
5.18 F.M. Jacobsen, O.E. Mogensen, N.J. Pedersen: in [1.7], 651-653 (1989)
5.19 F.M. Jacobsen, O.E. Mogensen: to be published
5.20 O.E. Mogensen, F.M. Jacobsen: Chem. Phys. **73**, 223-234 (1982)
5.21 Risoe work, unpublished
5.22 G. Bakale, W.F. Schmidt: Z. Naturforsch. **36a**, 802-806 (1981)
5.23 L. Onsager: Phys. Rev. **54**, 554-557 (1938)
5.24 C.L. Braun, T.W. Scott: J. Chem. Phys. **87**, 4776-4778 (1983)
5.25 R.A. Holroyd, T.K. Sham: J. Phys. Chem. **89**, 2909-2913 (1985)
5.26 G. Wikander, N.J. Pedersen, O.E. Mogensen: Chem. Phys. **86**, 213-225 (1984)
5.27 C.D. Jonah, M.S. Matheson, J.R. Miller, E.J. Hart: J. Phys. Chem. **80**, 1267-1270 (1976)
5.28 O.E. Mogensen, in [1.13], 418-441 (1988)
5.29 *Handbook of Physics and Chemistry*, ed. by R.C. Weast (Chemical Rubber Co., Cleveland, 1969), 50th edition.
5.30 O.E. Mogensen, N.J. Pedersen, B.D. Malhotra, R.A. Pethrick: Chem. Phys. **87**, 139-148 (1984)
5.31 C.Y. Leung, D.A.L. Paul: in [1.12], 409-416 (1967)
5.32 U. Sowada, J.M. Warman, M.P. de Haas: Phys. Rev. **25**, 3434-3437 (1982)
5.33 M. Tuomisaari, K. Rytsölä, P. Hautojärvi: in [1.7], 749-751 (1989)
5.34 O.E. Mogensen: in [1.9], 763-772 (1982)

6.1 P.E. Cade, A. Farazdel: J. Chem. Phys. **66**, 2598-2611 (1977)
6.2 A. Farazdel, P.E. Cade: J. Chem. Phys. **66**, 2612-2620 (1977)
6.3 O.E. Mogensen, N.J. Pedersen: Radiat. Phys. Chem. **28**, 33-48 (1986)
6.4 O.E. Mogensen: Chem. Phys. **37**, 139-158 (1979)
6.5 O.E. Mogensen, N.J. Pedersen, J.R. Andersen: Chem. Phys. Lett. **93**, 115-120 (1982)
6.6 J.R. Andersen, N.J. Pedersen, O.E. Mogensen: Chem. Phys. Lett. **63**, 171-173 (1979)
6.7 O.E. Mogensen, J.R. Andersen, N.J. Pedersen: in [1.9], 806-808 (1982)

6.8 O.E. Mogensen, N.J. Pedersen, J.R. Andersen: Chem. Phys. Lett. **98**, 405–407 (1983)
6.9 K.Y. Lam, J.W. Hunt: Int. J. Radiat. Phys. Chem. **7**, 317–338 (1975)
6.10 O.E. Mogensen, M. Eldrup, N.J. Pedersen: in [1.8], 756–758 (1985)
6.11 R.E. Weston Jr., H.A. Schwarz: *Chemical Kinetics* (Prentice-Hall, Englewood Cliffs, N.J., USA, 1972)
6.12 A.M. North: *The Collision Theory of Chemical Reactions in Liquids* (Methuen, London, 1964)
6.13 R.M. Noyes: Prog. React. Kinet. **1**, 129–160 (1961)
6.14 U.M. Gösele: Prog. Reac. Kinet. **13**, 63–161 (1984)
6.15 P.W. Atkins: *Physical Chemistry* (Oxford, U.K., 1989)
6.16 E.A. Moelwyn-Hughes: *Physical Chemistry* (Pergamon Press, Oxford, U.K., 1965)
6.17 S.A. Rice, P.R. Butler, M.J. Pilling, J.K. Baird: J. Chem. Phys. **70**, 4001–4007 (1979)
6.18 G. Czapski, E. Peled: J. Phys. Chem. **77**, 893–897 (1973)
6.19 S.R. Logan: Trans. Faraday Soc. **62**, 3416–3422 (1966)
6.20 S.R. Logan: Trans. Faraday Soc. **62**, 3423–3426 (1966)
6.21 J.M. Ziman: *Electrons and Phonons* (Clarendon, Oxford, U.K., 1960)
6.22 A.M. Stoneham: *Theory of Defects in Solids* (Clarendon, Oxford, U.K., 1975) chapter 14
6.23 C.D. Jonah, J.R. Miller, E.J. Hart, M.S. Matheson: J. Phys. Chem. **79**, 2705–2711 (1975)
6.24 C.D. Jonah, J.R. Miller, M.S. Matheson: J. Phys. Chem. **81**, 931–934 (1977)
6.25 L. Feldman, Z.B. Alfassi: J. Phys. Chem. **85**, 3060–3063 (1981)
6.26 J. Talamoni, G. Duplatre, J. Ch. Abbé, A. Haessler: Chem. Phys. **83**, 471–480 (1984)
6.27 J. Talamoni, J.Ch. Abbé, G. Duplatre, A. Haessler: Radiat. Phys. Chem. **21**, 431–438 (1983)
6.28 O.E. Mogensen: in [1.7], 552–554 (1989)

7.1 B. Levay, O.E. Mogensen: J. Phys. Chem. **81**, 373–377 (1977)
7.2 B. Levay, O.E. Mogensen: Chem. Phys. **53**, 131–139 (1980)
7.3 G. Wikander: Chem. Phys. Lett. **80**, 361–364 (1981)
7.4 D.A. Diehl, D.M. Schrader: J. Chem. Phys. **94**, 289–295 (1991)
7.5 J.M. Warman, K.D. Asmus, R.H. Schuler: J. Phys. Chem. **73**, 931–939 (1969)
7.6 P.P. Infelta, R.H. Schuler: J. Phys. Chem. **76**, 987–999 (1972)
7.7 C.A.M. van den Ende, J.M. Warman, A. Hummel: Radiat. Phys. Chem. **23**, 55–59 (1984)
7.8 L.H. Luthjens, H.C. deLeng, C.A.M. van den Ende, A. Hummel: in *Proc. 5th Tihany Symp. on Radiation Chemistry* (eds. J. Dobo, P. Hedvig, R. Schiller), 471–478 (1982)
7.9 O.E. Mogensen: Appl. Phys. **6**, 315–322 (1975)
7.10 M. Eldrup, V.P. Shantarovich, O.E. Mogensen: Chem. Phys. **11**, 129–142 (1975)
7.11 O.E. Mogensen: Chem. Phys. Lett. **65**, 511–514 (1979)
7.12 J.M. Warman, K.D. Asmus, R.H. Schuler: Am. Chem. Soc. Adv. Chem. Ser. **82**, 25–57 (1968)
7.13 C.D. Jonah, E.J. Hart, M.S. Matheson: J. Phys. Chem. **77**, 1838–1843 (1973)
7.14 H.T. Choi, S. Lipsky: J. Phys. Chem. **85**, 4089–4099 (1981)
7.15 K. Lee, S. Lipsky: J. Phys. Chem. **86**, 1985–1994 (1982)
7.16 H.T. Choi, F. Hirayama, S. Lipsky: J. Phys. Chem. **88**, 4246–4251 (1984)
7.17 J.Ch. Abbé, G. Duplatre, A.G. Maddock, J. Talamoni, A. Haessler: J. Inorg. Nucl. Chem. **43**, 2603–2610 (1981)
7.18 J.Ch. Abbé, G. Duplatre, A. Haessler: Radiat. Phys. Chem. **28**, 19–24 (1986)
7.19 G. Wikander: Chem. Phys. **38**, 181–197 (1979)
7.20 G. Wikander: Chem. Phys. **66**, 227–235 (1982)
7.21 G. Wikander: Chem. Phys. **49**, 153–163 (1980)
7.22 G. Wikander, O.E. Mogensen, N.J. Pedersen: Chem. Phys. **77**, 159–168 (1983)
7.23 G. Wikander: Chem. Phys. **39**, 309–323 (1979)
7.24 P. Jansen, O.E. Mogensen: Chem. Phys. **25**, 75–86 (1977)
7.25 G. Wikander: Chem. Phys. Lett. **77**, 120–126 (1981)
7.26 G. Wikander, O.E. Mogensen, N.J. Pedersen: Chem. Phys. **87**, 149–161 (1984)
7.27 I.L. Finar: *Organic Chemistry* (Longman, Harlow, U.K., 1973)

7.28 M.C.R. Symons: Pure and Appl. Chem. **53**, 223–238 (1981)
7.29 R.A. Holroyd, M. Allen: J. Chem. Phys. **54**, 5014–5021 (1971)
7.30 R.A. Holroyd, S. Tames, A. Kennedy: J. Phys. Chem. **79**, 2857–2861 (1975)
7.31 A. Henglein: Ber. Bunsen Ges. **79**, 129–135 (1975)
7.32 O.A. Anisimov, Yu.N. Molin: Khim. Vys. Energ. **9**, 539–540 (1975) (Engl. Transl. High Energy Chem. **9**, 471–472 (1976))
7.33 O.A. Anisimov, S.V. Vocel, Yu.N. Molin: Chem. Phys. **44**, 367–372 (1979)
7.34 O.A. Anisimov, S.V. Vocel, Yu.N. Molin: Chem. Phys. **53**, 123–129 (1980)
7.35 O.A. Anisimov, S.V. Vocel, Yu.N. Molin: Chem. Phys. **56**, 261–266 (1981)
7.36 Yu.N. Molin, O.A. Anisimov: in [1.2], 179–202 (1979)
7.37 Y. Ito, Y. Tabata: Radiat. Phys. Chem. **15**, 329–336 (1980)
7.38 Z. Zhicheng, Y. Ito, M. Washio, Y. Tabata: Radiat. Phys. Chem. **30**, 47–50 (1987)
7.39 Y. Ito, Y. Miyake, Y. Tabata: Radiat. Phys. Chem. **19**, 315–319 (1982)
7.40 Y. Ito, M. Hasegawa, Y. Miyake, Y. Tabata: Radiat. Phys. Chem. **21**, 217–223 (1983)
7.41 Y. Ito: in [1.1], 120–158 (1988)
7.42 B. Djermouni, H.J. Ache: J. Phys. Chem. **82**, 2378–2384 (1978)
7.43 G. Wikander: Chem. Phys. **39**, 21–32 (1979)
7.44 J.M. Warman, M.P. deHaas, E. Zador, A. Hummel: Chem. Phys. Lett. **35**, 383–386 (1975)
7.45 R.A. Holroyd, T.E. Gangwer, A.O. Allen: Chem. Phys. Lett. **31**, 520–523 (1975)
7.46 R.A. Holroyd: Ber. Bunsen Ges. **81**, 298–304 (1977)
7.47 L. Nyikos, C.A.M. van den Ende, J.M. Warman, A. Hummel: J. Phys. Chem. **84**, 1154–1155 (1980)
7.48 C.A.M. van den Ende, L. Nyikos, J.M. Warman, A. Hummel: Radiat. Phys Chem. **19**, 297–308 (1982)
7.49 N. Gee, G.R. Freeman, J. Chem. Phys. **90**, 5399–5405 (1989)
7.50 Yu.N. Molin, O.A. Anisimov: Radiat. Phys. Chem. **21**, 77–82 (1983)
7.51 Z.B. Alfassi, H.J. Ache: Ber. Bunsen Ges. **89**, 913–916 (1985)
7.52 R.E. Wild, H.J. Ache: J. Chem. Phys. **65**, 247–252 (1976)
7.53 P. Jansen, M. Eldrup, B.S. Jensen, O.E. Mogensen: Chem. Phys. **10**, 303–312 (1975)
7.54 A. Bisi, G. Gambarini, L. Zappa: Nuovo Cim. **B54**, 92–97 (1979)
7.55 G. Duplatre, A.G. Maddock, J.Ch. Abbé, A. Haessler: Chem. Phys. **28**, 433–440 (1978)
7.56 J.Ch. Abbé, G. Duplatre, A.G. Maddock, J. Talamoni, A. Haessler: J. Inorg. Nucl. Chem. **43**, 2603–2610 (1981)
7.57 A.G. Maddock, J.Ch. Abbé, G. Duplatre, A. Haessler: Chem. Phys. **26**, 163–167 (1977)
7.58 J.D. McGervey: in [1.12], 143–154 (1967)
7.59 G. Duplatre, C.D. Jonah: Radiat. Phys. Chem. **24**, 557–565 (1985)
7.60 D.A. Skoog, D.M. West: *Fundamentals of Analytical Chemistry* (Holt-Saunders, Tokyo, 1982)
7.61 J.H. Baxendale, R.S. Dixon: Z. Phys. Chem. **43**, 161–176 (1964)
7.62 L.J. Bartal, J.B. Nicholas, H.J. Ache: J. Phys. Chem. **76**, 1124–1129 (1972)
7.63 L.J. Bartal, H.J. Ache: Radiochim. Acta. **19**, 49–53 (1973)
7.64 H.J. Ache, R.E. Wild, L.J. Bartal: J. Phys. Chem. **81**, 941–943 (1977)
7.65 G. Duplatre, A. Haessler, J.Ch. Abbé: J. Phys. Chem. **89**, 1756–1760 (1985)
7.66 G. Duplatre, J.Ch. Abbé, J. Talamoni, A.G. Maddock: Chem. Phys. Lett. **100**, 553–558 (1983)
7.67 O.E. Mogensen, P.W. Percival: Radiat. Phys. Chem. **28**, 85–89 (1986)
7.68 G. Duplatre, J. Talamoni, J.Ch. Abbé, A. Haessler: Radiat. Phys. Chem. **23**, 531–536 (1984)
7.69 J.Ch. Abbé, G. Duplatre: Chem. Phys. Lett. **98**, 403–404 (1983)
7.70 O.E. Mogensen, N.J. Pedersen, J.R. Andersen: Chem. Phys. Lett. **98**, 405–407 (1983)
7.71 J. Talamoni, G. Duplatre, J.Ch. Abbé, A. Haessler: Chem. Phys. **83**, 471–480 (1984)
7.72 A.L.F. Lazzarini, A.F. Para, E. Lazzarini: J. Inorg. Nucl. Chem. **42**, 957–960 (1980)
7.73 A.F. Para, E. Lazzarini: J. Inorg. Nucl. Chem. **40**, 1373–1481 (1978)
7.74 A.L.F. Lazzarini, E. Lazzarini: J. Inorg. Nucl. Chem. **43**, 2599–2602 (1981)
7.75 E. Lazzarini: Radiat. Phys. Chem. **28**, 49–54 (1986)

7.76 F.A. Smith, M.K. Pickard: Chem. Phys. **118**, 445–455 (1987)
7.77 C.D. Beling, F.A. Smith, M.K. Pickard: Chem. Phys. **101**, 165–173 (1986)
7.78 F.A. Smith, S. Naidu: in [1.7], 561–563 (1988)
7.79 V.M. Byakov, V.I. Grafutin: Radiat. Phys. Chem. **28**, 1–18 (1986)
7.80 O.A. Anisimov, Yu.N. Molin: High Energy Chem. **9**, 473–475 (1975)
7.81 L.J. Bartal, H.J. Ache: J. Inorg. Nucl. Chem. **36**, 267–272 (1974)
7.82 A.L. Nichols, R.E. Wild, L.J. Bartal, H.J. Ache: Appl. Phys. **4**, 37–40 (1974)
7.83 J.B. Nicholas, R.E. Wild, L.J. Bartal, H.J. Ache: J. Phys. Chem. **77**, 178–182 (1973)
7.84 O.A. Anisimov: work at Risoe, unpublished
7.85 P. Jansen, M. Eldrup, O.E. Mogensen, P. Pagsberg: Chem. Phys. **6**, 265–271 (1974)
7.86 G. Wikander: Chem. Phys. **66**, 227–235 (1982)
7.87 V.L. Bugaenko, V.M. Byakov, V.I. Grafutin, O.V. Koldaeva, E.V. Minaichev: Radiat. Phys. Chem. **11**, 145–149 (1978)
7.88 V.M. Byakov, V.I. Grafutin, O.V. Koldaeva, E.V. Minaichev, O.P. Stepanova: High Energy Chem. **6**, 396–399 (1984)
7.89 B. Levay, P. Hautojærvi: Radiochem. Radioanal. Lett. **10**, 309–314 (1972)
7.90 B. Levay, S.J.G. Lund, O.E. Mogensen: Chem. Phys. **48**, 79–104 (1980)
7.91 F.M. Jacobsen: Chem. Phys. **109**, 455–464 (1986)
7.92 B. Djermouni, H.J. Ache: J. Phys. Chem. **82**, 2378–2384 (1978)
7.93 W. Brandt: Lett. Nuovo Cim. **33**, 499–501 (1982)

8.1 P. Jansen: Ph.D. thesis, Risoe-R-333
8.2 P.D. Burrow, A. Modelli, K.D. Jordan: Chem. Phys. Lett. **132**, 441–447 (1986)
8.3 E.C.M. Chen, W.E. Wentworth: J. Phys. Chem. **87**, 45–49 (1983)
8.4 R.B. Blaunstein, L.G. Christophorou: Radiat. Res. Rev. **3**, 69–118 (1971)
8.5 P. Kebarle, S. Chowdhury, Chem. Rev. **87**, 513–534 (1987)
8.6 V.M. Byakov, V.I. Grafutin, O.V. Koldaeva, E.V. Minaichev: J. Phys. Chem. **84**, 1864–1867 (1980)
8.7 W.J. Madia, A.L. Nichols, H.J. Ache: J. Am. Chem. Soc. **97**, 5041–5049 (1975)
8.8 E. Hall, W.J. Madia, H.J. Ache: Radiochem. Radioanal. Lett. **23**, 283–291 (1975)
8.9 I.B. Kevdina, V.P. Shantarovich: Khim. Vys. Energ. **9**, 477–479 (1975) (Eng. transl.: High Energ. Chem. **9**, 418–420 (1976))
8.10 E.S. Hall, H.J. Ache: J. Inorg. Nucl. Chem. **40**, 1690–1694 (1978)
8.11 V.I. Goldanskii, V.P. Shantarovich: Appl. Phys. **3**, 335–351 (1974)
8.12 V.I. Goldanskii, I.B. Kevdina, V.P. Shantarovich, K. Petersen: Dokl. Akad. Nauk **203**, 870–873 (1972)
8.13 Z.B. Alfassi, W.G.F. Filby, H.J. Ache: private information
8.14 W.J. Madia, H.J. Ache: J. Phys. Chem. **80**, 451–453 (1976)
8.15 Y.C. Jean: in [1.8], 202–204 (1985)
8.16 H. Nakanishi, Y.C. Jean: in [1.1], 159–192 (1988)
8.17 R. Forster: *Organic Charge-Transfer Complexes* (Academic Press, London, 1970)
8.18 *Spectroscopy and Structure of Molecular Complexes*, ed. by J. Yarwood (Plenum Press, London, 1973)
8.19 L. Levay, B. Levay, A. Vertes: Chem. Phys. **124**, 155–160 (1988)
8.20 O.E. Mogensen: unpublished
8.21 O.E. Mogensen: in [1.7], 552–554 (1988)
8.22 A.P. Buchikhin, V.I. Goldanskii, V.P. Shantarovich: Dokl. Akad. Nauk. **212**, 1356–1359 (1973)
8.23 Y. Kobayashi, Y. Ujihira: J. Phys. Chem. **85**, 2455–2458 (1981)
8.24 S. Rochanakij, D.M. Schrader: Radiat. Phys. Chem. **32**, 557–561 (1988)
8.25 I. Billard, J.Ch. Abbé, G. Duplatre: in [1.7], 540–542 (1988)
8.26 I. Billard, J.Ch. Abbé, G. Duplatre: J. Chem. Phys. **91**, 1579–1584 (1989)
8.27 R.S. Brusa, A. Dupasquier, S. Longano, S. Oss: Phys. Rev. **B43**, 12715–12722 (1991)

8.28 J.Ch. Abbé, G. Duplatre, A. Haessler, A.M. Netto, D.P. Veloso: J. Phys. Chem. **88**, 2071–2076 (1984)
8.29 Yu.N. Molin, K.M. Salikhov, K.I. Zamaraev: *Spin Exchange* (Springer, Berlin, 1980)
8.30 R.L. Wilson: Trans. Faraday Soc. **67**, 3008–3019 (1971)
8.31 V.P. Shantarovich, V.I. Goldanskii: Preprint in Collection of Preprints of the 2nd Int. Conf. Positron Annihilation, Kingston, Canada, 1971.
8.32 L.G. Aravin, V.I. Goldanskii, V.P. Shantarovich: Dokl. Acad. Nauk SSSR **230**, 106–109 (1976)
8.33 V.P. Shantarovich, P. Jansen: Chem. Phys. **34**, 39–45 (1978)
8.34 W.F. Magalhaes, J.Ch. Abbé, G. Duplatre: in [1.7], 531–533 (1989)
8.35 A.B. Doktorov, K.M. Salikhov, Yu.N. Molin: Dokl. Acad. Nauk SSSR **205**, 1385–1388 (1972)
8.36 K.I. Zamaraev, Yu.N. Molin, K.M. Salikhov: *Spinovui obmen* (Nauka, Sibirskoje Otd., Novosibirsk, 1977), §18, 292–303. Not included in the Engl. revised ed. [8.29].
8.37 A.P. Mills: J. Chem. Phys. **62**, 2646–2659 (1975)
8.38 A. Bisi, G. Consolati, G. Gambarini, L. Zappa: Nuovo Cim. **65B**, 442–454 (1981)
8.39 L.G. Aravin, B.I. Zaslavskii, V.P. Shantarovich: High Energy Chem. **21**, 455–460 (1988)
8.40 V.P. Shantarovich, B.V. Sobolev: Phys. Lett. **38A**, 391–392 (1972)
8.41 D.P. Kerr, A.M. Cooper, B.G. Hogg: Can.J. Phys. **43**, 963–965 (1965)
8.42 A. Bisi, G. Consolati, G. Gambarini, L. Zappa: Nuovo Cim. **1D**, 725–736 (1982)
8.43 V.I. Goldanskii, V.P. Shantarovich, A.V. Shishkin, A.O. Tatur: Dokl. Phys. Chem. **229**, 750–753 (1976)
8.44 V.M. Byakov, V.I. Goldanskii, V.P. Shantarovich: Elektrokhimiya **13**, 804–807 (1977)
8.45 V.P. Shantarovich, V.I. Goldanskii, Yu.N. Molin, V.P. Perminov, G.I. Skubnevskaya: Iz. Akad. Nauk. SSSR, Ser. Khim. **10**, 2188–2196 (1967)
8.46 A.L.F. Lazzarini, E. Lazzarini: J. Phys. **B22**, 541–552 (1989)
8.47 A.L.F. Lazzarini, E. Lazzarini: J. Phys. **B21**, 367–381 (1988)
8.48 A.L.F. Lazzarini, E. Lazzarini: J. Phys. **B20**, 183–197 (1986)
8.49 A.L.F. Lazzarini, E. Lazzarini: Chem. Phys. Lett. **113**, 380–383 (1985)
8.50 G. Trumpy: Phys. Rev. **118**, 668–674 (1960)
8.51 G. Duplatre, A. Haessler, A.M. Netto: J. Radioanal. Nucl. Chem. **82**, 219–228 (1984)
8.52 R. Zana, S. Millan, J.Ch. Abbé, G. Duplatre: J. Phys. Chem. **86**, 1457–1461 (1982)
8.53 V.I. Goldanskii, R.I. Zusman, Y.N. Molin, V.P. Shantarovich: Dokl. Akad. Nauk SSSR **188**, 1079–1081 (1969)

9.1 T. Goworek, in [1.7], 137–146 (1989)
9.2 G. Coussot: Thesis (Faculté des Sciences d'Orsay, Univ. of Paris, 1970)
9.3 T. Hyodo, Y. Takakusa. J. Phys. Soc. Japan, **42**, 1065–1066 (1977)
9.4 J. Kasai, T. Hyodo, K. Fujiwara: J. Phys. Soc. Japan **57**, 329–341 (1988)
9.5 O.E. Mogensen, M. Eldrup: J. Glaciology **21** (85), 85–98 (1978)
9.6 M. Eldrup, O.E. Mogensen, G. Trumpy: J. Chem. Phys. **57**, 495–504 (1972)
9.7 M. Eldrup: J. Chem. Phys. **64**, 5283–5290 (1976)
9.8 M. Eldrup, O.E. Mogensen, J. Bilgram: J. Glaciology **21** (85), 101–113 (1978)
9.9 J.A. Ghormley, C.J. Hochanadel: J. Phys. Chem. **75**, 40–44 (1971)
9.10 R.J. Douglas, M. Eldrup, L. Lupton, A.T. Stewart: in [1.10], 621–624 (1979)
9.11 T. Hyodo, A.T. Stewart: Phys. Rev. **B29**, 4164–4166 (1984)
9.12 S. Dannefær, D.P. Kerr, G.W. Dean, B.G. Hogg: Can.J. Phys. **56**, 1527–1530 (1978)
9.13 J.N. Sherwood (ed.): *The Plastically Crystalline State* (J. Wiley, London, 1979)
9.14 D. Lightbody, J.N. Sherwood, M. Eldrup: Mol. Cryst. Liq. Cryst. **96**, 197–210 (1983)
9.15 D. Lightbody, J.N. Sherwood, M. Eldrup: Chem. Phys. **93**, 475–484 (1985)
9.16 J. Bruce, J.N. Sherwood, N.J. Pedersen, M. Eldrup: in [1.8], 181–183 (1985)
9.17 P.C. Jain, M. Eldrup, N.J. Pedersen, J.N. Sherwood: Chem. Phys. **106**, 303–313 (1986)
9.18 M. Eldrup, N.J. Pedersen, J.N. Sherwood: Phys. Rev. Lett. **43**, 1407–1410 (1979)
9.19 D. Lightbody, J.N. Sherwood, M. Eldrup: Chem. Phys. Lett. **70**, 487–491 (1980)

9.20 S.J. Wang, Y.C. Yean: in [1.7], 809–811 (1989)
9.21 M. Eldrup, D. Lightbody, J.N. Sherwood: Chem. Phys. **63**, 51–58 (1981)
9.22 A.I. Kitaigorodsky: *Molecular Crystals and Molecules* (Academic Press, New York, 1973)
9.23 W. Brandt: private information
9.24 P.C. Jain, M. Eldrup, J.N. Sherwood: in [1.9], 674–676 (1982)
9.25 T. Goworek, C. Rybka, J. Wawryszczuk, R. Wasiewicz: Chem. Phys. Lett. **106**, 482–486 (1984)
9.26 T. Goworek, C. Rybka, R. Wasiewicz, J. Wawryszczuk: Phys. Stat. Sol. **b113**, K9-K13 (1982)
9.27 A. Bisi, G. Consolati, L. Zappa: Hyperfine Interact. **36**, 29–37 (1987)
9.28 B. Levay, M. Lalovic, H.J. Ache: J. Chem. Phys. **90**, 3282–3291 (1989)
9.29 M. Eldrup: Risoe work, unpublished
9.30 W.W. Walker: Phys. Lett. **59A**, 319–320 (1976)
9.31 S.Y. Chuang, S.J. Tao: in *Phase Transitions 1973*, ed. by L.E. Cross (Pergamon Press, Oxford, U.K., 1973) pp. 363–372
9.32 G. DeBlonde, S.Y. Chuang, B.G. Hogg, D.P. Kerr, D.M. Miller: Can. J. Phys. **50**, 1619–1622 (1972)
9.33 G. Consolati, F. Quasso: J. Phys. Cond. Matter **2**, 3941–3946 (1990)
9.34 S.Y. Chuang, S.J. Tao: Phys. Lett. **33A**, 56–57 (1970)
9.35 O.A. Anisimov, Y.N. Molin: High Energy Chem. **9**, 331–332 (1975)
9.36 R. Lenc: Chem. Phys. **131**, 443–453 (1989)
9.37 T. Goworek, C. Rybka, R. Wasiewicz, J. Wawryszczuk: Chem. Phys. Lett. **92**, 202–205 (1982)
9.38 W.W. Walker: in [1.9], 727–729 (1982)
9.39 M. Milosevic-Kvajic, O.E. Mogensen, G. Kvajic, M. Eldrup: J. Chem. Phys. **56**, 2567–2571 (1972)
9.40 V.P. Shantarovich, O.E. Mogensen, V.I. Goldanskii: Phys. Lett. **9**, 485–486 (1970)
9.41 Y. Jean, H.J. Ache: J. Phys. Chem. **82**, 656–659 (1976)
9.42 M. Eldrup, E. Lund-Thomsen, O.E. Mogensen: J. Chem. Phys. **56**, 4902–4906 (1972)
9.43 Y. Ito, Y. Tabata: Bull. Chem. Soc. Japan **48**, 808–812 (1975)
9.44 A.F. Wells: *Structural Inorganic Chemistry* (Clarendon, Oxford, U.K.,1985)
9.45 R.K. Wilson, P.O. Johnson, R. Stump: Phys. Rev. **129**, 2091–2095 (1963)
9.46 W. Brandt, J. Wilkenfeld: Phys. Rev. **B12**, 2579–2587 (1975)
9.47 M. Eldrup, A. Vehanen, P.J. Schultz, K.G. Lynn: Phys. Rev. Lett. **51**, 2007–2010 (1983)
9.48 M. Eldrup, A. Vehanen, P.J. Schultz, K.G. Lynn: Phys. Rev. Lett. **53**, 954 (1984)
9.49 M. Eldrup, A. Vehanen, P.J. Schultz, K.G. Lynn: Phys. Rev. **B32**, 7048–7064 (1985)
9.50 O.E. Mogensen: Phys. Lett. **118**, 357–362 (1986)
9.51 P.C. Rice-Evans, D.L. Smith: Phys. Lett. **A141**, 201–203 (1989)
9.52 M. Eldrup, O.E. Mogensen, L. Kevan: Chem. Phys. Lett. **10**, 379–380 (1971)
9.53 P. Sferlazzo, S. Berko, K.F. Canter: Phys. Rev. **B35**, 5315–5318 (1987)
9.54 M. Tuomisaari, R.H. Howell, T. McMullen: Phys. Rev. **B40**, 2060–2069 (1989)
9.55 M. Tuomisaari, R.H. Howell, T. McMullen: Helsinki Univ. Techn. Lab. of Phys. report 199 (1990)
9.56 H. Ikari, K. Fujiwara, J. Phys. Soc. Japan **46**, 92–96 (1979)
9.57 E.M. Gulliksen, A.P. Mills, Jr.: Phys. Rev. Lett. **57**, 376–379 (1986)
9.58 E.M. Gulliksen, A.P. Mills, Jr., E.G. McRae: Phys. Rev. **B37**, 588–591 (1988)

10.1 R.A. Pethrick, F.M. Jacobsen, O.E. Mogensen, M. Eldrup: J. Chem. Soc. Faraday Trans. 2 **76**, 225–232 (1980)
10.2 B.D. Malhotra, R.A. Pethrick: J. Chem. Soc. Faraday Trans. 2 **78**, 297–306 (1982)
10.3 O.E. Mogensen, N.J. Pedersen, B.D. Malhotra, R.A. Pethrick: Chem. Phys. **87**, 139–148 (1984)
10.4 A.Z. Varisov, Y.N. Kuznetsov, E.P. Prokopiev, A.I. Filipev: Russ. Chem. Rev. **50**, 991–1008 (1981)

10.5 J.R. Stevens: "Positron Annihilation" in *Methods of Experimental Physics*, ed. by R.A. Fava, Vol. **16A** (Academic Press, New York, 1980) pp. 371–403
10.6 A.E. Hamielec, M. Eldrup, O.E. Mogensen, P. Jansen: J. Macromol. Sci. Rev. Macromol. Chem. **C9**, 305–337 (1973)
10.7 L. Breitenhuber, P. Kindl, H. Sormann, W. Puff: Proc. Int. Conf. on Amorphous Systems Investigated by Nucl. Meth., Balatonfüred, Hungary, **1**, 327–337 (1981)
10.8 P. Kindl, H. Sormann, W. Puff: in [1.9], 685–687 (1982)
10.9 G. Reiter, P. Kindl: Phys. Stat. Sol. **a118**, 161–168 (1990)
10.10 P. Kindl, G. Reiter: in [1.7], 806–808 (1989)
10.11 J. Serna, J.Ch. Abbe, G. Duplatre: Phys. Stat. Sol. **a115**, 389–397 (1989)
10.12 D.P. Kerr: Can. J. Phys. **52**, 935–939 (1974)
10.13 G. Consolati, F. Quasso: Appl. Phys. **A50**, 43–48 (1990)
10.14 C. Dauwe, M. Kwete: in [1.9], 663–665 (1982)
10.15 B.S. Malhotra, R.A. Pethrick: Polymer Comm. **24**, 165–167 (1983)
10.16 A.P. Mills, E.M. Gullikson, L. Pfeiffer, W.S. Rockward: Phys. Rev. **B33**, 7799–7802 (1986)
10.17 I.K. MacKenzie, P.Z. Ghorayshi: Solid State Commun. **55**, 125–127 (1985)
10.18 B.D. Malhotra, R.A. Pethrick: Macromolecules **16**, 1175–1179 (1983)
10.19 Y.C. Jean, T.C. Sandreczki, D.P. Ames: J. Polymer Sci. **B24**, 1247–1258 (1986)
10.20 A.D. Kasbekar, P.L. Jones, A. Crowson: in [1.7], 799–802 (1989)
10.21 P. Kindl: Phys. Stat. Sol. **a81**, 293–298 (1984)
10.22 G. Brauer, W. Hertle, A.G. Balogh: Phys. Stat. Sol. **a105**, K7–K12 (1988)
10.23 G. Duplatre, F. Didierjean, J.Ch. Abbe: in [1.14], 72–75 (1990)
10.24 T. Suzuki, Y. Ito, K. Endo, S. Fujita: Appl. Radiat. Isot. **40**, 397–402 (1989)
10.25 W. Brandt, I. Spirn: Phys. Rev. **142**, 231–237 (1966)
10.26 A. Bisi, G. Gambarini, L. Zappa: Nuovo Cim. **2D**, 1465–1479 (1983)
10.27 M. Bertolaccini, A. Bisi, G. Gambarini, G. Padovini, L. Zappa: Appl. Phys. **12**, 93–96 (1977)
10.28 G.M. Sessler (ed.): *Electrets*, Topics of Appl. Phys., Vol. 33 (Springer, Berlin, 1980)
10.29 A. Bisi, G. Gambarini, L. Zappa: Lett. Nuovo Cim. **31**, 58–60 (1981)
10.30 A. Cornaz, J. Brunner, E. Cartier: in [1.7], 793–795 (1989)
10.31 S. Millan, U. Geckle, B. Levay, H.J. Ache: Ber. Bunsenges. Phys. Chem. **94**, 781–785 (1990)
10.32 P.C. Jain: in [1.18], 143–166 (1982)
10.33 P.C. Jain: in [1.13], 507–523 (1987)
10.34 G.D. Cole, W.W. Walker: J. Chem. Phys. **39**, 850–851 (1963)
10.35 G.D. Cole, W.W. Walker: J. Chem. Phys. **42**, 1692–1694 (1965)
10.36 E. Mueller, W.W. Walker: Phys. Lett. **44A**, 320 (1973)
10.37 W.W. Walker: Mol. Liq. Cryst. **102**, 129–132 (1984)
10.38 M. Szuszkiewiez: Acta Phys. Pol. **32**, 365 (1969)
10.39 J.B. Nicholas, H.J. Ache: J. Chem. Phys. **57**, 1597–1603 (1972)
10.40 J.K. Nash, S. McKenney, P.J. Boatman, J.D. McNutt: in [1.11], 74–85 (1976)
10.41 C. Kaur, J. Kumar, P.C. Jain: in [1.7], 803–805 (1989)
10.42 P.C. Jain: private information
10.43 Y.C. Jean, H.J. Ache: J. Am. Chem. Soc. **99**, 7504–7509 (1977)
10.44 Y.C. Jean, H.J. Ache: J. Am. Chem. Soc. **100**, 6320–6327 (1978)
10.45 B. Djermouni, H.J. Ache: J. Phys. Chem. **83**, 2476–2479 (1979)
10.46 L.A. Fucugauchi, B. Djermouni, E.D. Handel, H.J. Ache: J. Am. Chem. Soc. **101**, 2841–2844 (1979)
10.47 A. Boussaha, B. Djermouni, L.A. Fucugauchi, H.J. Ache: J. Am. Chem. Soc. **102**, 4654–4658 (1980)
10.48 E.D. Handel, H.J. Ache: J. Chem. Phys. **71**, 2083–2087 (1979)
10.49 K. Kano, A. Romero, B. Djermouni, H.J. Ache, J.H. Fendler: J. Am. Chem. Soc. **101**, 4030–4037 (1979)

10.50 J. Serrano, R. Reynoso, R. Lopez, O. Olea, B. Djermouni, L.A. Fucugauchi: J. Phys. Chem. **87**, 707–714 (1983)
10.51 S. Millan, R. Reynoso, J. Serrano, L.A. Fucugauchi: in [1.8], 628–630 (1985)
10.52 Y.Y. Wang, A.J. Hancock, Y.C. Jean: J. Am. Chem. Soc. **105**, 7272–7276 (1983)
10.53 Y.C. Jean, Y.Y. Wang: in [1.8], 845–847 (1985)
10.54 S.R. Choudhury, M. Zaboli, A.N. Maitra, P.C. Jain: in [1.7], 821–823 (1989)
10.55 G. Bakale, G. Beck, J.K. Thomas: J. Phys. Chem. **85**, 1062–1064 (1981)
10.56 M.B. Perkal, W.B. Walters: J. Chem. Phys. **53**, 190–198 (1970)
10.57 Y. Ito, M. Hirose, Y. Tabata: Appl. Phys. **A50**, 39–42 (1990)
10.58 Y. Ito, T. Takano, M. Hasegawa: Appl. Phys. **A45**, 193–201 (1988)
10.59 M. Debowska, J.Ch. Abbe, G. Duplatre: Phys. Stat. Sol. **b146**, 91–96 (1988)
10.60 A.M. Habrowska, E.S. Popiel: J. Appl. Phys. **62**, 2419–2423 (1987)
10.61 W.M. Meier: Pure Appl. Chem. **58**, 1323–1328 (1986)
10.62 R. Paulin, G. Ambrosino: J. de Phys. **29**, 263–270 (1968)
10.63 T. Chang: in [1.7], 150–152 (1989)
10.64 T. Hyodo, M. Kakimoto, T.B. Chang, J. Deng, T. Akahane, T. Chiba, B.T.A. McKee, A.T. Stewart: in [1.7], 878–880 (1989)
10.65 A.P. Mills Jr., E.D. Shaw, R.J. Chichester, D.M. Zuckerman: Phys. Rev. **B40**, 2045–2052 (1989)
10.66 G.W. Ford, L.M. Sander, T.A. Witten: Phys. Rev. Lett. **36**, 1269–1272 (1976)
10.67 W. Brandt, R. Paulin: Phys. Rev. Lett. **21**, 193–195 (1968)
10.68 S.Y. Chuang, S.J. Tao: Appl. Phys. **3**, 199–204 (1974)
10.69 V.I. Goldanskii, A.D. Mokrushin, A.O. Tatur, V.P. Shantarovich: Appl. Phys. **5**, 379–382 (1975)
10.70 M. Kakimoto, T. Hyodo, T.B. Chang: J. Phys. **B23**, 589–597 (1990)
10.71 O.E. Mogensen, M. Eldrup, S. Moerup, J.W. Oernbo, H. Topsoee: in [1.8], 980–982 (1985)
10.72 Y.Y. Wang, Y.C. Jean: in [1.1], 353–378 (1988)
10.73 G. Graf, E.D. Handel, P.L. McMahon: in [1.2], 67–87 (1979)
10.74 D.R. Gustafson: Biophys.J. **10**, 316–322 (1970)
10.75 K. Diem, C. Lentner: (eds.): *Documenta Geigy Scientific Tables* (Ciba-Geigy, Basle, 1970)
10.76 F.H. Hsu, B.G. Wen, J.F.R. Kuck, N.T. Yu: Phys. Stat. Sol. **a102**, 571–575 (1987)
10.77 J. Dworakowski, W. Hendrich, J. Wesolowski: Photosynthetica **8**, 344–352 (1974)
10.78 J.M. Warman, M. Eldrup: Biopolymers **25**, 1865–1874 (1986)

Index of Chemicals

The letters f or t appearing after a page number indicate that the subject is dealt with on that page in a figure or a table.

Subject Index

The letters f or t appearing after a page number indicate that the subject is dealt with on that page in a figure or a table.

Springer Series in Chemical Physics

Editors: Vitalii I. Goldanskii Fritz P. Schäfer J. Peter Toennies

1 **Atomic Spectra and Radiative Transitions** By I. I. Sobelman
2 **Surface Crystallography by LEED** Theory, Computation and Structural Results. By M. A. Van Hove, S. Y. Tong
3 **Advances in Laser Chemistry** Editor: A. H. Zewail
4 **Picosecond Phenomena** Editors: C. V. Shank, E. P. Ippen, S. L. Shapiro
5 **Laser Spectroscopy** Basic Concepts and Instrumentation By W. Demtröder 3rd Printing
6 **Laser-Induced Processes in Molecules** Physics and Chemistry Editors: K. L. Kompa, S. D. Smith
7 **Excitation of Atoms and Broadening of Spectral Lines** By I. I. Sobelman, L. A. Vainshtein, E. A. Yukov
8 **Spin Exchange** Principles and Applications in Chemistry and Biology By Yu. N. Molin, K. M. Salikhov, K. I. Zamaraev
9 **Secondary Ion Mass Spectrometry SIMS II** Editors: A. Benninghoven, C. A. Evans, Jr., R. A. Powell, R. Shimizu, H. A. Storms
10 **Lasers and Chemical Change** By A. Ben-Shaul, Y. Haas, K. L. Kompa, R. D. Levine
11 **Liquid Crystals of One- and Two-Dimensional Order** Editors: W. Helfrich, G. Heppke
12 **Gasdynamic Laser** By S. A. Losev
13 **Atomic Many-Body Theory** By I. Lindgren, J. Morrison
14 **Picosecond Phenomena II** Editors: R. M. Hochstrasser, W. Kaiser, C. V. Shank
15 **Vibrational Spectroscopy of Adsorbates** Editor: R. F. Willis
16 **Spectroscopy of Molecular Excitions** By V. L. Broude, E. I. Rashba, E. F. Sheka
17 **Inelastic Particle-Surface Collisions** Editors: E. Taglauer, W. Heiland
18 **Modelling of Chemical Reaction Systems** Editors: K. H. Ebert, P. Deuflhard, W. Jäger
19 **Secondary Ion Mass Spectrometry SIMS III** Editors: A. Benninghoven, J. Giber, J. László, M. Riedel, H. W. Werner
20 **Chemistry and Physics of Solid Surfaces IV** Editors: R. Vanselow, R. Howe
21 **Dynamics of Gas-Surface Interaction** Editors: G. Benedek, U. Valbusa
22 **Nonlinear Laser Chemistry** Multiple-Photon Excitation By V. S. Letokhov
23 **Picosecond Phenomena III** Editors: K. B. Eisenthal, R. M. Hochstrasser, W. Kaiser, A. Laubereau
24 **Desorption Induced by Electronic Transitions DIET I** Editors: N. H. Tolk, M. M. Traum, J. C. Tully, T. E. Madey
25 **Ion Formation from Organic Solids** Editor: A. Benninghoven
26 **Semiclassical Theories of Molecular Scattering** By B. C. Eu
27 **EXAFS and Near Edge Structures** Editors: A. Bianconi, L. Incoccia, S. Stipcich
28 **Atoms in Strong Light Fields** By N. B. Delone, V. P. Krainov
29 **Gas Flow in Nozzles** By U. G. Pirumov, G. S. Roslyakov
30 **Theory of Slow Atomic Collisions** By E. E. Nikitin, S. Ya. Umanskii
31 **Reference Data on Atoms, Molecules, and Ions** By A. A. Radzig, B. M. Smirnov
32 **Adsorption Processes on Semiconductor and Dielectric Surfaces I** By V. F. Kiselev, O. V. Krylov
33 **Surface Studies with Lasers** Editors: F.R. Aussenegg, A. Leitner, M.E. Lippitsch
34 **Inert Gases** Potentials, Dynamics, and Energy Transfer in Doped Crystals. Editor: M. L. Klein
35 **Chemistry and Physics of Solid Surfaces V** Editors: R. Vanselow, R. Howe
36 **Secondary Ion Mass Spectrometry, SIMS IV** Editors: A. Benninghoven, J. Okano, R. Shimizu, H. W. Werner
37 **X-Ray Spectra and Chemical Binding** By A. Meisel, G. Leonhardt, R. Szargan
38 **Ultrafast Phenomena IV** By D. H. Auston, K. B. Eisenthal
39 **Laser Processing and Diagnostics** Editor: D. Bäuerle

MIX
Papier aus verantwortungsvollen Quellen
Paper from responsible sources
FSC® C105338

If you have any concerns about our products,
you can contact us on
ProductSafety@springernature.com

In case Publisher is established outside the EU,
the EU authorized representative is:
Springer Nature Customer Service Center GmbH
Europaplatz 3, 69115 Heidelberg, Germany

Printed by Libri Plureos GmbH
in Hamburg, Germany